VITAL RAILS

VITAL RAILS

*The Charleston & Savannah
Railroad and the Civil War in
Coastal South Carolina*

H. DAVID STONE, JR.

THE UNIVERSITY OF SOUTH CAROLINA PRESS

© 2008 University of South Carolina

Published by the University of South Carolina Press
Columbia, South Carolina 29208

www.sc.edu/uscpress

Manufactured in the United States of America

17 16 15 14 13 12 11 10 09 08   10 9 8 7 6 5 4 3 2 1

Library of Congress Cataloging-in-Publication Data

Stone, H. David., 1966–
  Vital rails : the Charleston & Savannah Railroad and the Civil War in coastal South Carolina /
H. David Stone, Jr.
    p. cm.
  Includes bibliographical references and index.
  ISBN-13: 978-1-57003-716-0 (cloth : alk. paper)
  ISBN-10: 1-57003-716-7 (cloth : alk. paper)
  1. South Carolina—History—Civil War, 1861–1865—Transportation. 2. Charleston and
Savannah Railroad—History—19th century. 3. Railroads—South Carolina—History—19th
century. 4. United States—History—Civil War, 1861–1865—Transportation. I. Title.
  E545.S76 2008
  973.7'1—dc22
                                            2007026233

To Anne

CONTENTS

List of Illustrations *ix*
Acknowledgments *xi*

Introduction *1*
ONE Birth of a Railroad *6*
TWO Construction, Connection, and the Convention *24*
THREE A Confederate Railroad *40*
FOUR Opening Guns *59*
FIVE The Most Formidable Earthwork *80*
SIX Secessionville to Pocotaligo *99*
SEVEN The Business of War *120*
EIGHT Singletary's Inheritance *133*
NINE Unbroken Lines *142*
TEN The Squeaky Wheel *158*
ELEVEN Unrealized Gains *173*
TWELVE Foster Tries Charleston *186*
THIRTEEN Under Siege *200*
FOURTEEN Honey Hill *210*
FIFTEEN Sherman's Neckties *226*
SIXTEEN Running the Gauntlet *246*
SEVENTEEN One More River to Cross *267*
EIGHTEEN Postwar Debts *281*
NINETEEN Terminus *301*

Notes *317*
Bibliography *339*
Index *349*
About the Author *370*

ILLUSTRATIONS

FIGURES

Brigadier General Thomas Fenwick Drayton, C.S.A. *10*
Advertisement for Charleston & Savannah Railroad stock, 1854 *20*
Charleston & Savannah Railroad tracks crossing a swamp, 1862 *25*
State guaranty bond for the Charleston & Savannah Railroad, 1856 *28*
Green Pond Drive station *31*
Schedule of the Charleston & Savannah Railroad, April 1861 *57*
Notice of Charleston & Savannah stockholders' resolutions, 1861 *65*
Colonel Charles Jones Colcock, C.S.A. *66*
General Robert E. Lee, C.S.A. *71*
Henry Stevens Haines *72*
Brigadier General Roswell Sabine Ripley, C.S.A. *81*
Notice to civilian shippers, 1862 *83*
William Joy Magrath *87*
Brigadier General Isaac Ingalls Stevens, U.S.A. *94*
Brigadier General William Stephen Walker, C.S.A. *96*
Brigadier General Stephen Elliott, C.S.A. *96*
Major General Ormsby MacKnight Mitchel, U.S.A. *111*
The Forty-eighth New York Volunteer regiment attacking a Confederate train *118*
General Pierre G. T. Beauregard, C.S.A. *121*
Charleston & Savannah notices to passengers crossing the Ashley River, 1862 *123*
Charleston & Savannah notice of rice-storage rates, 1862 *125*
Charleston & Savannah notice of fare increases, 1863 *134*
African American soldiers in battle *143*
Major General Quincy Adams Gillmore, U.S.A. *144*
Brigadier General Truman Seymour, U.S.A. *175*
Brigadier General John Porter Hatch, U.S.A. *189*
Major General John Gray Foster, U.S.A. *191*

Train transporting prisoners from Savannah to Charleston *204*
Lieutenant General William Joseph Hardee, C.S.A. *206*
Major General Gustavus Woodson Smith, C.S.A. *213*
Corporal Andrew Jackson Smith *220*
Major General William Tecumseh Sherman, U.S.A. *227*
Sherman's troops destroying the rails *241*
Evacuation of Savannah, December 1864 *253*
Pocotaligo Depot, winter 1865 *262*
Sherman's XV Corps crossing the South Edisto River *265*
Historical marker for Charleston & Savannah rail line *282*
Causeway originally built for tracks of the Charleston & Savannah Railroad *290*
Railroad spikes found along the causeway near Stono Ferry *297*
Causeway leading to Frampton Creek *306*
Present-day CSX Transportation railroad tracks at Coosawhatchie *314*

MAPS

Route of the Charleston & Savannah Railroad in 1856 *8–9*
Confederate railroads *48–49*
The Broad River and its tributaries *77*
Troop movements during the First Battle of Pocotaligo, May 29, 1862 *92*
Confederate defenses at Charleston, spring 1862 *100*
The Battle of Secessionville, June 16, 1862 *103*
Route of the Second Battle of Pocotaligo, October 21–22, 1862 *113*
The Battles of Second Pocotaligo and Coosawhatchie, October 22, 1862 *115*
Union assaults planned for July 9, 1863 *148*
Confederate and Union positions around Charleston, June 1863 *159*
Proposed defenses for the area between the Ashepoo and Combahee rivers, November 4, 1863 *170*
Sherman's march through Georgia, September–December 1864 *211*
The Battle of Honey Hill *222*
Union navy movements around Gregorie's Neck, December 6–9, 1864 *233*
Union and Confederate positions outside Savannah, December 1864 *249*
Confederate railroads in the Bentonville Campaign, January–March 1865 *273*

ACKNOWLEDGMENTS

The concept for *Vital Rails* evolved over time from a search for my ancestral roots into what might best be termed a calling. My quest began in 1992 at the old South Carolina Archives building on Bull Street in Columbia as I researched the Civil War service of two of my great-great-grandfathers: Private William H. Stone of Company A, Third South Carolina Cavalry, and Private Isaac Oliver Polk of Company K, Eleventh South Carolina Infantry. Both men were from Colleton County, and both spent a significant portion of their Confederate military service in camps along the line of the Charleston & Savannah Railroad. In fact Private Polk was on an ill-fated train that Yankee troops ambushed near Coosawhatchie and was seriously wounded in the battle. Intrigued, I continued to seek information about the railroad, hoping to learn more about my ancestors in the process. Though I found no further mention of them, I did, however, become fascinated with the Charleston & Savannah Railroad, the men who ran it, and the role it played in the Civil War. Given its importance to the region, I began to wonder why no one had written a history of the railroad. The more I searched and the more information I accumulated, the more I became convinced that I should tell its story. To borrow the words of Robert Black, author of *Railroads of the Confederacy*, I wrote *Vital Rails* "primarily because [I had] long desired such a volume, and no one [had] seen fit to produce it for [me]" (xix).

My interest in railroads stems from my early childhood watching trains pass in front of my grandfather's seafood market in Mullins, South Carolina, on what was formerly the old Wilmington & Manchester line, which became part of the Seaboard Coastline system. Even as a boy, I appreciated the history of my hometown and the fact that it grew up around a railroad depot. My fascination with the history of my home state broadened as I read about the adventures of Caw Caw, Little Possum, and King Salkehatchie in my favorite elementary school text of all time, *South Carolina from the Mountains to the Sea*, by Mary Simms Oliphant and Mary C. Simms Oliphant Furman. When my third grade class traveled to Charleston on a field trip at the end of that year, I was hooked.

A number of years later, I was inspired by my history professor at Furman University, Dr. Newton Bond Jones, and by my wife's grandmother Rosa Waring Thornton. Mrs. Thornton was the person who—with her volumes of research on the

Waring, Thornton, and Gaillard families—piqued my curiosity as to the origins of my own family. Dr. Jones was the only history professor I had during my undergraduate years at Furman, and it was during a classroom discussion of the *Star of the West* incident that he posed to me the question, "Where . . . is your *pride*, Mr. Stone?" He made me question who I was, where I came from, and what it meant to be American, Southern, and South Carolinian. It is a disappointment to me that neither of them is around to read my finished work, but I hope that they would both be proud of it.

In the process of researching this book I was blessed to encounter many helpful librarians, whose assistance and cooperation should not be overlooked. I extend a big "thank you" to the staffs of the South Carolina Historical Society, College of Charleston Library, the Citadel Library, the South Carolina Room at the Charleston County Public Library, and the Charleston Library Society—all in Charleston; the Georgia Historical Society and the Kaye Kole Genealogy and Local History Room at the Bull Street branch of the Live Oak Public Libraries in Savannah; the Thomas Cooper and South Caroliniana Libraries at the University of South Carolina and the South Carolina Department of Archives and History in Columbia; the University of North Carolina Library in Chapel Hill, especially those working with the Documenting the American South Collection; the Virginia Tech Library in Blacksburg; the Henry B. Plant Library in Tampa, Florida; and the South Carolina Room of the Beaufort County Public Library. Certain individuals stood out along the way: Mike Coker, who went above and beyond the call in assisting me with the illustrations for *Vital Rails*, and Gloria Beiter, who provided valuable research assistance, at the South Carolina Historical Society; Bruce Pencek at Virginia Tech; Henry Fulmer and Beth Bilderback at the South Caroliniana Library; Jane Yates at the Citadel Archives and Museum; Kristin Veline at the Henry B. Plant Library; and Kenneth Johnson, of the photoduplication service at the Library of Congress in Washington, D.C.

When I was unable to leave town to research, I was fortunate to have two excellent libraries in Florence at my disposal. I would like to thank the staff of the James A. Rogers Library at Francis Marion University, especially those who work in the microfilm room, for tolerating the many hours I spent occupying the microfilm readers scouring the pages of the *Charleston Mercury* and *Daily Courier*. I would also like to recognize the helpful staff of the Zeigler South Carolina Room at the Doctors Bruce and Lee Foundation Library in Florence as well as the reference librarians who assisted with interlibrary loan materials. I also received help from a distance from William C. "Bill" Floyd, whose expertise in researching at the Library of Congress, the Valentine Museum in Richmond, Virginia, and the National Archives branches in Washington, D.C., and Maryland, was an incredible boon to my project.

There are several other individuals deserving of special mention for their assistance. Lucy Gross and Steve Haines were gracious enough to provide me with helpful information and a photograph of their grandfather Henry S. Haines. Greg

Whitman at the Links at Stono Ferry toured me around the areas of the golf course that were pertinent to the railroad and allowed me to photograph the area over which the railroad once ran. I was also able to attend a tour—sponsored by the University of South Carolina, Beaufort—of the battle sites along the western end of the railroad. Through the efforts of JoAnn Kingsley, the tour's organizer, and Doctors Stephen Wise and Larry Rowland, experts on the history of the region, we toured the sites at Port Royal Ferry, Pocotaligo, Coosawhatchie, and Honey Hill. This tour brought past events to life much more effectively than could any map, table, or book.

I must also take this opportunity to thank two gentlemen who helped me bring my project to its completion. Philip Grose graciously agreed to read my manuscript and advised me to go forward with the project. His advice and encouragement has meant more to me than he will ever realize. Alexander Moore, acquisitions editor at the University of South Carolina Press, believed that the story of the Charleston & Savannah Railroad was worth telling. He has shepherded me through the writing and publication process, patiently answering questions and helping me to solve many problems—both complicated and mundane—along the way.

Finally I would like to thank my family and friends, who for the last four years have had to listen to me talk about a railroad as if it were a living, breathing human; my parents, Kathryn Stone and David Stone, Sr., and grandparents, who instilled in me a sense of doing my best and of "remembering where I came from"; and my children, H.D., Daniel, Anna Caroline, and William, who had to endure their daddy being gone a number of weekends over the last few years to do research, as well as his often being preoccupied with writing his book even when he *was* at home. Last, but certainly not least, I owe an untold debt of gratitude to my wife, Anne. She has endured nights of silence while I sat in our den putting thoughts onto paper, tolerated my many research trips, and kept our household running despite her husband's near obsession with locomotives, boxcars, Civil War battles, and the South Carolina lowcountry. In addition to her busy routine, she often served as my sometimes-reluctant reader, providing helpful grammatical advice, even though the work was not in her preferred genre.

I have thoroughly enjoyed my trip on the Charleston & Savannah Railroad, and I am richer for having taken it. My sincere hope is that I have done justice to those who projected the railroad, constructed it, managed it, and fought in battles adjacent to it.

Introduction

From its onset the railroad movement in South Carolina met with stiff resistance. Charleston merchants encountered resistance in the late 1820s when they financed and built a railroad from Charleston to Hamburg, a South Carolina town just across the Savannah River from Augusta, Georgia. At the time of its organization neither South Carolina planters nor the planter-dominated state legislature showed much interest in purchasing stock in the venture. Because of the tight-fisted attitude of their fellow Carolinians, the railroad's president, William Aiken, Sr., and Alexander Black, one of its key promoters, went to Washington, D.C., to persuade the U.S. Congress to purchase $250,000 worth of the company's stock, arguing that the road would connect two states and promote national defense and stressing that the U.S. government's investment in the project would give South Carolina a fair share of Federal appropriations.[1]

Though the railroad was meant to protect Charleston's stake in the Savannah River trade, Aiken and Black's petition provoked strong opposition when the South Carolina legislature convened in the latter part of 1829. Led by Robert Barnwell Rhett and William Campbell Preston—nullifiers, who believed that states had the right to prevent the enforcement of Federal laws within their borders—felt that it was hypocritical for the state to fight the traditional American governmental philosophy on one hand and seek its benefits on the other. They also feared that the petition would have the effect of dividing the state and corrupting its political opinions. With this line of reasoning Rhett and Preston were able to secure a House resolution requesting South Carolina lawmakers to "oppose with all their zeal" internal improvement projects and "particularly all such appropriations for the benefit of South Carolina or any of her citizens."[2]

Opposition to the railroad was rooted in the desire to preserve of South Carolina's agrarian way of life. The state's lowcountry planters have been described as "a leisure class dedicated to achieving the exclusiveness and refinement displayed by English country gentlemen." During the so-called sickly season from May to November, wealthy land barons fled the mosquito-borne diseases of the intracoastal swamps, leaving their plantations in the hands of trusted overseers and gangs of slaves. Many planters escaped to Charleston, where they could live in opulence and further their various cultural pursuits. The absentee planters strove to create their own identity, and the South's "peculiar institution" was its underpinning. Not only did slave owning enable slaveholders to accrue and maintain their wealth, it also freed them to develop their aristocratic culture.[3]

With its predominantly agrarian orientation Charleston was, according to William W. Freehling, "as close to an anti-entrepreneurial city as any enterprising city could be."[4] Rather than investing in banks and factories, most planters poured their capital into land and slaves. Consequently in the early nineteenth century foreigners from the North and from Europe held sway over Charleston's East Bay Street countinghouses. Northerners also owned a large portion of Charleston's shipping and insurance companies and took their profits home with them when they left to escape the summer heat and disease. Even though these merchants and bankers were often wealthy and cultivated, they were usually unable to overcome the planters' contempt for Yankees and Yankee endeavors, so they rarely attained the upper echelon of Charleston society. Charleston was the planters' city, and even those Charlestonians who were not planters served agricultural interests.

This mindset played a prominent role in the reluctance to invest in the railroad from Charleston to Hamburg. During the legislative session of 1829, the railroad renewed its plea for a state appropriation, and the legislature responded with an inadequate loan of one hundred thousand dollars payable over seven years. Furthermore, the state would disburse the loan in installments of ten thousand dollars to match each thirty thousand dollars paid (not simply subscribed) by private investors, and the state required that the entire railroad property be assigned as a security.[5] The mercantile community in Charleston promptly organized a public meeting and renounced the state loan. They urged the company to reapply for a congressional appropriation, angering radicals throughout the state. Lowcountry planters felt that the request gave South Carolina's sanction to the principle that property was held "not by the tenure of the Constitution, but by the mere permission of the majority in Congress." Newspapers in the Midlands declared that the city of Charleston was "a colony of Yankee speculators, cherishing not a spark of Southern feeling." The company decided that an enthusiastic nationalist presentation would increase the chances of congressional approval; therefore, they asked Senator Daniel Webster of Massachusetts to present it before Congress. The angry and embarrassed South Carolina delegation was able to enlist enough opposition to defeat the petition, and the company eventually had no choice but to accept the state loan.[6]

The controversy demonstrated how Henry Clay's nationalism served the economic interests of the Charleston mercantile community and railroad promoters better than John C. Calhoun's sectionalism. Any Northern interference—whether it be in the form of industrial progress, entrepreneurial activity, or financial assistance from the Federal government—was dangerous politically and economically because Yankees were likely to oppose Southern radicalism and be sympathetic to antislavery sentiment. Also many in the community felt that commercial and industrial activity would threaten the city's ambience and disturb its social order.[7] Southern opposition to a strong central government and Southern resistance to industrialization were major contributors not only to the dissolution of the United States but also to the inefficiency of the Southern railroad system and the fragility of the economy of the Confederate States of America.

Such reluctance was not as typical across the Savannah River in neighboring Georgia. Even though agriculture was the foundation of Georgia's economy, the planter elite were a very small part of the white population, and—as in South Carolina—the "old wealth" was concentrated predominantly along the coast. Most Georgia planters were self-made men who planted cotton because it was more profitable than other economic pursuits. Many saw the wisdom of diversification and were alert to other means of turning a profit. Kenneth Coleman describes them as "shrewd opportunists" who fully utilized the credit facilities of local merchants and invested in Georgia's rapidly developing industries, creating an expansion in the number of tanneries, quarries, turpentine distilleries, lumber mills, and especially cotton mills.[8]

When South Carolina chartered the railroad from Charleston to Hamburg, Georgia leaders saw a threat to Savannah's seaport and reacted quickly. The Georgia legislature chartered railroads from Athens to Augusta (the Georgia Railroad) and Savannah to Macon (the Central of Georgia) in the mid-1830s, and—under the direction of prorailroad governor Wilson Lumpkin—it also chartered a state-owned road, the Western & Atlantic, in December 1836. In 1847 Savannah interests received a charter for the Savannah & Albany Railroad and later expanded their plan for a Savannah, Albany & Gulf Railroad extending all the way to Pensacola, Florida. A downturn in the economy hampered construction of these roads, but the plans for them remained in place. With the end of the depression in 1843 and with ample leadership, there was no doubt that the Southern railroad race was on.

Nevertheless times were changing in South Carolina, and the state's economy changed too. With the defeat of nullification and with Northern abolitionist sentiment on the rise in the mid-1830s, Southern incendiaries such as George McDuffie tried to convince planters to throw off their "prejudice against the mercantile character" and to consider it patriotic rather than demeaning to enlist in such activities. Consequently successful planters began to invest funds in Charleston's mercantile community. With little work to do on their plantations and with the medical and legal professions saturated, many wealthy planters steered their sons into the business arena, creating a bridge between planter wealth and commercial activity in the Port City.[9]

In a few short years Charlestonians such as James Hamilton, Jr., became leaders in the city's mercantile efforts. Hamilton and his contemporaries George A. Trenholm and Henry W. Conner hoped to extend their power base by chartering steamship companies, thus securing a share of the shipping industry. The state legislature also put forth a concerted effort to tap into the trade of the Midwest and to foster politico-economic ties with that region by granting generous support to the Louisville, Cincinnati & Charleston Railroad. Public and private banking capital expanded rapidly, and several new textile mills began operation in the upper part of the state.

The economy of the 1840s was much stronger than it had been two decades earlier, and Charleston became the manufacturing center of South Carolina. Charleston businessmen lobbied for and received appropriations for harbor improvements, and Charleston's increasingly merchant-dominated city council invested large amounts of public money in railroad projects. The city's tributary railroads shrank transportation costs. A robust banking system helped to stave off bankruptcies and kept currency in circulation, and improved agricultural techniques led to increased profits for Carolina planters. Those bound to the ways of the Old South, however, still cast a suspicious eye on what they considered Yankee enterprises.

Although the railroads powered an economic surge in the early 1850s, Southern railroads still had difficulties. Railroads built in staple-producing areas such as South Carolina were dependent on the prices of local goods in international markets. The self-sufficiency of cotton plantations posed a problem: there was very little demand for commercial goods since most daily necessities could be produced on the plantation. What was in high demand was a way to get produce to market. Railroads prospered when cotton was in demand at good prices, but there was very little to transport when it was not. For a time the frontier railroads had served Charleston well, bringing in agricultural products for exportation; however, the manufacturing capacity and retail markets in the regions they served were inadequate to sustain the railroads' growth.

In the mid–nineteenth century commercial and industrial activity in Charleston was on the rise, but it was not keeping pace with that of the South's other major population centers. Although its economy was somewhat more diverse, the city's failure to promote economic ties with other southeastern cities put its commercial future at risk. Railroad promoters in Wilmington tried to persuade Charleston investors to join them in constructing a railroad between the two cities, but the shortsighted Charlestonians refused. The North Carolinians proceeded with their venture, extending their line from Wilmington through Florence, South Carolina, all the way to the hamlet of Manchester on the Wateree River. Later the South Carolina Railroad completed its Camden branch, which intersected with the Wilmington & Manchester line, and in time passengers and freight from the Wilmington & Manchester could be shunted westward through Augusta. After the completion of other, rival railroads, commerce typically bound for Charleston could easily find its way instead to Mobile, New Orleans, and Pensacola. In Savannah new railroads began to branch westward, extending the city's reach in the direction of the Gulf of Mexico. Unless Charleston's leaders took action, their city would become isolated.

Talk of economic issues centered on this much-feared isolation soon dominated political discussions throughout the South Carolina lowcountry. The focus shifted from banking to the promotion of agriculture and commerce through railroad construction. Economic leaders and slaveholding politicians from the middle and upper parishes wooed planters, merchants, and other property owners between Charleston and Savannah with the same tactics used to rally support for nullification and states' rights. They held dinner parties at the homes of influential leaders and hosted gatherings in conjunction with state and national observances. Promoters loudly touted the economic opportunities that would come to local farmers and planters, but they were more select about mentioning the advantages that would accrue to land speculators and merchants.

At its heart, however, Charleston was a city deeply rooted in tradition and the influence of patriarchal wealth. In the minds of many of her most prominent citizens, that distinctive character was all defining and all encompassing—something to be not only embraced but nurtured and preserved. To some it meant gentility and refinement, a romantic sort of notion, an appreciation of its aristocratic European heritage. To others it referred to an agrarian lifestyle—stately plantation houses, bales of long- and short-staple cotton, bushels of rough rice, and slave labor. Any incursion of outside—that is, Northern—interests into the region would threaten the institution of slavery, and without it the economic system—and Charlestonians' whole way of life as they knew it—would crumble. For that reason many Charlestonians feared that the anticipated fiscal explosion to be brought by the railroad would threaten the very thing that made the city what it was, the thing that made it Southern and eventually made it Confederate.

Efforts to expand the railroads pitted those who held firm to the Palmetto City's cherished past against those who feared being crippled by it. In the early 1850s the state legislature and Charleston's mercantile-dominated city council came forward with funds, and the efforts of an odd marriage of lowcountry planters, slave-owning politicians, and capitalistic merchants resulted in the birth of a railroad. It was a generation, however, before their original goal of an east-west connection was fully realized.

ONE

Birth of a Railroad

The sun was just beginning to set as the locomotive approached the station. The state flags of South Carolina and Georgia were suspended overhead, and a host of onlookers lined the tracks near St. Andrews Station, across the Ashley River from Charleston. As the train passed beneath the banners, the guns of the Marion Artillery burst forth with a twenty-one gun salute. Hearty applause greeted the passengers as they exited the coaches, and they were escorted to the wharf to await the ferry ride into Charleston. Among them were Thomas Fenwick Drayton, Edward M. Manigault, and Savannah's delegates to the Democratic National Convention, which was being held that week in Charleston. The date was April 21, 1860, the day of the first two trips on the Charleston & Savannah Railroad between the Savannah River and Charleston. The celebration marked the arrival of the Charleston-bound train.

Although the mood this day took on the air of a Fourth of July gala, a striking turn of events in the coming weeks and months had a profound effect on the lives of those in attendance. Within days the delegates aboard the train had participated in the Democratic National Convention, during which many Southern delegates withdrew to form a separate political party. The split—the result of a dispute over the party's stance on slavery—proved irreparable and helped to accelerate the nation down a path toward disunion. On April 12, 1861, the artillerists who had welcomed the train not quite a year earlier were firing fieldpieces from Battery Trapier toward Fort Sumter, and later they were involved in other engagements defending the railroad. Indeed, the Charleston & Savannah Railroad became vital to the defense of the South Carolina coast, participating in its own defense—as did many of the men who projected, built, and operated it.

The citizens of Charleston and Savannah had long anticipated the completion of the railroad. Through it the two cities shared not only commerce and culture but also one another's fate. The railroad was their iron link. Their successes were its successes. Their failures were its failures. Their war was its war.

Between 1830 and 1860 approximately one thousand miles of railroad were constructed in South Carolina, in large part owing to ambitious plans to solidify Charleston's trade base. The initial line was intended to attract trade from northwestern South Carolina, but attention soon turned toward the midwestern and southwestern

United States. Development of railroads in the state occurred in different phases. The initial phase was the pioneer railroad from Charleston to Hamburg, South Carolina (the South Carolina Railroad). Next were the railroads to the mountains and foothills and branches of the South Carolina Railroad to Columbia and Camden. Subsequently construction was begun on branch lines from Columbia to every district in the Piedmont, and services were expanded to the northeastern section of the state. Finally a connection was sought between Charleston and Savannah—the last attempt to use rail transportation to achieve Charleston's dream of acquiring the trade of competing ports.[1]

A group of venture capitalists in New York was interested in creating a railroad along the shortest seaboard route from New York to Norfolk, Virginia. Contemporary thought was that they might plan to continue the line south through Wilmington and Charleston, possibly using the Charleston & Savannah line in the project. If this road were continued to Pensacola, Florida, which was considered the best harbor on the Gulf of Mexico, the distance between New York and Pensacola on this route would be considerably shorter than the distance between New York and New Orleans.[2] If Pensacola were made a packet station for steamers running to different routes across the Isthmus of Tehuantepec, cargo and passengers from California, South America, and the West Indies could be channeled through Charleston and Savannah.[3] The initiative for connecting these two cities with that route was undertaken by several well-known citizens of the lowcountry, prominent among whom were Charles Jones Colcock and Thomas Fenwick Drayton.

Charles Jones Colcock was born in Barnwell District in April 1820 and lived until adulthood with his grandfather Judge Charles Jones Colcock. The younger Colcock was a planter of Sea-Island cotton, residing at Bonnie Doone Plantation on the Okatie River near Grahamville and later owning a plantation near Foot Point at the junction of the Colleton and Broad rivers. Having an enterprising mind and a magnetic personality, Colcock exerted a great deal of influence in social, commercial, and political circles. He often used this influence in aid of public endeavors, especially those involving personal friends. He served South Carolina as a director of the Bank of the State and was a director of the Memphis & Charleston Railroad. After his second marriage in 1851, he joined the cotton firm of Fackler, Colcock, and Company, a branch of the factorage of Bradley, Wilson, and Company of New Orleans. He later established the factorage firm of Colcock, McCauley, and Molloy, which did a considerable business importing cotton from north Alabama, Georgia, Tennessee, and areas of South Carolina.[4]

In those days the only mode of transportation connecting Charleston to Savannah was a single daily steamboat. Seeing an opportunity in this inconvenience, Colcock held a dinner party on June 25, 1853, at his home near Grahamville. With the proposed railroad as the chief topic of conversation, this meeting was an attempt to persuade the local gentry that the railroad would bring even more prosperity to the region. It would theoretically open new markets for their agricultural products as

Route of the Charleston & Savannah Railroad in 1856.
Courtesy of South Carolina Historical Society

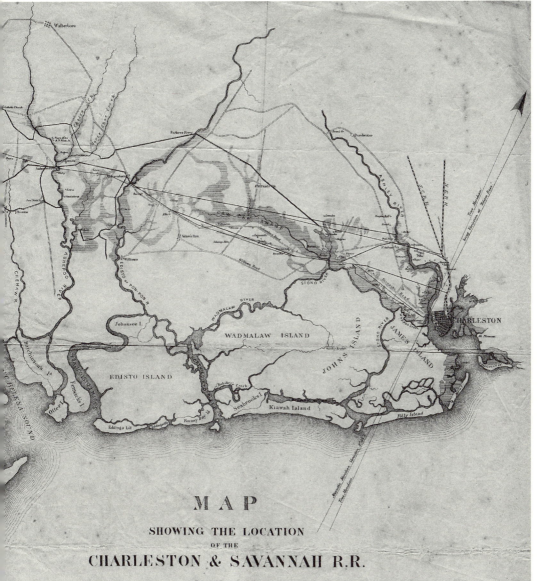

Brigadier General Thomas Fenwick Drayton, C.S.A., president of the Charleston & Savannah Railroad, 1854–1862, photograph by George S. Cooke. Courtesy of South Caroliniana Library, University of South Carolina, Columbia

well as increase property values in the area. Colcock and other early backers of the road knew of the proposed north-south coastline rail network, and they wanted the cities of Charleston and Savannah to gain their share of the economic benefits that the network would bring. They also knew the mindset of their neighbors and that there would be more than a little opposition to such an endeavor. It was hoped that local landowners would "see the necessity of acting liberally, and if the meeting at Grahamville attain[ed] this single object, they [would] find that they [had] made a great advance towards its progress; that they [had] acted wisely."[5]

Before adjourning for dinner those present at the meeting passed resolutions advocating the establishment of a railroad on the most direct and available route between Charleston and Savannah, the promotion of the line's early commencement, and the calling of a convention to take up the matter. Involving delegations from all interested parishes and districts in the region, the convention was planned for November 8, 1853, in Charleston, South Carolina. The delegation from Beaufort District was appointed from among those assembled that evening. In addition to Colcock it included prominent citizens such as Robert J. Davant (future president of the Port Royal Railroad), Thomas Hutson Gregorie, Julius G. Huguenin, Robert L. Tillinghast, John H. Webb, John Richardson, John E. Screven, Paul Pritchard, James A. Strobhart, J. J. Strong, R. H. Kirk, and Thomas Fenwick Drayton.[6]

Thomas Fenwick Drayton was born in South Carolina in 1807. His father, William Drayton, was a U.S. congressman from 1825 to 1833 and a staunch supporter of President Andrew Jackson. The younger Drayton graduated from the U.S. Military Academy at West Point in 1828, in the same class as future Confederate president Jefferson Davis. After his service in the army, Drayton was a civil engineer employed as assistant surveyor for the Louisville, Cincinnati & Charleston Railroad for two years. He returned to Saint Luke's Parish, South Carolina, where he became a planter and served as a captain in the South Carolina militia.[7]

Drayton felt that countries lacking adequate railroads and telegraph lines would be at a disadvantage in the event of an internal or external conflict. Without interior lines of communication, there would be no means of mobilizing personnel from remote locations in the North or South or of efficiently receiving and transmitting information. It concerned him that the population and wealth of the lowcountry of South Carolina and Georgia had been gradually redistributed over the years to enhance the power and resources of other states. Drayton also knew of the planned seaboard route running northeast to southwest, and he felt that a railroad between Charleston and Savannah was the ideal path on which it should travel. Drayton helped Colcock organize the dinner meeting at Grahamville and served as its chairman. Around this time he began a correspondence with William Brown Hodgson of Savannah to enlist his support in the enterprise.[8]

Hodgson was a scholarly and well-connected gentleman. Specializing in foreign languages and cultures, he had served the U.S. State Department in several Asian and African countries. He had retired to Savannah after his marriage to Margaret Telfair in 1842, and he took an active part in managing her plantations and vast financial holdings.[9] Of particular value to Drayton were Hodgson's connections to the Bank of the State of Georgia and his interest in railroad transportation and other means of shipping. With Hodgson's backing, Drayton thought, he could enlist the support of other enterprising men in Savannah and the surrounding area.

The idea of a seaboard rail line was received favorably by those present at Colcock's dinner party. Before they could organize their November 1853 convention in

Charleston to discuss promotion of the planned rail connection between Charleston and Savannah, another meeting caught Drayton's attention. A group of planters from the upper part of St. Peter's Parish, Beaufort District, met in Lawtonville near the end of June 1853 with designs on their own route. This area produced roughly twenty thousand bales of cotton per year. If a planter wished to sell his cotton in Charleston, he first had to ship it to Savannah on a steamer and then reship it to Charleston at added expense. The upstarts proposed that the new railroad should join the South Carolina Railroad at a point between Charleston and Lowery's Turn-Out, terminating at Purrysburg and connecting with Savannah by steamer if necessary. They believed a coastal route would be more expensive because of the bridges and embankments required. Others of the group did not care which route was constructed as long as it ran through the upper part of the parish. The road they envisioned would send their cotton to Charleston—an idea much more favorable to partisan South Carolinians. This group did not think that the citizens of Savannah would lend much assistance to a railroad favoring Charleston, proclaiming that they were too "wide awake to [their] own interests."[10]

Drayton notified Hodgson of this "injudicious" meeting, making sure to accentuate the offensive remarks directed toward Savannah. Drayton felt that the meeting was too intertwined with politics and that its were organizers self-serving. He assured Hodgson that these parish meetings would not have any effect on the final location of the railroad. This would be decided by the stockholders and not until the engineers had completed their survey. In reply Hodgson reassured the people of Charleston that Savannah would do everything in its power—short of subscription of stock—to make the railroad answer to its intended purpose. He said that there was no reason for one city to fear competition from the other, since communication between the two cities would give one city a connection to the southwest and the other an avenue to the north.[11]

As time passed, however, momentum grew for a route through St. Peter's Parish. Some felt that a route from Branchville on the South Carolina Railroad to Savannah would eventually be the "great thoroughfare" from New York to the Gulf of Mexico. Even if the coastal route were constructed, St. Peter's Parish and Barnwell District would still be prime ground for a railroad. If this line were built, Savannah would still be on the seaboard route, but Charleston would be excluded. One solution was to build a rail line paralleling the Savannah River to a point opposite Branchville and then run it along the shortest route between there and Charleston. This route would appease the planters in that area, send the area's cotton to Charleston, and keep both Charleston and Savannah on the seaboard railroad.

As planned the previous summer, railroad advocates from the various parishes and districts between Charleston and Savannah convened at the Hibernian Hall in Charleston on November 8 and 9, 1853. Thomas Drayton was quietly optimistic that the group assembled would further his plan of connecting Charleston to the seaboard rail line. In the group were his old friends Charles J. Colcock and Christopher

Gadsden of St. Luke's Parish. He also expected support from William E. Martin and Thomas P. Huger of Charleston. He was less certain of Alexander J. Lawton and his St. Peter's delegation.

The meeting was called to order, and officers were elected. William Ferguson Colcock was elected president of the convention, and a seven-man executive board was appointed: Solomon Cohen, George A. Trenholm, A. R. Johnston, Alfred P. Aldrich, Reuben Stevens, Burwell McBride, and John Richardson. On the motion of Thomas Drayton, one person from each delegation was appointed to a committee to review and report on the resolutions being considered. The delegates on this committee and their respective districts were Edmund Rhett of St. Helena's, William E. Martin of Charleston, William B. Hodgson of Savannah, Thomas F. Drayton of St. Luke's, Burwell McBride of Prince William, Dr. Benjamin W. Lawton of Barnwell District, Dr. John S. Lawton of St. Peter's, and Nathaniel Barnwell Heyward of St. Bartholomew's.[12]

The committee deliberated into the afternoon of the ninth and eventually submitted two resolutions to the assembly. The first called for the president to appoint a committee to memorialize the legislatures of South Carolina and Georgia for the charter of a railroad connecting the cities of Charleston and Savannah, specifying that it be built by the "shortest practicable route." The second charged the committee appointed by the president to commission a survey of the proposed routes.[13]

Hodgson rose in support of the resolutions and accentuated the importance of a short and direct route that would be one segment of the "great chain that would link the St. Lawrence with the Gulf of Mexico" and might lead eventually to "transoceanic" travel.[14] Alfred P. Aldrich, though not claiming prejudice toward any region or route, objected to the use of the word *shortest* to describe the route. Since there would be other factors in their decision—economy of construction, time required for travel, and prospective revenues—he proposed that the word *most* should replace *shortest*. Drayton spoke up next in favor of the original resolutions, restating the importance of short, direct routes.

John S. Lawton gave a concise review and his opinions of the three potential routes for their project. The seaboard route was 105 miles through swampy land and across several streams and rivers. He objected to this route because of the difficulty of construction, the lack of timber on the route, and the small amount of freight and passenger travel that the railroad would pick up from intermediate stations along the way. The middle route ran from George's Station through Walterboro, Gillisonville, and Grahamville—a total of 125 miles. This route had similar problems to the seaboard route and was removed from important planting and producing areas. The upper route started at George's Station or any preferred point on the South Carolina Railroad, went to Duck Branch in Barnwell District, then to Purrysburg, and across to Savannah, covering 135 miles. Lawton advocated this route mainly, it appeared, because of his allegiance to St. Peter's Parish. He supported the Aldrich amendment.

William E. Martin opposed the amendment and urged direct and immediate connection of the cities. He knew that it would be impossible to satisfy every individual delegation, and his main concern was the linkage with the seaboard route. He appealed to everyone to unite on the wording "shortest practicable route," pointing out that the main route could serve as a trunk line from which branches might be built to serve sections whose population warranted them. Edmund Rhett also supported the original wording, contending that building the most direct route was the best way to attract capital investment.

When the amendment was brought to a vote, the Savannah delegates abstained to keep out of the South Carolinians' squabble. The contingent from St. Peter's and Barnwell voted for the amendment; the St. Bartholomew's delegates were split; and the remaining four delegations voted against it, thus defeating the measure. After the vote Solomon Cohen addressed the group, alluding to the pending connection of Savannah by rail to Mobile and Pensacola on the Gulf of Mexico. When this line was finished, the idea of a direct route from Charleston to Savannah and the Gulf would become a reality. After Cohen's remarks, the original two resolutions were again voted on and decisively passed.

On returning home the angry delegates from St. Peter's Parish hastily organized a community meeting. Citizens from both St. Peter's and Barnwell District attended the gathering, chaired by Joseph A. Lawton, to discuss a Branchville to Savannah railroad connection. The views expressed there by A. M. Martin reflected the displeasure of the St. Peter's planters. Since the Charleston convention had "rejected uniting with that area," he felt that their "duty to Charleston had been discharged." They were prepared to cooperate with the City of Savannah and with the Wilmington & Manchester Railroad in building the Branchville-Savannah route.[15] They seem to have been using the two cities as means to an end. They preferred shipping to Charleston because that would be the "South Carolinian" thing to do, but if that city did not consider them important enough to run its rail line through St. Peter's Parish, they would ply their wares in Savannah.

Colonel John Lartigue next introduced a resolution that the group request a charter for a railroad running from Owen's Crossroads to Savannah along the lines of Barnwell and Beaufort districts. The terrain in this area was better suited to railroad construction, and the region's cotton would provide a richer source of income than lowcountry rice. He believed that a lowcountry railroad would sacrifice the interests of St. Peter's for the those of Charleston. He did not want to be subservient to Charleston. He wanted Charleston to serve the interests of the St. Peter's planters.

Sentiment favoring the Branchville-to-Savannah line could not be ignored. Many felt that the South Carolina state government had given too much aid to Charleston over the years. Some considered it hypocritical that a state that so strongly advocated states' rights, free trade, and equal privileges would object to a charter for this railroad simply because Charleston opposed it. Many were upset that the taxpayers had been pumping tens of thousands of dollars into a treasury that was used to fund the build up of Charleston's economy through railroads and other projects.[16] The Lawtonville group was also angry that they were getting little, if any, assistance from Savannah in the Georgia state legislature. Drayton hoped his Savannah connections would continue to afford the Charleston & Savannah Railroad a more favorable position among that city's leaders.

The toils of Colcock, Drayton, and Hodgson finally came to fruition on December 20, 1853, when Robert F. W. Allston, president of the Senate of South Carolina, and Representative James Simons, speaker of the South Carolina House, signed an act chartering the Charleston & Savannah Railroad. The railroad received a charter from the state of Georgia on February 18, 1854, and the opening of subscription was advertised in the *Charleston Daily Courier* on February 23. In an attempt to entice the masses to purchase stock in the railroad, a related article reminded citizens of the potential commercial isolation of Charleston. The steamers arriving from Wilmington were nearly deserted by this point, and Charleston was receiving only about 10 percent of the travelers from the Wilmington & Manchester Railroad and the Augusta branch of the South Carolina Railroad. If the Branchville & Savannah Railroad were finished before the Charleston & Savannah, Charleston would again be isolated from the principal routes of commerce. The projectors also alluded to recent difficulties with mail delivery to convince potential subscribers that the seaboard route would help alleviate these problems as well.[17]

Stock subscription was opened on March 6, 1854, at the office of the Charleston Insurance and Trust Company with future Charleston & Savannah board members Otis Mills, Lorenzo T. Potter, and William C. Bee serving as commissioners. The stock was offered at one hundred dollars per share, and a payment of five dollars per share was required at the time of subscription. Subscribers were to remit the remainder in five-dollar-per-share installments. To make the stock offering more attractive, the directors also declared that all subscriptions would bear 6 percent interest per year from the date of payment.[18]

Shortly thereafter plans began for construction of the railroad. A survey of the line was funded by the Charleston City Council and directed by John McRae. His assistants included John Johnson, who surveyed the lands between Charleston and the Salkehatchie Bridge and Combahee Ferry, and H. Lee Thurston, who surveyed from that point to Savannah.[19] Every stake placed by the engineers was under the scrutiny of the planters of St. Peter's Parish and lower Barnwell District.

McRae initially favored the lower route since much of the land passed over viable pine forests and was firmer than he had expected. He had spent some time in Savannah and had detected a generally supportive attitude toward the project there. He also heard a rumor that the Georgia Central Railroad might offer the Charleston & Savannah road a portion of its land in Savannah for building a depot. In order to appease the Lawtonville interests, the Charleston City Council ordered McRae to survey a route through Lawtonville, so that an objective comparison could be made. The preliminary work was completed in May 1854, and the report was submitted to Charleston mayor Thomas Leger Hutchinson on June 24.

While McRae's men surveyed potential routes, a city meeting was held in Charleston to discuss municipal aid to the railroad. William Gregg pointed out to those in attendance that cities such as Boston had generously funded projects that expanded their regional communications and had seen huge economic benefits. He enthusiastically recommended that Charleston undertake the whole enterprise if that was what it took to keep Charleston on the seaboard route. The meeting ended with a resolution authorizing Mayor Hutchinson to present the council's approval for any amount of aid necessary to complete the Charleston & Savannah Railroad.[20] The councilmen later approved a subscription of $260,000.

The politically savvy Lawtonvillle planters were not idle during this time. They organized their own meeting, at which they pledged a two hundred thousand dollar subscription to the Charleston & Savannah Railroad on one condition: that the road be constructed through their region. It was becoming increasingly apparent that they not only wanted a railroad, they wanted to control it. It mattered not to them whether the railroad fed Charleston, Savannah, or Macon as long as it was to their benefit and on their terms.

The company's charter called for a course that should travel the "shortest practicable route" between Charleston and Savannah. In the survey there were two possible routes from which to choose. An upper route was surveyed from the depot of the Northeastern Railroad in Charleston running to the depot of the Central Railroad of Georgia in Savannah. This route spanned 117 miles, and the estimated cost included "grading, grubbing, clearing, etc., $346,956.14; bridging and trestlework (total length 31,500 feet), $359,900.00; and superstructure (117 miles with 60 lb. rail at $9,060 per mile), $1,060,020.00; coming to a total of $1,766,876.14."[21]

The proposed lower route began at the intersection of Meeting and Spring streets, where the South Carolina Railroad and the Northeastern Railroad joined. From this point it paralleled the coast, traveling as directly as possible through pine forests, swamps, and plantations and passing through small towns such as Jacksonboro, Adams Run, and Pocotaligo until it crossed the Savannah River near Hutchinson Island. After that point it deflected to the left and ran parallel to the river, joining the Central Railroad of Georgia where it crossed the Augusta Road three-fourths of a mile from the depot.[22] The lower route measured only 105 miles and crossed the Edisto, Ashepoo, Combahee, Pocotaligo, and Coosawhatchie rivers a short distance below the head of schooner navigation; thus it would require the construction of drawbridges across these streams. The estimated cost of the lower route was "grading, grubbing, clearing, etc., $262,616.79; bridging and trestlework (total length 34,470 feet), $536,000; and superstructure (104 3/4 miles with 60 lb. rail at $9,060 per mile), $949,035; coming to a total of $1,748,251.79." The Lawtonville route followed the line of the upper route to the Salkehatchie Bridge. It then followed a circuitous path to Lawtonville and turned back toward Savannah, running through Robertville and Purrysburg. It joined the other lines at New River Bridge and from there followed the other lines into Savannah. The route covered approximately 150 miles with an estimated cost of "grading, grubbing, etc., $404,388; bridging and trestlework, $365,182; and superstructure (150 miles at $9,060 per mile), $1,359,000; coming to a total of $2,128,570."[23]

In estimating the construction costs for the railroad McRae was careful to point out that he had "made no estimate for the [purchase of] right of way, as it is impossible to tell, even approximately, what it will come to. No doubt in some cases the Company will have to pay very heavily, but it is to be hoped that a work which is so essential to the prosperity of the city and country will be treated with the greatest liberalities by the landed proprietors."[24] This issue eventually put a significant strain on the financial health of the company.

The estimated cost of the bridging on the lower route was higher than that on the upper route. Except for the region between Coosawhatchie and the Great Swamp, the lower route lacked suitable timber, and on the upper route the Pocotaligo and Coosawhatchie rivers did not require bridges. The most expensive undertaking on the lower route would be the crossing of the Ashley and Savannah rivers. A bridge over the Ashley would be more expensive because of its greater width and depth. McRae suggested that the location of the bridge on the lower route could be changed to that of the upper route by lengthening the road by approximately four miles, actually lowering the cost of construction by $75,719.[25]

The lower route fit the needs of the charter by being the least expensive and the most direct route between the two cities. It also was found to be the shortest route that would allow a future branch line to be constructed through St. Peter's Parish if a connection were later desired between Charleston and Macon, Georgia, via the Georgia Central Railroad at Millen. This connection would put Charleston into direct communication with the richest portion of the cotton belt between the Atlantic Ocean and the Mississippi River. Even if this connection were not affordable immediately, it could be of great benefit if undertaken at some point in the future. This suggestion actually scared some of the railroad's proponents in Savannah. If Charleston were eventually connected to a point seventy-five miles above Savannah, that city would then be excluded from the main line. This connection would have the same effect on Savannah that the Branchville & Savannah Railroad would have on Charleston. The citizens of Savannah could not and would not tolerate this. The most politically palatable solution was to preserve a direct route from Florence to Savannah.

With the completion of the Northeastern Railroad, freight and passengers could travel from Florence to Savannah via Charleston, thus cutting 119 miles from the previous route from Florence to Savannah via Kingville, Branchville, and Augusta, which had spanned 323 miles. Transportation of goods and passengers from Washington, D.C., to New Orleans would be more favorable on a route through

Charleston and Savannah than on the more interior routes because of the gentler grade of the land and fewer curves and obstructions. This would allow travel at higher speeds with less risk, thus helping to make up for the slightly longer distance by this route. The Charleston to Savannah route would also decrease the number of times passengers, baggage, and freight would have to be transferred to other cars, and it would reduce the need to connect arriving trains with other companies' trains running in other directions. Once the Charleston to Savannah route was finished, rail travel could be shunted toward Charleston and Savannah from any region south and west of Augusta, Georgia.[26]

McRae believed that this route would profit from the commercial trade in the port cities of Charleston and Savannah and from transporting cotton and rice as well as mail and passengers. A railroad in the lowcountry would decrease the cost of taking agricultural goods to market and might also spur the introduction to the area of corn as a cash crop, or possibly cattle farming. There would be a steady stream of people, who would bring new energy and enterprise to the region. Hotels would be filled with guests and retail stores filled with customers. This rejuvenation of the economy was bound to increase real-estate values in the Combahee, Ashepoo, and Savannah areas. Of course these dreams would all evaporate if the Charleston & Savannah were not favorably positioned on a proposed north-south seaboard line.

From a purely economic standpoint this plan was sound. As with the other railroads in the state, the plans for the Charleston & Savannah were intended to maximize the transport of goods and passengers to and through Charleston. Although preliminary steps had been made to update South Carolina's military equipment, there was no thought at this point how to protect the railroad in the event of military conflict with the North. Had there been, greater focus would have been placed on the bridge crossing the Ashley River and the fact that the lower route crossed many of the lowcountry waterways near or below the head of navigation.

In June 1854, after the necessary amount of stock was subscribed, the stockholders met to elect a board of twelve directors. One group of stockholders nominated William Gregg for the presidency. Though he had spoken in favor of the road, he had not invested himself in the project as much as Thomas Drayton had. At their meeting on July 12, 1854, the stockholders honored Drayton by electing him as the railroad's first president. Its first board of directors included Charleston mayor T. Leger Hutchinson, William Kirkwood, William Brown Hodgson, Lorenzo T. Potter, Otis Mills, James Butler Campbell, Edward Frost, John Bradley, Nathaniel Barnwell Heyward, William Ferguson Colcock, Christopher G. Memminger, and Daniel Heyward.

Not long afterward McRae's survey was published in the Charleston newspapers, and the supporters of the Lawtonville route set out to antagonize him. They said that McRae had "acted as witness, judge, and attorney" in making a case against the route.[27] They criticized the methodology of his survey and his commentary about the benefits of such a route. They claimed that McRae's survey of the Lawtonville route was inaccurate and too circuitous. They countered that the distance from Lawtonville to the Georgia Central was 22 miles rather than 42, and that the distance of the straightest version of a road through their region would be only 123 miles rather than McRae's estimate of 140–150.

The critics also commented on McRae's apparent ignorance of the economic benefits of the region. Many of the region's planters were evidently promising to grant rights-of-way freely through their lands, which would reduce the cost of construction. Timber was found abundantly along the proposed route, which would obviate the need for long-distance transport of wood for cross ties and other uses. Furthermore a railroad along this path would also have access to the rich cotton plantations of St. Peter's Parish and Barnwell District. This last point was possibly the strongest argument, considering that railroads of the day got most of their income from freight transport. A less valid argument was that the proposed path toward Lawtonville would be Charleston's only possible route to the Pacific. Apparently they were forgetting about Charleston's connection with the South Carolina Railroad.

The Lawtonville planters were an impediment to the Charleston & Savannah Railroad at every step of the way. They wanted the main route, not a branch line, to come to them. Drayton wondered how could they argue so fervently that their route would benefit Charleston, while asserting that being on a branch line off the main route would be unacceptable. If one would be of benefit, surely the other would too. The Lawtonville group warned that if their proposal was neglected or ignored it would "convert those who are [the Charleston & Savannah Railroad's] best friends into her worst and bitter enemies."[28]

The opponents of the Charleston & Savannah Railroad threatened that if St. Peter's Parish were disregarded, they would build the Branchville & Savannah Railroad before construction on the Charleston & Savannah was completed. The reactionaries, led by the Lawtons, called for a railroad meeting in Lawtonville. At this meeting they agreed to subscribe liberally to the Charleston & Savannah road if the Charleston City Council adopted the Lawtonville route. This route would be acceptable to them if it came to within ten miles of Lawtonville Church, but they considered a branch line "impracticable." At the close of the meeting, they appointed a committee—B. L. Willingham, B. F. Buckner, Henry Smart, A. M. Ruth, and John Lartigue—to raise the conditional subscriptions to the railroad.[29]

Stock sales for the Charleston & Savannah road were progressing more slowly than hoped. The directors decided to go back to the public and reopened subscription at numerous points along the proposed line of the railroad. Commissioners for the stock sale included, among others, W. H. Lowndes of Ashepoo, Wilson Ferebee of Grahamville, John Frampton of Whippy Swamp, and Nathaniel Barnwell Heyward of Blue House—all strong supporters of the railroad. Drayton knew that there had been problems with disease, storms, and economic hardship toward the close of 1854. He maintained that it was not a good time to seek venture capital—especially

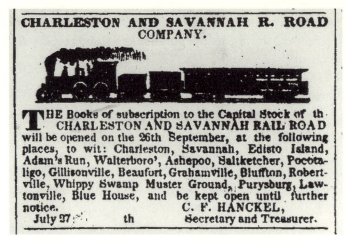

Advertisement for Charleston & Savannah Railroad stock. From the *Charleston Daily Courier,* July 27, 1854

not for a large public-works project. Still they must do something, and he hoped that, with the return of more prosperous times, people would have renewed interested in the project.

By January of 1855 the railroad's treasurer, Charles F. Hanckel, reported that a total of 3,167 shares of stock had been pledged; 2,600 shares by the City of Charleston itself.[30] It was also noted that the largest individual stockholder at that time was a prominent Savannahian, presumably Hodgson. The state had pledged $270,000, conditional on the sale of a certain amount of stock.

At a meeting in late November 1854, the board of directors had given McRae the go-ahead to organize two teams to survey the final route of the line, but he was not destined to complete the project. Before the winter was over McRae had resigned his position. Whether this was because of the constant criticisms over the Lawtonville route or the constant indictment of his professional abilities, will forever be a matter of speculation. To replace him the directors hired James S. Williams, a native of Savannah. Williams had been the chief engineer on the Cairo & Fulton Railroad in Arkansas, and his hiring was welcome news to the Georgians involved in the project.

It was not long before the Lawtonville group was again up in arms over the surveys. Williams had not been instructed to survey the Lawtonville route. The Branchville & Savannah supporters were angry that the railroad would be traveling through the lower part of Beaufort District, which they felt did not deserve a railroad. There was growing support for the route to pass through Lawtonville and connect with the Georgia Central Railroad at Millen. With increasing individual subscriptions, support pledged from the state, and a controlling interest held by the City of Charleston, it was also becoming apparent that the road would not travel through the Lawtonville area. The next obstacle for the railroad's directors and engineers was the Ashley River terminus.

In setting the final location of the roadway Williams primarily took into consideration the physical characteristics of the land. Edward M. Manigault was the assistant engineer conducting the survey from the Ashley River to the Salkehatchie. Charles O. Davis was entrusted with the task west of the Salkehatchie. The line began at the edge of John H. Steinmeyer's land, 1,400 feet below New Bridge Ferry, and then ran parallel to the road from New Bridge to Rantowles. Along that route it passed through the lands of Thomas Grange Simons, Joseph Prevost, George J. Crafts, Christian Hanckel, William Ravenel, Edward Sebring, Mrs. Caroline Geddes, Dr. Edward M. Haig, John Wilkes, Andrew Milne, and Samuel Barker. After a slight turn, the road passed through the lands of Benjamin Fuller and James O'Hear, and then crossed the Stono Marsh at its narrowest point. From there it traversed Benjamin Chaplin's and A. B. Williman's lands and crossed the Wiltown Road about fifty feet below its intersection with the Jacksonboro Road.

It passed through more than a quarter mile of rice land, and then entered the Berry Hill tract of George William Logan. From there, it traversed a mixture of swampland and pine ridges before entering the "high swamp" near a place called The Wilderness and owned by Dr. Smith. From The Wilderness the route crossed to a point adjacent to Oak Lawn, home of William Elliott, and then encountered the steepest grade on the line—twenty-six feet per mile—as it found the high ground on Major Hawkins King's Oakvilla tract on its way to Charles Baring's plantation.

The route struck the public road at Saxby's Hill, then another thousand feet of swamp on its way to higher ground owned by Dr. Joseph Edward Glover. From there it ran on a line crossing the branches of Deer Creek Swamp, which would require a mile-long trestle. After passing through the lands of Elliott, Glover, John Lewis, and Charles Warley, the route crossed the Ashepoo River and followed a straight line through Charles Myers's land to the Pynes Plantation of William Henry Heyward. From there it entered White Hall, the plantation owned by Nathaniel Barnwell Heyward, and crossed the lands of James B. Heyward, Charles Heyward, Daniel Heyward, and A. M. Peeples. From there the road passed through four miles of Daniel Blake's Board House Plantation (known today as Cherokee Plantation) to the Salkehatchie Bridge.

The line of the roadway from the Salkehatchie Bridge passed a mile to the west of Pocotaligo and crossed the Coosawhatchie a tenth of a mile above the highway bridge. After a very slight curve, it ran on a straight course through Gopher Hill, near Grahamville, and through Ferebeeville and Hardeeville to John Francis Pelot's plantation, ending at the Savannah River swamps near Union Landing.

The location of the Charleston terminus was still undecided. Initially there were three potential options. The first provided for a bridge across the Ashley at Bee's Ferry with the road connecting with the Northeastern Railroad at its depot. The second route, which crossed the Ashley at a lower point near Hacker & Riker's wharf, also terminated at the Northeastern depot. The final option called for the line to end at a depot-wharf complex on the western side of the Ashley River on the lands of John H. Steinmeyer. The Hacker & Riker line was quickly eliminated because of the expense and the depth of the water where it would cross the Ashley

River. This left a choice between the Bee's Ferry line, which was favored by the city council, or the Steinmeyer terminus, favored by the engineer and directors of the railroad.

In the final survey Williams objected to the construction of a permanent bridge across the Ashley because of its excessive cost. He also objected to the building of a temporary, perishable pile bridge across such a wide river in an area that was prone to strong storms. His final recommendation was to terminate the railroad at a wharf built near Steinmeyer's Point on the Ashley River, with the remainder of the trip into Charleston being made via steamer.[31] Even with the purchase of two steamboats, this proposal was nearly one hundred thousand dollars cheaper than any other plan. The land was situated just across from the heart of Charleston on high ground, providing more room for repair shops, turntables, and engine houses.

Some Charlestonians objected to this route because it ended outside the city limits and might cause the city to lose commercial business. The city council was upset because the city's subscription in the railroad was predicated on the fact that it would terminate inside the city. They authorized the mayor to protest the recommendation as "most injurious to the city, and inconsistent with the design and intent" of their original goals.[32] The Georgia Central had offered to allow its depot to be used as the Charleton & Savannah's Savannah terminus for the sum of ten thousand dollars per year for three years, then fifteen thousand dollars per year for three years after that. This might save the company money that could be used in defraying the cost of establishing the other terminus within the bounds of Charleston.

Williams argued that freight could be delivered from the Steinmeyer terminus to wharves at East Bay less expensively than from a terminus within the city because transport by water was cheaper than by dray. Supporters of the Steinmeyer route also thought that building warehouses west of the Ashley would be cheaper than doing so in the city and that the price of weighing and storing cotton in the west Ashley warehouses—and of shipping it from there—could be done more cheaply than it could from the city. They anticipated that cotton factoring and shipping businesses would move west of the Ashley in a few years because the terminus and warehouse were located there.

Williams also contended that the new location would lessen congestion in the streets of Charleston by reducing the number of carts and drays moving freight through the city. He pointed out that it would also decrease railroad congestion, since a road joining the Charleston & Savannah with the Northeastern's depot would have to cross the lines of both the South Carolina Railroad and the Northeastern Railroad before its termination. His proposal would lessen traffic in the city limits and still keep trade coming into the city from the same planters to the same factors and consigners—but at less expense.[33] Many taxpayers and the city council held the view that the railroad should terminate as near to the business district as possible. They and the mayor held firmly to their belief that the Bee's Ferry route should be chosen. In August, Alderman James M. Eason proposed a compromise that would give preference to the Steinmeyer route as long as a cost-effective bridge could be built from that point into the city. This proposal was undesirable to both parties and was defeated.[34]

Finally, and with reluctance, the board abandoned the Steinmeyer location under the pressure of city council resolutions and adopted the Bee's Ferry route. The directors sent Manigault to survey and stake this route in preparation for adding it to the proposed line. Before grading started, the city council and the Bank of Charleston rescinded the resolution from the previous summer that called for the Bee's Ferry route. The directors of the railroad reconvened on January 14, 1856, and voted not to construct the line from Rantowles to Bee's Ferry and the Northeastern Railroad. As the Steinmeyer route once again received their endorsement, Drayton lamented that construction on it had been delayed a full five months because of the whims of the city council.[35]

In the end Williams and the board won their argument. The company had their direct route from Charleston to Savannah. It would not run through Lawtonville, and it would not cross the Ashley River into Charleston. For a corporation looking for the most economical solution to a problem, the Steinmeyer terminus was a good business decision. For a group looking for the most advantageous way to join a seaboard railroad stretching from the northeast to the Gulf of Mexico, it seemed a little shortsighted. In a few short years the gap separating Steinmeyer's Point from the eastern bank of the Ashley would seem to stretch for miles.

TWO

Construction, Connection, and the Convention

As with most Southern railroads of the day, the lion's share of the physical work was placed on the shoulders of slaves. Thomas Drayton favored the use of slave labor not only because it was cheaper and "could be kept under better discipline,"[1] but also because teams could be worked throughout the year and would not create the "annoyances" that came with indentured white labor from the North. To ensure that the railroad was built as cheaply as possible, the Charleston & Savannah stipulated in its contracts that the work be performed "exclusively with slave labor."[2] These contracts also meticulously spelled out the responsibilities of food, shelter, clothing, and medical care for the slaves. The majority of slaves performed hard and menial tasks, but some eventually filled positions of skill and responsibility, becoming mechanics, firemen, and brakemen. Blacks working for railroad companies during this period seemed to have received better treatment than most members of their race.

Some companies, such as the South Carolina Railroad, owned their own slaves; but the Charleston & Savannah Railroad followed the more common practice of hiring their slave labor. Slaves were often rented individually or in gangs from their masters and returned to them in time for harvest season. Some planters contracted for the work on the portion of the railroad that ran through their own land, as was the case with one unnamed planter west of the Salkehatchie who oversaw six miles of construction performed by his own slaves.[3] The average cost of purchasing a slave at the time was $1,000, and the cost of hiring a "prime Negro man" for labor was around $180 per year.[4] The Charleston & Savannah paid approximately 58 cents per day for the hire of an individual slave. Basing his calculations on productivity estimates, Drayton estimated that the cost of slave labor to the railroad would be 5.5 cents per mile excavated, about one-third the usual contract price.[5] This was a bargain for Drayton, who felt that a slave was just as productive as—if not more productive than—a white laborer. Despite the risks of disease and injury to their slaves, planters were often eager to accept payment for their slaves' labor in cash or in stocks and bonds of the company.

Building a railroad in the antebellum South was a grueling endeavor. Technology was not only primitive but also labor intensive. The longest and most burdensome

Engraving from *Harper's Weekly,* March 15, 1862. Courtesy of South Carolina Historical Society

aspect of the job was grading the roadbed. Guided by notched stakes placed by surveyors, workers used picks, shovels, and wheelbarrows to build up, break down, and reshape the earth to "make the grade" and form the most even foundation upon which to lay the cross ties. Whether in suffocating heat and humidity or in bitter cold, a black slave could in a typical day exhume twelve cubic yards—nearly fourteen tons—of the boggy mixture of silt, clay, and sand indigenous to the coastal plain between Charleston and Savannah. On the rare occasion that they encountered rockier ground, they used gunpowder to blast the rock into pieces before hauling it away to use as dirt fill. Over the next two years companies of slaves dutifully graded, grubbed, and bridged the 102-mile stretch of swamps, forests, and marshes that lay between the two ports. They would commence laying track nearly a year after their job was begun.[6]

In Beaufort District much of the push for a railroad through the South Carolina lowcountry came from areas that did not already possess a ready commercial avenue to Charleston by water. Thomas Drayton was able to enlist support from entrepreneurial planters such as George Parsons Elliott, Charles Jones Colcock, and Julius G. Huguenin, but some residents of St. Peter's Parish had pressed for a route farther inland. Even though the shortest route between Charleston and Savannah ran through the lower portion of St. Bartholomew's Parish, there were few activists among the planter elite of this area. Nathaniel Barnwell Heyward and William Henry Heyward were early backers; however, many of the planters already had good access to Charleston by water and did not need the economic boost of a railroad to develop their land. Rather men such as Senator Lewis O'Bryan, Sheriff Lawrence W. McCants, and D. S. Henderson—economic leaders from the middle and upper sections with strong political ties—initiated much of the organized activity in the parish.[7] There was also strong support in Savannah, which needed a northern rail connection, and Charleston, where the mercantile-oriented city government had invested more than three million dollars in various railroad projects to keep the city on a par with other key commercial centers in the South.

With wealthy and improving cities at each end of their line and a rich country and population in between, the railroad's directors were confident that locating the road on the shortest route between New York and the Gulf of Mexico would attract enough business to profit producer and proprietor alike. Not everyone shared this optimism. Almost from the outset Drayton and the directorship were confronted by the prejudice and apathy of the tideland gentry, who were set against the extension of any railroad to the wharves of Charleston. Forbidden to use private roads, contractors building the railroad were forced to haul in timber for trestles and cross ties from remote areas.[8] This caused further delays on the sections requiring extra trestlework. There were also those who underhandedly pitted the citizens of the two cities against one another by instilling the idea that one city would benefit from the railroad at the expense of the other. These shortsighted adversaries of the project stirred up many who felt the railroad to be not only an annoyance but also a danger to their livelihood. Against these unfriendly ideas, Drayton argued that "no section advances which bestows no thought upon the interests of those around them, whose prosperity is and must be inseparable from their own."[9]

Enmity toward the railroad was not exclusive to individual planters. Some municipalities passed ordinances prohibiting railroads from entering their boundaries, failing to see the future benefits that such entry might bring. The citizens of Grahamville were so adverse to the railroad bringing its smoke and clamor into their town that they insisted the track be "laid no nearer the village than one full mile." The railroad's depot was also built to their specifications: "westward at a point where the track crossed the public road leading from Euhaw's to Sister's Ferry."[10] Sometimes this antagonism backfired. In the years after the completion of the railroad, the population of Grahamville began to decline, and the village surrounding the depot, Gopher Hill, became the commercial center of the area—the present-day city of Ridgeland.

The proprietary sentiment displayed during the early stages of the South Carolina railroad movement again came to the fore as disputes developed between the railroad and neighboring landowners over the cost of right-of-way. Commissioners and juries granted excessive awards to the property owners—some amounting to one hundred times the true value of their land—resulting in the payment of fifty thousand dollars for one ten-mile stretch on the line. Many of the assessments were felt by the railroad's management to be weighted excessively in favor of the landowners, possibly because the railroads were viewed as nuisances rather than vehicles for opening up avenues of trade. At least one planter, Andrew Milne, donated the right-of-way through his plantation, because he felt the road was requisite to Charleston's future prosperity.

Some of these disputes were never resolved. Daniel Blake was awarded damages for right-of-way through his lands amounting to $40,000 in March 1859. The railroad appealed and had the judgment reduced to $4,725 in May 1860. Blake appealed that ruling and was awarded another trial. When war broke out, more pressing concerns prevented the case from being reheard. Though Richard DeTreville—the company's attorney in the matter—continued to negotiate a compromise throughout the war, one was never achieved. Blake finally dropped the case in 1869.[11]

When construction of the railroad began in February 1856, Thomas Fenwick Drayton remained its president. The board of directors was dominated by a mixture of politicians and merchants from Charleston. It consisted of Charleston mayor William Porcher Miles, William Kirkwood, William Ferguson Colcock, Otis Mills, Christopher G. Memminger, Lorenzo T. Potter, James Butler Campbell, William E. Martin, Edmund Rhett, and Henry Gourdin—all from South Carolina—and Georgians William Brown Hodgson and Richard W. Bradley. They contracted with Etienne and Gerald B. Lartigue for the work between Rantowles Bridge and the Edisto River. William Phelan began his section between the Ashley River and Rantowles in June of 1856, and Robert H. Hunter of the firm Potter & Hunter supervised the work on the wharf and trestle on the Ashley. Additional grading and trestlework was contracted out to Luder F. Behling and F. Nims, and the track laying was supervised by the firm of Greig & Jones. Later that same year the state legislature authorized the endorsement of $510,000 of the railroad's 6 percent bonds, which were issued for purchasing iron and equipment.[12]

Grading and construction of the road were halted briefly with the sudden collapse of the U.S. economy in the summer of 1857. The Panic of 1857 was sparked by failure of the New York branch of the Ohio Life and Trust Company following a massive embezzlement. Confidence in the economy was further shaken when British investors began to pull funds out of U.S. banks and when a hurricane sank the SS *Central America* off the East Coast, sending thirty thousand pounds of U.S. gold to the ocean floor. A backlog of manufactured goods led to massive layoffs, and widespread railroad failures caused the collapse of land-speculation ventures, plunging thousands of investors into ruin. Leaders in the North blamed the crisis on protective tariffs and Southern slavery while their Southern counterparts argued that their labor pool was better than the "wage slaves" of the North.

Since the cotton trade was not as severely affected as some other sectors of the economy, the South was not as hard hit as other regions of the country, and work on the Charleston & Savannah Railroad was able to be resumed later in the year. By this time construction had been progressing so slowly that the company was running dangerously low on funds. In order to complete the railroad the company was granted permission on January 1, 1858, to issue up to one million dollars in bonds

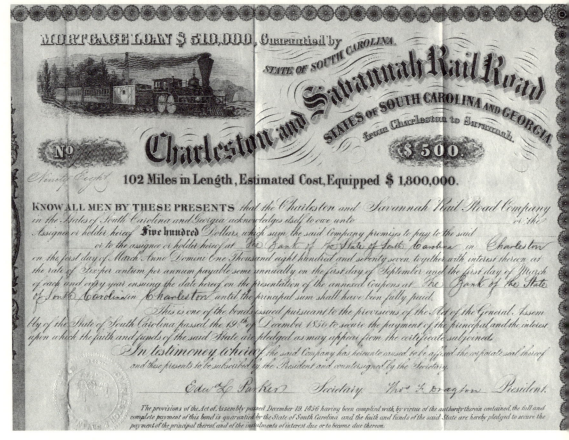

State guaranty bond for the Charleston & Savannah Railroad, 1856. Collection of the author

on a first mortgage (second lien), the interest on which would be paid semiannually at a rate of 7 percent. To secure the bonds an indenture was made, naming Isaac W. Hayne, Edward Sebring, and Erastus M. Beach trustees of the mortgage and deeding to them the railroad, its right-of-way, estates, rolling stock, and all other property owned by the company.[13]

Tired of the delays in construction, a member of the company's board of directors decided it was time for action. On December 24, 1857, John S. Ryan contracted for the remaining construction of the road from the Salkehatchie Bridge to the Savannah River except for one short stretch of track through the lands of William Henry Heyward, where Drayton had assigned the construction to Heyward himself. Ryan's contract provided for him to receive one-third of his payment in company stock at par value and two-thirds in the company's 7 percent first-mortgage bonds. One half of these bonds were to be redeemed by the company at par value one year after the completion of the contracts. Because of the company's unsteady financial condition, Ryan protected himself in the event the company folded. If he failed to complete the work for which he contracted, the only penalty stipulated by his contract was forfeiture of a portion of the stock payment. Ryan wasted little time in getting workers back on the line, and the railroad's wharf on the Ashley River was completed by the end of the month. W. S. Hudson, with whom Ryan had subcontracted, began work west of the Salkehatchie in February with a crew of fifty slaves hired from planters along the route. The company's first locomotive—fittingly named the *Ashley*—was put into service on March 29, 1858, carrying cross ties and rails for the tracklayers.[14]

In addition to their conflicts with local landowners, Hudson and the other railroad contractors found themselves in a fierce contest with nature, battling topography, weather, and disease. Grading of the segment of the railroad between the Edisto and Ashepoo rivers was delayed because of the perpetual sinking of the soft soil in the area. This was particularly evident between Adams Run and Jacksonboro, where the soil was composed of a type of clay that was tenacious to moisture. As a result the interior of the embankments in this area remained in a semifluid state, and drainage by any economical means would be impossible. The settling of the embankments would complicate track maintenance, but Drayton was confident he could alleviate the problem by gradually adding sand or gravel under the track, which should eventually solidify the ground and provide a stable base for the superstructure.[15] Wet weather plagued the workers that year, causing rice fields and other low-lying areas to remain under water for an exasperatingly long time, bringing much of the grading and earthwork to a complete halt. Flooding also made it difficult for crews to obtain sufficient earth for embankments and necessitated the construction of more trestlework than was originally intended.

The prevalence of yellow fever in the summer of 1856 severely depleted Phelan's workforce between the Ashley and Rantowles, but by November he had one hundred hands back on the line. The threat of the "sickly season" caused many Midlands planters to recall their slaves that same year, slowing the work of the Lartigues' crew between Rantowles and the Edisto. Outbreaks of malaria also hindered progress during the early phases of construction and likely would have done further damage if not for the expertise of Dr. Henry W. DeSaussure. He was called out to treat one of the contractors during work on the section of the road between the Ashepoo and Combahee rivers and gave the man quinine, at the time a revolutionary treatment for the ailment. Preventative treatment for the disease was in its infantile stage at that point, but Dr. DeSaussure advised that all railroad hands, white or black, should receive daily doses of quinine. Only one more case of malaria was reported during the remainder of construction.[16]

At the end of 1858 Drayton and the railroad's directors feared that potential investors in the Charleston & Savannah were still too apathetic. The newly appointed treasurer of the railroad, Edward L. Parker, reported in January 1859 that stock purchases were proceeding more slowly than expected, much slower than needed for the continued operation of the railroad. If the current stockholders could not coax others to purchase stock in the road and show more consistent support for operations the consequences could be dire.[17] The planters owned the land, the slave labor, and the lumber for cross ties. If they would part with them on reasonable terms, certainly the port cities of Charleston and Savannah had adequate monies to provide the iron, rolling stock, and any buildings and shops needed for the railroad. The lethargy of the stockholders and any further delays in construction would require the railroad to borrow more money and issue discounted bonds, which would make it more difficult to realize a profit even with generous receipts from freight and passengers. The delays would also retard completion of the planned seaboard line from Weldon, North Carolina, to the Gulf of Mexico. As long as a gap existed between Charleston and Savannah, commercial wealth and prestige could not and would not be drawn to the Carolina lowcountry.

A frustrated Drayton lamented in the company's annual report that "no matter what the circumstances, when a railroad is contending for a fair judgment of the right-of-way, the juries were too apt to lean to the idea that corporations are rich, careless, and insolent, and can afford to be found guilty and learn a lesson through the pocket of humility and propriety." Bickering with local landowners, prodding contractors, and raising capital were beginning to wear a little on Drayton. He had for quite some time neglected the operation of his plantation in Beaufort County and had not spent enough time with his wife, the former Elizabeth Pope, and their eight children. He expressed these sentiments in a letter to his longtime friend Jefferson Davis, maintaining his belief that railways would be the most efficient means for uniting the Southern states and preventing them from becoming overly dependent on the North. Drayton was, however, growing tired of working so hard "to serve a people who sternly frown down and oppose everything new, however faultless, useful, and beneficial, for no other reason than that its existence can not be traced back to their boyhood. They all want to ride, but not push."[18] Drayton was beginning to feel alone in his passion for the railroad and was beginning to wonder if he would have the patience and stamina to see it to its completion.

Later that summer Drayton was at least temporarily lifted out of his doldrums when the state came through with its sizable stock subscription. In December 1854 the state of South Carolina had pledged a $270,000 subscription in the railroad, contingent on the company's having sold $750,000 in stock. Because of the sluggish pace at which their stock was selling, they had been unable to receive this aid from the state. Finally, however, with the aid of large subscriptions by the City of Charleston and the Central Railroad and Banking Company of Georgia, the sale of the railroad's capital stock topped $750,000 on August 21, 1858. The state subsequently transferred $200,000 of stock in the Wilmington & Manchester Railroad, $50,000 of stock in the Kings Mountain Railroad, and $20,000 of stock in the South Carolina Railroad to the Charleston & Savannah. Although these assets were not immediately made available, it was thought that the company would receive them before construction of the road was completed. In the meantime the Charleston & Savannah maintained a substantial voting interest in the Wilmington & Manchester Railroad. With this advantage it would have a voice in electing directors friendly toward the Charleston & Savannah and could potentially quash any attempts by rival

Green Pond Drive station, near the Combahee River, sketch by Alfred R. Waud. Courtesy of Library of Congress, DRWG/US Waud No. 783

railroads to encroach on their business. Certainly the state's investment was welcome news, but it was news that Drayton had hoped to have heard much earlier.

John Ryan's difficulties continued. Late in 1859 W. S. Hudson abandoned his portion of the project, leaving Ryan liable for a large portion of unfinished track. The company agreed to save him from financial straits and paid him an additional Eighty-five thousand dollars to finish the job. To complicate matters William Henry Heyward had also been unable to fulfill his contract, further burdening Ryan and his subcontractors.[19] In need of additional hands, Ryan procured the services of Robert Legare Singletary, a native of Marion County, South Carolina, and Henry M. Drane, from Wilmington, North Carolina, who together had just completed the construction of the Fernandina & Cedar Keys Railroad in Florida. Drane and Singletary's fresh crew began work in October, and by November 29 the drawbridge over the Stono River was finished. They built a similar bridge across the Edisto, and by December 21 they had completed work on the second twenty-mile section of the road. When they assumed the contract in October, the railroad owned only two locomotives, two boxcars, and thirteen platform cars. Two more locomotives, the *Andrew Milne* and the *Stono*, were put into service in the latter part of 1858, and by January 19, 1859, there was a daily train carrying passengers and freight the twenty-nine miles between Jacksonboro and the Ashley River.[20]

Work progressed more efficiently under the direction of Drane and Singletary, and soon twenty-mile sections were being finished at a much faster pace. With the completion of each section, the state endorsed an additional one hundred thousand dollars of the company's bonds. Trees were felled; pilings were driven; and miles of track were laid with an energy not previously exhibited along the line. Miles of tidal swamps previously considered impervious were now bridged. Only eighteen months after Drane and Singletary were hired, the only barriers preventing a through connection between Charleston and Savannah were the rivers at the terminal ends of the road. The railroad's completion to the Combahee afforded Charles Heyward his first trip on the new railroad. He recorded in his diary on November 10, 1859, that he sent his carriage home from Charleston ahead of him, compelling him "to leave (his) door in a public carriage the first time in [his] life." Heyward's first trip took nearly four hours. The train was drawn by a "quaint little locomotive burning wood and with a smokestack nearly as large as the engine itself."[21] This mode of travel was a big change for Heyward, but he must have been pleased with it judging from his favorable dealings with the road.

By the spring of 1860 work had begun on the Savannah River Bridge. The bridge was to be similar to the bridges crossing the Pee Dee and Santee rivers, which were composed of a series of sunken iron cylinders. The Savannah bridge required two-and-one-half miles of trestlework on the Georgia bank and one-and-one-half miles on the South Carolina side. As originally proposed, a bridge across the Savannah River at Hutchinson Island would have required two separate bridges—one from the South Carolina bank to the island and the other from the island to Savannah—each requiring its own drawbridge. Instead, engineers chose a new site sixteen miles above the city, where the terrain was more favorable. An engineer with the Charleston & Savannah Railroad, M. P. Muller, supervised the work, with the able assistance of W. S. Smith, an engineer and consultant with the Trenton Locomotive and Machine Manufacturing Company. As work crews struggled to sink the iron cylinders into place, work was progressing well on the thirteen miles of the railway that eventually connected the line west of the Savannah River Bridge with the Central Railroad of Georgia three miles above the depot.[22]

Spurred by the added incentive of a twelve thousand dollar bonus, Drane and Singletary worked their hands at a furious pace, and on the afternoon of April 20, 1860, all the rails and trestles were in place between the Ashley and the Savannah rivers. The ceremonial spikes for the final rail were hammered into place midway between Grahamville and Hardeeville by Thomas Drayton and board members William Colcock, Lorenzo T. Potter, and Otis Mills. Drane and Singletary, who had overseen a major portion of the construction, were present that day and provided those present with a "handsome collation on the ground." A daily passenger service between St. Andrews Station and the Savannah River was put into operation the following day, but steamers were still be required at both termini because the Savannah River bridge remained unfinished and one was not yet planned for the Ashley.[23]

The first through trips between Charleston and the Savannah River occurred on April 21, 1860. Aboard the locomotive *Combahee* when it left St. Andrews Station at 7:40 A.M. were conductor John Strohecker and engineer Jesse Howell, who guided the train on its first of many trips to the Georgia port. When they arrived at the station near Grahamville, they met the train making its maiden trip from Savannah to

Charleston. After the groups exchanged congratulations, those aboard the *Combahee* left Grahamville, stopped at the place where the final spike had been driven the previous day, and christened the rails with a bottle of wine. After the brief ceremony the excursion continued on until it reached the wharf at the Savannah terminus. Here the passengers boarded the steamship *Teaser* to complete their journey into the city.[24]

The first train to Charleston departed Savannah at 9:00 A.M. and arrived at the railroad's wharf on the Ashley River at just before 6:30 P.M. As it approached St. Andrews Station, its riders cheered the state flags of Georgia and South Carolina and the Stars and Stripes suspended above the tracks. As the locomotive passed beneath the banners, the Marion Artillery rocked the air with a twenty-one-gun salute. Festivities in Charleston had begun at around noon that day, with salutes fired by many of the ships in Charleston harbor. The harbor was aglitter with the signals and colors from all the ships in port, and the air was filled with the music of Gilmore's Brass Band from Boston. One hundred and fifty passengers made the nine-hour trip that day, among whom were Thomas Drayton, Edward M. Manigault, and Savannah's delegates to the National Democratic Convention. After their ferry ride into Charleston, many of those who made the trip were entertained at a party given by John S. Ryan, whose entrepreneurial spirit had made the day's events possible.[25]

Daily passenger service was immediately established between the two cities, a train departed from Charleston at 7:40 A.M. and arrived at the Savannah River station at 4:35 P.M. while another train left Savannah at 9:00 A.M. and reached St. Andrews Station on the Ashley River at 6:20 P.M.— approximately nine hours of travel each way, including the time spent on sidings waiting for the train passing in the opposite direction. Nearly all the railroad's 104 miles traversed a belt of coastal pine forests intersected by large swamps, which bordered rather deep rivers. Though relatively new, the trestles and high wooden drawbridges were frail looking, and passengers held their breath as they crossed them for the first time. About every five or ten miles, locomotives stopped at a way station consisting of a high platform and a small station house. Some also had wood racks for replenishing the engine's fuel supply.

From the train the countryside looked almost uninhabited, and passengers were struck by the natural beauty of the woods in the spring, when various shades of green and black decorated the rivers and swamps. As described by Duncan Clinch Heyward, there was "so little sign of human habitation that it was hard to realize that only a few miles downriver lay many of the largest and most prosperous rice plantations in the world." During the day passengers often saw flocks of wild turkeys feeding in the open woods and deer crossing in front of the train. When the train ran through a pasture, cattle congregated on the tracks, especially on summer nights, and the engineer or fireman had to drive them off.[26]

In the beginning the railroad's chief engineer, Edward M. Manigault, restricted the speed of the locomotives because of the newness of the tracking, which had not established its strength and durability in its roadbed. A promotional fare of three dollars each way was in effect for the week of the Democratic Convention and the week following. The usual fare for passengers was three cents per mile. Freight was transported three days a week between Charleston and Grahamville as well as the stations in between. Packages and parcels could also be sent to Savannah and way stations via Adams' Express Company's southern express service, which had negotiated an arrangement with the railroad.[27]

Because the railroad was considered a domestic—that is, Southern—enterprise, it was finding more favor among Charlestonians. Such demonstrations of sectional loyalty were beginning to meld into an attitude that was later branded "Southern nationalism." Such loyalty did not refer to a particular style of governance but to an emotional bond, a sense of "Confederateness." Some scholars believe that the Confederacy functioned as a nation only in a technical, organizational sense and that an imperfectly formed sense of nationalism was the eventual undoing of its quest for independence. From this perspective only slavery gave the South its own identity, and many Southerners became Confederates not because they shared a sense of nationhood but because they feared a life without slavery.[28] This sentiment was gaining momentum as the Democratic delegates descended on the South Carolina Institute Hall in Charleston for their 1860 convention.

Despite the enthusiasm of the railroad celebration in Charleston that day, others were in a more somber mood. The roots of sectionalism ran deep in the soil of the Cotton States, and they were about to crack the weak foundation of the Democratic Party. A prelude to the fiery dogma about to be unleashed on the convention—and the entire nation—could be found in that day's *Charleston Mercury*: "Shall not the South be as united and firm in the maintenance of their rights, as the North is in denying and overthrowing them? Must justice always succumb to injustice—right to wrong? And is it consistent with the vaunted dignity of human nature, and of our far-famed free institutions, that the oppressed shall not only cling to the oppressor, but like captives behind the car of a Roman conqueror, minister to the pomp of his August triumph? The Charleston Convention may fill an important page in history. It may tell of the rebound of a great and free people in the maintenance of their rights, or of their final submission and downfall."[29]

Unity in the Democratic Party at that time was fragile at best; the situation had been exacerbated by Southern Democrats' opposition to the nomination of Stephen A. Douglas for president. The party's weakness was about to be exploited by extremists led by Senator William L. Yancey of Alabama and *Charleston Mercury* publisher Robert Barnwell Rhett. Yancey and his allies were looking forward to a floor fight over the party's platform—especially with regard to how to handle slavery in the territories. Douglas felt that it would be impossible for any Democratic candidate to get any Northern votes if the party platform favored forcing slavery upon a territory even if it were unwanted there. A few Southern leaders had convinced themselves that a slave code was essential to the preservation of Southern security, but for most the issue was simply a political maneuver to block Douglas's nomination.[30]

Fanning the flames for the secessionists was the mercurial newspaperman Rhett. In an editorial he called for the slave code to be adopted into the party platform and urged Southern delegates to withdraw from the convention and nominate their own candidate if this was not done. This was the position adopted by the Yancey contingent. Rhett even went as far as stating that if that candidate did not win election, "the next legislature should recall the state's members from Congress and invite the cooperation of other Southern states on the matters affecting the common safety."[31] Through establishing their alliances and inciting public opinion, the secessionists felt that they could define what Southerners would consider an unsatisfactory outcome in the Democratic convention, the national election, or both.

On April 27 the platform committee brought its reports to the convention floor. The core of each report was based on the party's platform of 1856; the assertion that Congress lacked authority to interfere with Southern slavery; endorsement of the Compromise of 1850, which allowed California to enter the union as a slave state while leaving open the question of slavery in the new territories of Utah and New Mexico; and affirmation of the doctrine of popular sovereignty, which asserted that the people in a territory should decide for themselves whether their territory should enter the Union as a slave state or a free state. Douglas supporters backed a minority report affirming that the Supreme Court should decide the issue of slavery in the territories. Instead of deferring this issue to the Supreme Court, the majority report, engineered by a coalition of cotton-producing states, declared that Congress had no power to abolish slavery in the territories and that the territorial legislatures could neither abolish nor interfere with slave ownership in the territories.[32]

By April 30 many of the out of town attendees had started to return home, and the galleries of the convention hall were populated by overzealous South Carolinians. Without much ado the delegates brought the platform issue to a much-anticipated —and by some a much-dreaded—vote. The Douglas platform was accepted by a vote of 165 to 138, and in a conciliatory gesture Douglas lieutenants offered to drop the deferment of the popular-sovereignty issue to the Supreme Court. The unmoving delegates in the Cotton State alliance, cheered on by the fire-breathing locals, announced their withdrawal from the convention. Later that evening dissenters led by an ecstatic Yancey announced the seceding delegates' formation of the Constitutional Democratic Convention. Those delegates who remained at the Democratic Convention tried to carry on the business of the party, but the damage inflicted by the split was evident. On May 3, after twenty-three unsuccessful ballots, the convention finally adjourned with plans to reconvene in Baltimore in June.

Writing of the Democratic Party's demise, Robert Barnwell Rhett prophetically editorialized: "The last party, pretending to be a national party, is broken up, and the antagonism of the two sections of the Union has nothing to arrest its fierce collisions."[33] The Baltimore convention resulted in the nomination of Douglas for president and Senator Benjamin Fitzpatrick of Alabama for vice president. The Constitutional Democratic Party—with a platform of Southern states' rights and protection of slavery—nominated John Breckinridge for president and Joseph Lane for vice president. The division of the Democratic Party made it much easier for the more unified Republicans, who nominated Abraham Lincoln for president and Hannibal Hamlin for vice president.

A month after the opening of the Charleston & Savannah Railroad—with rumblings from the Charleston convention beginning to die down—officials of the railroad could once again concentrate on the mundane issues of day-to-day operation. At the company's monthly meeting in May of 1860, the board of directors decided to consolidate the positions of superintendent and chief engineer, naming Edward M. Manigault to fill the new post. A Charleston native, Manigault had been educated at South Carolina College and had an extensive background in the classics as well as mathematics. He served in the army as an infantry officer during the Mexican War and later as an ordnance officer.[34] He had extensive knowledge of the railroad, having surveyed a large proportion of the line as an assistant engineer.

In addition to promoting Manigault, the board addressed the company's lack of available cash. They voted to issue another five hundred thousand dollars worth of equipment bonds to aid in finishing construction of depots, wharves, machine shops, and the bridge across the Savannah River. Also on the minds of several directors was clarifying the role they would play in the development of the Port Royal Railroad. This upstart railroad had been projected in January by a combination of early supporters of the Charleston & Savannah and some of the original Branchville & Savannah proponents from the Barnwell area. Apparently the division between the two groups was beginning to heal. The originators of the new railroad hoped to connect the cities of Beaufort, South Carolina, and Augusta, Georgia, intersecting the Charleston & Savannah Railroad adjacent to the Salkehatchie River at or near Pocotaligo. When subscription was opened, it was touted as a second connector between Charleston and the West. After having some time to consider it, the directors of the Charleston & Savannah Railroad voted later that summer to subscribe $150,000 to the new road, contingent on its completion. Of that amount the directors pledged $50,000 in the form of the loan of equipment and transportation of supplies during construction and the remaining $100,000 in stocks of the Charleston & Savannah Railroad. It would be dispersed at the rate of $18.75 per mile of road constructed in exchange for stock in the Port Royal Railroad.[35]

During the first week of August 1860 Drayton was invited to represent the directors at an organizational meeting of the Port Royal Railroad held at Barnwell Courthouse. There he was asked to address the attendees, who included some of the same men who had vehemently opposed his efforts to build his railroad and even the idea of building a branch line through their region. Tonight he and aid from his company were welcomed heartily. The conciliatory Drayton was met with great applause. He informed the projectors of the subscription package proposed by the Charleston & Savannah Railroad. Afterward he offered wisdom that he had gained while shepherding his own road to completion. His advice was simple. Raise as much money as possible early in the process, and build on the *most direct* line possible between Salkehatchie and Augusta.³⁶ When the delegates selected the officers of the Port Royal line, Drayton and William E. Martin were named to its board of directors. The Port Royal gained much needed financial assistance to begin construction, and the Charleston & Savannah gained an active voice in the future plans of the potentially profitable connector. The key word here was *potential*, since this road was not completed until after the Civil War.

At long last, on October 26, 1860, the final rail was laid on the track, and the final spike was driven to complete the connection between Charleston and the city of Savannah. While the line between the Savannah River and Charleston had been completed the previous spring, the final section from the river to the city, including the Savannah River bridge, took another six months. Finishing the Charleston & Savannah Railroad had taken much longer and was a much more expensive undertaking than had originally been expected. Both time and expense would have been far less had the company had cash instead of stocks and bonds to pay its contractors and had there not been such staunch resistance to the railroad both in the cities and in rural areas. The belief of many that the railroad was too risky and impractical had discouraged supporters of the railroad and gave the opponents excuses not to subscribe. The cost of the railroad at its completion had ballooned from the initial estimate of $1.75 million to just short of $3 million, a sum that hampered the railroad in the years to come.³⁷ Drayton had expected the cost to be high, but he winced at the final amount. The discrepancy between the estimated and actual costs was felt to have come from several unexpected expenditures. Also to blame were the excessive time required for the railroad's completion, the sale of discounted bonds, the high cost of right-of-way, and problems encountered at the terminal ends of the railroad.

Because it had taken almost twice as long to complete as initially expected, the railroad accrued extra engineering costs of approximately $24,000 and extra construction costs and overtime bounty for working around the clock of another $100,000. This made it necessary to raise additional cash by selling stocks and bonds owned by the corporation. These securities were either sold at a great discount or given to contractors at par value in exchange for work. The loss on the sale of the stocks and bonds alone came to $318,000, and the interest paid on those bonds added an additional $70,000.³⁸

The excessive cost of the right-of-way had been a problem for Drayton since the onset of grubbing and grading on the first section of the railroad. The first nine miles of the railroad, cut through the land of Thomas Grange Simons, had cost the company $3,746 per mile. Drayton balked at the excessive prices for land, citing as an example of what other railroads were paying for their land the Wilmington & Manchester's payment of only $8,500 for 171 miles of roadway. Even the private sale of land adjacent to the tracts bought by the Wilmington & Manchester brought lower prices. Eventually commissioners were needed to mediate some of the disputes between landowners and the Charleston & Savannah Railroad. The directors felt that even these arbitrations were weighted in favor of the landowners and appealed many of the judgments.

Even the landowners who had dealt favorably with the railroad desired concessions from the company. When the company obtained two and three-fourths miles of right-of-way through Charles Heyward's plantation in Colleton District, he parted with the land for a total of only five dollars. In exchange for this low rate the railroad agreed not to allow any freight trains to stop on any portion of track adjacent to Heyward's Calf-pen and Rogerson tracts. The railroad also allowed passenger trains to stop at the intersection of the tracks with the plantation's main avenue to pick up Heyward or members of his family—with sufficient notice, of course. When all its right-of-way was finally purchased, the railroad had paid more than $111,000 for 102 miles of track, an average of almost $1,100 per mile.[39] The decision to locate the Charleston end of the railroad west of the Ashley River had also cost the Charleston & Savannah dearly. When the company decided to terminate the road at St. Andrews, they purchased 206 acres of land from John H. Steinmeyer for $20,000. The wharf at the Steinmeyer terminus and the steamers for transporting passengers and freight into the city cost the railroad almost $90,000.[40] Whether or not this was actually an "extra" cost could be debated, however, since the railroad was avoiding the high cost of bridging the Ashley River.

The Savannah River also presented a problem at the close of construction. The chief engineer, Edward M. Manigault, had felt that the planned bridge over the Savannah River would take between one and a half and two years to construct completely. The iron caissons for the proposed cylinder bridge were shipped to Savannah from New Jersey in November of 1859, and their placement was fraught with difficulty. The company hired engineer C. C. Martin to place and sink the cylinders to specified heights in the silt beds of the river using the latest in vacuum technology—a long and tedious task. A work crew accidentally dropped one of the cylinders overboard and could not retrieve it. When contractors suspended construction of the bridge in July because of illness among the work crew, the directors opted to build a temporary bridge until the permanent structure could be completed.

The firm of McDowell and Callahan was contracted to furnish the material and labor necessary to complete the task in ninety days. When completed, the bridge spanned more than 1,800 feet, originating on the South Carolina bank with approximately one thousand feet of trestling and piling, which was followed by one 150-foot span of the standard Howe-truss design, a 190-foot pivot draw affording 80 feet of passage on either side, and 500 additional feet of trestling on the Georgia side. Under the watchful eyes of Manigault and Muller, the wooden structure was completed on schedule at a cost of $20,000. The permanent iron bridge was never brought to completion, and the iron cylinders were eventually melted down and used for ammunition by the Confederate army.[41] The total of the "extraordinary expenses" was $746,302. If this price is subtracted from the final cost of the railroad, the result is much closer to the average cost of most Southern railroads of the day—$21,000 per mile. Regardless of their hindsight, the directors found themselves on tenuous financial footing at the conclusion of construction. They went to the Charleston City Council requesting its assistance, either through endorsement or guaranty of the purchase of the railroad's bonds. They returned from their meeting empty-handed.[42]

THREE

A Confederate Railroad

In November 1860 the Republicans profited from the political conflict that split the Democratic Party. Republican candidate Abraham Lincoln was elected the nation's sixteenth president on the sixth. Yancey, Rhett, and their allies had achieved their goal of destroying the Democratic Party and moving the South closer to secession. News of Lincoln's election arrived in Charleston just days before the opening of the entire length of the Charleston & Savannah Railroad was to be celebrated there, heating the already simmering kettle of secessionist feeling to a full boil.

On All Saints Day, November 1, 1860, through rail communication had been opened between Charleston and Savannah. The *Charleston Daily Courier* hailed the figurative obliteration of the Savannah River from the map, stating that it existed "only as a convenient line of demarcation between neighbors and friends."[1] To observe the occasion the mayor and aldermen of Savannah had invited their Charleston counterparts to participate in an opening celebration in Savannah, but because November 1 was set aside as a day of thanksgiving in Charleston, the ceremony was postponed until the following day. On November 2 Thomas Fenwick Drayton, Charleston mayor Charles Macbeth, and eighty Charleston dignitaries boarded cars coupled behind the locomotive *Mayor Macbeth* for the grand opening trip to Savannah. As members of the delegation enjoyed bountiful refreshments and a few hours of lively discourse, the train sped easily along the older, more settled portions of the railroad, but the engineer eased it more cautiously over the more recently laid tracks at the western end of the line. It was not long before the train arrived at "the magnificent stretch of trestle-work" on the approach to the Savannah River, "thundered over the new bridge," and was soon "within sight of the massive masonry, which [marked] the railroad structures of Savannah."[2] A number of locals were milling about the depot at the road's junction with the facilities of the Georgia Central line, anticipating the train's appearance.

When it finally pulled into the station, Savannah mayor Charles Colcock Jones and the city council warmly greeted the Charleston delegation. After a few brief remarks, Jones led them in a procession to the Pulaski House, and the gentlemen spent the day touring Savannah or settling into their accommodations. Later that evening the Georgians treated their guests to a dinner at the hotel, during which the hosts toasted the railroad, the cities of Charleston and Savannah, and the uniting of the neighboring states. In addition to Jones and Macbeth, speakers that night included Thomas Fenwick Drayton, Alfred Huger, former South Carolina governor Robert F. W. Allston, and William Ferguson Colcock. In what was perhaps a foreshadowing of future events, Colcock posed to the audience the question of how Georgia would respond if Abraham Lincoln were elected on November 6. In response Captain Francis Stebbins Bartow of Savannah, hailed the union created by the railroad, observing that South Carolina and Georgia likewise were "united in heart, feeling, hopes, institution, by blood, by everything that can unite a people" and submitting that if the United States of America should be divided, "there [could] never be any division between Georgia and South Carolina." After several more speeches, the guests retired to their rooms at a late hour, exhausted as much from the protracted festivities as they were from the six-hour train ride that day.[3]

After a night's rest at the Pulaski House, the Charlestonians departed Savannah at ten o'clock the next morning. As they retraced the previous day's trek, they decided to host their Savannah counterparts at an equally festive gathering in Charleston the following week. The purpose of the gala was to celebrate the railroad, but for some something much more intriguing lay beneath the surface. The thoughts expressed by Captain Bartow the previous evening had drawn a rousing ovation, and with the presidential election approaching South Carolina separatists might soon put Georgia's loyalty to the test. A meeting honoring the railroad that brought together South Carolina hard-liners with like-minded Georgians might present such an occasion. They hoped, as *Mercury* editor Robert Barnwell Rhett, Jr., wrote, that the meeting would "strengthen the bonds of union between Georgia and South Carolina, and make Charleston and Savannah sister cities, indeed, in heart and purpose."[4]

The railroad celebration was set for November 9 in Charleston, and was planned by a special committee appointed by Mayor Macbeth and the Charleston City Council. The committee included councilmen—John S. Riggs, William Ravenel, and Dr. Robert Lebby—and three private citizens—Samuel Y. Tupper, Edward P. Milliken, and Robert N. Gourdin. Gourdin was a key member of the 1860 Association, a group of incendiary secessionists who met every Thursday in Charleston. The mission of this political organization was to shape public opinion and spur political activism by whatever means were at their disposal. They often sponsored political rallies and distributed pamphlets advocating the protection and preservation of slavery and Southern rights. The group's activities in conjunction with the opening of the railroad drove the state to the threshold of secession.

On Wednesday, November 7, 1860, news of Abraham Lincoln's election sparked intense excitement in the streets of Charleston. Crowds congregated in the business district as the *Mercury* staff posted news of the resignations of Federal officials such as Judge Andrew Gordon Magrath, U.S. District Attorney James Conner, and collector of the port of Charleston William Ferguson Colcock. Shouts rang out at noon when a states' rights flag was unfurled from the upper window of the *Mercury*'s office and stretched across Meeting Street. The sight of the red flag with the palmetto tree and lone star gave those assembled a sense of solidarity and state pride as the movement for Southern self-government gained momentum. Later that afternoon the 1860 Association met and resolved to assemble a large body of citizens at the Institute Hall to urge the state legislature to call for a state convention as soon as practical. They reserved the hall for the evening of November 9 and sent special invitations to Judge Magrath, James Conner, Daniel H. Hamilton, and William F. Colcock. As the cast was assembled for this rally, the city council's special committee busily choreographed the fete for the Charleston & Savannah Railroad.[5]

While the denizens of Charleston—and the rest of the slaveholding South—reacted to the election of Lincoln, Charleston & Savannah Railroad president Thomas Drayton was busy brokering an arms deal with the United States War Department. On November 5, 1860, South Carolina governor William H. Gist had authorized Drayton to purchase ten thousand rifles from Secretary of War John B. Floyd, and, if able, to attain from acting secretary of state William H. Trescot, a South Carolinian, any intelligence he could regarding the Buchanan administration. Drayton worked covertly to acquire the guns from the government, even enlisting the aid of a Southern-sympathizing New York financier to help disguise the purchase. His presence was required in Washington in due time, but his duties to the railroad kept him in Charleston through the gala at the end of the week.[6]

Meanwhile in Columbia, Governor Gist's office was a flurry of activity. Dispatches periodically announced the resignation of yet another South Carolinian from his Federal post, and unofficial communiqués from other Cotton State officials called for quick and decisive action by South Carolina. When the General Assembly had convened on November 5 to select South Carolina's delegates to the Electoral College, Gist had requested that they wait for the election results, and, if the voters elected Lincoln, pass legislation calling for a state convention. When the results were confirmed two days later, they obliged his request.

Most South Carolina legislators appeared to be of one accord toward the issue of the convention; however, there developed a division in the ranks as to the appropriate time for the meeting. Debate centered on calling an early convention versus delaying the convention in order to enlist the cooperation of other like-minded states. Radicals backed bills sponsored by Edmund Rhett and Robert Barnwell Rhett, Jr., which called for a convention as early as December 17. A coalition of moderates and conservatives supported George A. Trenholm's resolution, which was later modified to specify January 15 as the convention date. As night fell on November 8, it was the Trenholm resolution that seemed to garner the most favor. The Rhetts and their radical backers considered the later date unacceptable, fearing that it would dampen secessionist fervor and that other states might interpret the delay as a blunting of South Carolina's resolve. Never an opponent of separate action, Trenholm argued that his intent was to foster unity among the Charleston delegation and the entire legislature as well as to allow time "to gain the support of Georgia for the secession movement."⁷ Very soon, courtesy of a seven-hour train ride from Savannah to Charleston, both sides got their wish.

Before the festivities began on the afternoon of November 9, Major (later colonel) Peter F. Stevens, superintendent of the Citadel Academy, assembled a battalion line of cadets from the academy and the Washington Light Infantry and led them down Meeting Street on their way to Southern Wharf to greet the guests from Savannah. As they marched, they halted underneath the "States Rights Resistance" flag suspended from the offices of the *Charleston Mercury* and the State Bank. Major Stevens then announced, "By request of this command, and with my own hearty accord, I propose three cheers for Secession." Three lusty cheers rang through the streets, and they continued on to the wharf.⁸

Charleston harbor was again colored with signals, ensigns, national standards, and state flags displayed on every ship in port that day. A fifteen-gun salute by the Washington Artillery greeted the Savannah delegation as their train rumbled into St. Andrews Station shortly after 2:30 on the afternoon of November 9. A committee of the city council met the visiting dignitaries at the depot and escorted them to the railroad's wharf, where they boarded the steamer *Carolina* bound for Charleston. As they came down the Ashley River and into the harbor, the visitors received hearty salutes from many of the ships anchored there. After arriving at the wharf, the assembly, dampened by several rain showers that afternoon, proceeded up East Bay Street to Broad Street, past the railroad's offices, then onto Church, Chalmers, and Meeting streets to the Mills House hotel, where a banquet awaited them.⁹

That afternoon the room was abuzz with talk of Abraham Lincoln's election and of the resistance rally planned for later in the evening. News had also begun to circulate that earlier in the day Governor Gist had received a letter from Governor Joseph E. Brown of Georgia urging South Carolina to call an immediate state convention. After dinner, pleasantries were exchanged by Mayor Macbeth of Charleston and Mayor Jones of Savannah, and toasts were offered honoring the railroad and the union of the two cities. What happened next more closely resembled a political rally than the celebration of a railroad.

Mayor Jones was first to stir emotions when at the end of his remarks he stated that "the first rude blast which rustles your Palmetto, will vibrate among the sturdy oaks on our shore; and I tell you now that your Palmetto has been rustled, and that our oak is sending out its arms in a brawny defense." Several moments later Colonel Thomas Y. Simons toasted "the state of South Carolina: the night of her bondage has passed; the day of her deliverance has dawned."¹⁰ Edward P. Milliken summed up the effect and purpose of the meeting as he proposed a toast to "Georgia and South Carolina—Bound together by the tie of a common destiny; let no sacrilegious hand ever dare to break the links which unite them forever." Secessionist sentiment seemed to gain momentum each minute as glasses were raised in honor of "a Southern Confederacy," "the conservative Slave States of the South," and "a great Southern Republic"—provoking thunderous applause.¹¹ Further fanning the flames of secession, a number of Savannahians—including Captain Francis Stebbins Bartow, Henry R. Jackson, Samuel P. Hamilton, and Judge John M. Millen—responded to the toasts, expressing Georgia's solidarity with the Palmetto State and reaffirming their governor's call for firm resolve and prompt action. Enthusiasm coursed from the banquet hall to Institute Hall as the Georgians and their Carolina brethren ambled the short distance from the Mills House to join the public meeting that had assembled two blocks away.[12]

The meeting, a grand resistance rally organized by the 1860 Association, was one of the most imposing events Robert Barnwell Rhett, Jr., had ever chronicled in the pages of the *Mercury*. Hundreds of the city's most ardent separatists gathered at Institute Hall, which was "filled to every inch of its surface of floor and gallery." At seven o'clock John Henry Honour took his seat as chairman and called the meeting to order. After a slate of officers was elected and seated on the stage, Leonidas W. Spratt took the floor and presented two resolutions. The first resolution implored the General Assembly to call a state convention "to meet at the earliest possible moment, and sever our communication with the present government." The second ordered that copies of the first resolution be telegraphed to Charleston's legislative delegation in Columbia with the request that they be submitted to both houses of the legislature. With that "a wild storm seemed suddenly to sweep over the minds of men," and the revolutionary spirit that had taken hold of the railroad meeting soon permeated the hall. Next to take the stage was the popular judge Andrew Gordon Magrath, who eloquently affirmed his conviction that the time for deliberation had passed and the time for action had arrived. The 1860 Association could not have scripted it any better. The ovation that followed Magrath's remarks shook the rafters of the building, and the speakers who followed him were similarly received. Charlestonians Michael P. O'Connor, Thomas Y. Simons, and William Ferguson Colcock followed Magrath to the podium, and later in the evening, several of the guests from Georgia—Henry R. Jackson, Judge John M. Millen, Francis Stebbins Bartow, and Thomas M. Norwood—addressed the crowd. All were for immediate secession.[13]

The meeting of Charleston and Savannah citizens was one of the defining moments of the Southern secession movement. Its implications were so great that Robert Barnwell Rhett, Jr., later suggested that, had it not occurred, there may not have been a Southern Confederacy.[14] Whether or not it was, as Rhett described it, a "strange coincidence of circumstances," there is no denying its connection with the railroad gala with respect to its planning, its leadership, and its rhetoric. A prominent member of the 1860 Association, Robert N. Gourdin, helped plan and host the railroad gala and may have had a hand in its temporal and geographic proximity to the resistance rally. Of the fifty vice presidents elected at the rally, ten had past, present, or future ties to the Charleston & Savannah Railroad, including William Ferguson Colcock, Charles V. Chamberlain, John S. Riggs, William Porcher Miles, Thomas Y. Simons, Thomas Ryan, Edward McCrady, James H. Taylor, William C.

Bee, and John H. Steinmeyer. It was probably not coincidental that many of the speakers addressed both meetings, helping to transform a small strip of Meeting Street into the epicenter of disunion.

Energized by the resolve of its leadership and buoyed by Georgians' pledge of cooperation, the convention overwhelmingly passed Spratt's resolutions and appointed a special committee—Magrath, Colcock, and Conner—to travel to Columbia the next morning by special train to confer with Charleston's legislative delegation. At 10:30 that evening the committee sent a telegraph to the delegation in Columbia to inform them of the meeting's outcome and to announce their plans to leave for the capital city the next morning.[15]

The news spread quickly in Columbia, and by the time the train bearing the three emissaries arrived on the afternoon of November 10, a host of locals had gathered at the depot. Signs hailing "Magrath, Colcock, and Conner" and calling for "Immediate Separate State Action" decorated the city streets. The head of the Charleston delegation, Henry Buist, met them at the depot and escorted them to the Congaree Hotel, where they spent a few hours conferring with Charleston senators and representatives. Later that evening, Magrath, Colcock, and Conner addressed an impassioned gathering outside the hotel congratulating their fellow citizens "upon the great measure which had that day been so happily consummated."[16]

Inspired by the public sentiment in Charleston, the state legislature reconsidered its position. At the urging of the Charleston legislators, a House committee—which had been ready to present a compromise bill to the full House—amended a bill to call for a state convention on December 17, 1860, instead of the original date of January 15, 1861. The amendment sparked some debate from cooperationist upstate legislators, who were concerned that their constituents had not heard of the recent events in Charleston, but the promise of Georgia's support alleviated their fears. When called to a vote after its first reading, the bill passed by 91 to 14. After a second reading the vote was unanimous, 117 to 0. The Senate considered the bill during a special evening session, where it was passed without debate by a vote of 42 to 0. This vote was regarded at the time as South Carolina's decision to take the vanguard in the Southern secession movement. All that remained was for the citizens to stand by the actions of their representatives.[17]

After much political maneuvering by both unionists and secessionists, representatives from every district in the state convened at the First Baptist Church in Columbia on December 17, 1860. There was a concern that the secessionist juggernaut was losing steam. South Carolina seemed to be the only state prepared to take the initial step, and even some of her prominent political leaders—including Wade Hampton and Senator James H. Hammond—were less than enthralled with the idea. Columbia was not considered a secessionist stronghold, and many of the delegates preferred an environment more sympathetic to disunion. The size of the church and the rumor of an outbreak of smallpox provided the impetus the delegates needed to move the convention to Charleston.[18]

Courtesy of the South Carolina Railroad, the delegates traveled by way of a special train to Charleston on December 19 to continue their deliberation at Institute Hall. Their arrival was greeted by an enthusiastic crowd and a salute from the Marion Artillery. These men finished what had been set in motion in the Port City a mere eight months earlier. On December 20, 1860, South Carolina's Ordinance of Secession was adopted unanimously, and the fracturing of the Union was complete. While most cheered, a handful of intuitive South Carolinians could foresee difficult times ahead.

As many Southern political leaders were preparing speeches to lead their constituents toward secession, the railroads south of the Mason-Dixon Line had few established central, or trunk, lines. Many of the railroad companies were formed to serve as feeders to a particular waterway or as routes to local centers of trade. The Charleston & Savannah certainly fit into this mold as it was initially projected with designs on attracting cotton trade to the port of Charleston. Even in instances where promoters had attempted connections with the West (such as the Charleston & Savannah's attempt to join the Seaboard line), the roads were in actuality not much more than a combination of smaller roads with separate organizational and operational structures.

Despite the lack of a Southern system of rails, there were two routes running southwest to northeast that could in a loose sense be called the major railroads of the Confederacy. The first line began in Louisiana and traveled to Richmond, Virginia, via Corinth, Mississippi; Chattanooga, Tennessee; and Bristol, Tennessee. The second route was not quite complete in the first months of 1861; it ran from Pensacola, Florida, to Richmond through Montgomery, Alabama; Atlanta, Georgia; Augusta, Georgia; Wilmington, North Carolina; and Petersburg, Virginia.[19] The Charleston & Savannah Railroad was situated such that it helped to provide an alternative route to the east of Atlanta. If need be, a train could travel from Atlanta to Savannah via the Macon & Western and the Central Railroad of Georgia, then to Charleston via the Charleston & Savannah. After a steamer ride across the Ashley River to connect to the Northeastern Railroad, freight or passengers could travel to Florence, South Carolina. At Florence the Northeastern connected back to the main route by joining the Wilmington & Manchester and then continued toward Richmond.

While Northern roads were already taking advantage of interchange and cooperation, the advantages of through traffic were underemphasized in the South. Adjoining railroads often did not have connecting rails, and when they did, officials often refused to allow their rolling stock to travel on a neighboring company's tracks.[20] Freight, if traveling long distances, had to be unloaded and reloaded several times and at times required transport via wagon or boat between trains. At the Ashley River terminus of the Charleston & Savannah Railroad, goods had to be unloaded at the company's wharf, loaded onto a steamer, and transported to the city, where they had to be reloaded onto trains of the South Carolina Railroad or Northeastern Railroad for transport to other areas.

Southern Democratic politics usually dictated a strong resistance to any Federal investment in public works and internal improvement projects. South Carolina politicians balked at any potential Federal encroachment, not wanting to lose control of their agrarian economy. Neither did they want to grant any Northern interests a voice in the issue of slavery. State legislatures provided vital capital for the railroads and consequently had a significant amount of control over their management. This state involvement also had the effect of preventing the creation of a strong interstate railroad system. Railroads connecting one state to another were seen as potential conduits for the efflux of a state's economic resources. Railroads that stayed within their own state's borders and served their adjacent interests well—such as the Central of Georgia—thrived. Those that attempted a connection across several states—such as the Louisville, Cincinnati & Charleston—met with much resistance and ultimate failure.

A contrarian attitude also existed between neighboring railroads within the same state. Interline cooperation in the Confederacy existed mainly in the form of payments to adjacent railroad companies for trackage rights.[21] Occasionally railroads shared short stretches of track, such as the Charleston & Savannah's connection to the Central Railroad of Georgia several miles east of Savannah. This cooperation, more the exception than the rule in the secessionist South, was obtained for just over four thousand dollars per mile per year. That Thomas Drayton was able during the mid-1850s to promote and build a railroad marketed as part of an interstate seaboard railway connecting two major port cities in different states is a testament to his political dexterity. He had masterfully woven together a coalition of merchants, planters, South Carolina legislators, and Charleston city councilmen into the largest public works project seen in lowcountry South Carolina at that time.

Even though the South lacked resources for capital equipment, the railroads in the Confederate states were owned primarily by Southerners. Despite this local focus and Southern control, they did not refuse Northern capital. In many ways they were, in fact, dependent on it. The Charleston & Savannah contracted for more than two thousand tons of rails from the Lackawanna Iron Company of Pennsylvania. Spikes for the rails were obtained from the Troy Iron and Nail Foundry, and chairs for the rail connections came from J. B. Green and Company, both in New York. Even the locomotives of the road came from north of the Mason-Dixon Line—from the Rogers Locomotive and Machine Works in Paterson, New Jersey.

Southern planters were wealthy but often not enthusiastic about subscribing to the railroads, which could potentially benefit them so much. Many times, their failure to invest was because they lacked liquid assets. Because the planters possessed little cash and because the railroads needed investors, railroad promoters allowed many investments in Southern railroads to be made in kind. In several cases stock in the railroad was exchanged for land, cotton, and even slave labor for grading and grubbing the roadbeds. This policy was one of the contributing factors to the

Confederate railroads. From *Railroads of the Confederacy*, by Robert C. Black, © 1952 by the University of North Carolina Press; renewed 1980 by Robert C. Black III. Reproduced by permission of the publisher

NUMERICAL KEY TO RAILROADS

1. Baltimore & Ohio
2. Alexandria, Loudoun & Hampshire
3. Orange & Alexandria
4. Winchester & Potomac
5. Virginia Central
6. Richmond, Fredericksburg & Potomac
7. Richmond & York River
8. Richmond & Petersburg
9. Richmond & Danville
10. South Side
11. Norfolk & Petersburg
12. Petersburg R.R.
13. Seaboard & Roanoke
14. Virginia & Tennessee
15. Piedmont R.R.
16. Raleigh & Gaston
17. Roanoke Valley
18. Wilmington & Weldon
19. Atlantic & North Carolina
20. North Carolina
21. Western North Carolina
22. Western R.R.
23. Atlantic, Tennessee & Ohio
24. Wilmington, Charlotte & Rutherford
25. Wilmington & Manchester
26. Cheraw & Darlington
27. Charlotte & South Carolina
28. Kings Mountain
29. South Carolina R.R.
30. Greenville & Columbia
31. Spartanburg & Union
32. Laurens R.R.
33. Blue Ridge R.R.
34. Northeastern
35. Charleston & Savannah
36. Georgia R.R.
37. Augusta & Milledgeville
38. Western & Atlantic
39. Etowah R.R.
40. Rome R.R.
41. Central R.R. of Georgia
42. Macon & Western
43. Upson County
44. Macon & Brunswick
45. Southwestern R.R.
46. Muscogee R.R.
47. Augusta & Savannah
48. Savannah, Albany & Gulf
49. Atlantic & Gulf
50. Brunswick & Florida
51. Atlantic & West Point
52. Florida, Atlantic & Gulf Central
53. Florida R.R.
54. Pensacola & Georgia
55. Tallahassee R.R.
56. Alabama & Florida R.R. of Fla.
57. Alabama & Florida R.R. of Ala.
58. Montgomery & Eufaula
59. Montgomery & West Point
60. Tuskegee R.R.
61. Mobile & Girard
62. Mobile & Great Northern
63. Spring Hill R.R.
64. Mobile & Ohio
65. Mississippi, Gainesville & Tuscaloosa
66. Memphis & Charleston
67. Wills Valley
68. Nashville & Chattanooga
69. Winchester & Alabama
70. McMinnville & Manchester
71. Tennessee & Alabama
72. Nashville & Northwestern
73. Louisville & Nashville
74. Memphis, Clarksville & Louisville
75. Edgefield & Kentucky
76. East Tennessee & Georgia
77. East Tennessee & Virginia
78. Knoxville & Kentucky
79. Rogersville & Jefferson
80. Memphis & Ohio
81. Northeast & Southwest
82. Alabama & Mississippi Rivers

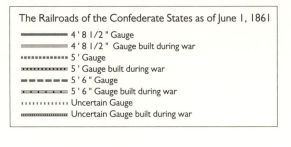

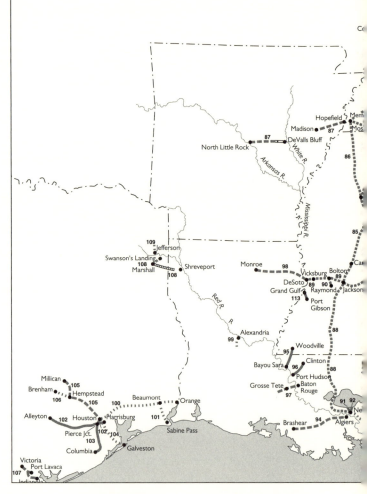

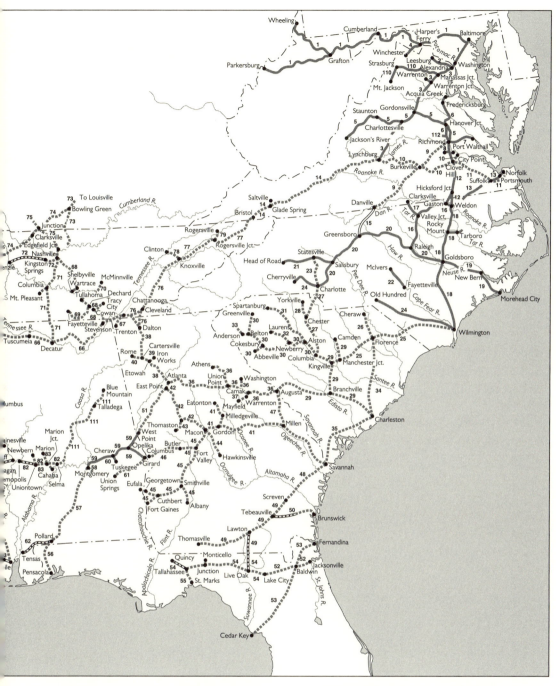

83. Cahaba, Marion & Greensboro
84. New Orleans & Ohio
85. Mississippi Central
86. Mississippi & Tennessee
87. Memphis & Little Rock
88. New Orleans, Jackson & Great Northern
89. Southern R.R. of Mississippi
90. Raymond R.R.
91. Jefferson & Lake Pontchartrain
92. Pontchartrain R.R.
93. Mexican Gulf R.R.
94. New Orleans, Opelousas & Great Western
95. West Feliciana R.R.
96. Clinton & Port Hudson
97. Baton Rouge, Grosse Tete & Opelousas
98. Vicksburg, Shreveport & Texas
99. Alexandria & Cheneyville
100. Texas & New Orleans
101. Eastern Texas R.R.
102. Buffalo Bayou, Brazos & Colorado
103. Houston Tap & Brazoria
104. Galveston, Houston & Henderson
105. Houston & Texas Central
106. Washington County R.R.
107. San Antonio & Mexican Gulf
108. Memphis, El Paso & Pacific
109. Southern Pacific
110. Manassas Gap
111. Alabama & Tennessee Rivers
112. Hungary Branch
113. Great Gulf & Port Gibson

burdensome debt experienced by many Confederate railroads, including the Charleston & Savannah.

By trading in kind the Charleston & Savannah, and many other Southern railroads of the day, could achieve two goals. The railroad gained investors and also acquired cheap labor for the construction of the road. As Scott Reynolds Nelson has stated, the "planters neatly folded the structure of plantation labor into the hierarchy of Southern railway construction."[22] From the banks of the Ashley and Stono to the flood plains of the Savannah, many lowcountry planters became some of the largest stockholders of the Charleston & Savannah Railroad.

In the early stages of the railroad movement, sectional sentiment—especially in South Carolina—kept many companies from requesting aid from the Federal government. That same philosophy did not exist with regard to public assistance from state governments. In a general sense this method of financing was a microcosm of Henry Clay's "American System."[23] It was based on the tenet that public agencies should lend aid to private business endeavors that were considered desirable for the public good. Though cash subsidies were occasionally offered, more often a state purchased stocks or bonds of a new carrier. Many times payment was made in securities owned by the state. Other ways in which states offered public assistance were by donating the costs of surveys, granting monopoly franchises, offering degrees of tax exemption, and awarding government contracts. By the beginning of the Civil War, South Carolina had extended roughly five million dollars in various sorts of aid to its dozen railroad companies.

In December of 1854 the state subscribed to the Charleston & Savannah Railroad $270,000 worth of stock of the South Carolina, Kings Mountain, and Wilmington & Manchester railroads. In December of 1856 the state legislature authorized the endorsement of $510,000 of the Charleston & Savannah's 6 percent bonds used for iron and equipment. The City of Charleston gave aid to the railroad in the form of a $260,000 cash subscription to help the directors raise the required $300,000 to obtain its charter. The city was also awarded $23,400 in company stock in exchange for surveys, contractors, right-of-way, and real estate. Of the remaining stock private citizens owned $390,700, and the Central Railroad and Banking Company of Georgia owned $75,000. More than half the company actually belonged to the State of South Carolina and the City of Charleston.[24]

Following the example of other Southern states, the government of South Carolina became an agent between railroads and both Northern and foreign investors by guarantying or endorsing railroad stocks and bonds to make them more marketable. This process was known as *hypothecation*. Railroad directors persuaded the government to buy railroad stock with state bonds or simply to place the state seal on railroad stock. The railroad companies could then sell these stocks—now converted into state bonds—abroad. The seal of a Southern state lent some solidity to these stocks, which allowed them to sell for more in foreign markets. This helped Southern railroads in their competition with their Northern counterparts for English capital. In exchange for this conversion, states obtained voting rights at stockholders' meetings and often liens on a portion of a railroad's capital equipment. They also took on a part of the risk since the states became the venture capitalists in the deal. Southern politicians who favored strong state governments disliked this practice. Men such as Senator James H. Hammond felt that South Carolina was mortgaging its good name and credibility as a state to English investors in the hope that state-financed railroads would succeed.

At the beginning of 1861—although the South could boast more railroad mileage than Great Britain—its 8,783 miles was only one-fifth the total mileage of the entire United States. Distribution of the rails between Texas and Washington, D.C., was uneven. Virginia led all Southern states with 1,800 miles, followed by Georgia with 1,400 miles of track. Despite its small area South Carolina contained 1,000 miles of railroad snaking through the lowcountry, Sandhills, and Piedmont. Variability was also found in the amount of capital investment made by the different states. Virginia led the South with an investment of $70 million. Tennessee, because of its difficult terrain, came next with an outlay of $31 million. Georgia, which had the second highest railroad mileage, had invested only $27 million. The average amount of capital investment in Southern railroads was easily less than $35,000 per mile, while three Northern states—Massachusetts, New York, and Pennsylvania—had invested more than $45,000 per mile.[25] Though the South had initiated a valiant attempt to capitalize on the benefits of the railroads, the effort was found wanting when compared to the better-organized and better-managed railroads of the North.

Because of their inferior construction, most Southern railroads of the time could sustain only light traffic. Heavy grading was avoided if at all possible, and most tracks were laid directly on thin earthen embankments. At certain points along the road, the clay-enriched soil used for embankments retained so much water that the constant addition of sand and gravel was required for preservation of the grade. The flat terrain between Charleston and Savannah helped to reduce the amount of grading and grubbing on the Charleston & Savannah line, but rivers, streams, and swamps hindered its construction and operation. Railroads situated along coastal plains required trestles and pile bridges to cross the many creeks, streams, and swamps. Bridges across larger rivers were mostly wooden trestles that were susceptible to floods and fire. Building such bridges also added to the considerable expense of railroad construction.[26]

The maintenance of rails and cross ties was also a source of consternation for railroad companies of the Confederacy. Pine, oak, and gum trees lined the Southern rails in abundance, but the companies with access to cypress trees had the best quality cross ties. Though they had some limited access to cypress from the bays south of Charleston, the Charleston & Savannah Railroad used "the best yellow or pitch pine timber" found along the line for their cross ties. Because the excessive heat and humidity in the South hastened the deterioration of the wood, cross ties lasted approximately five years there as compared with eight to ten years for ties in the more temperate North. The cross ties of the Charleston and Savannah were heavily weathered and ready for replacement by early 1863, a time when resources were needed in other areas.[27]

At the start of the War Between the States, Confederate rails were of the rolled, wrought iron "T" variety. They ranged in length from eighteen to twenty-four feet and in weight from thirty-five to sixty-eight pounds per yard. The rails used by the Charleston & Savannah Railroad were fifty-six pounds per yard and cost approximately fifty-five dollars per ton. They were not as strong as modern rails, but they could be rerolled and reused if necessary. Even with this advantage, heavier traffic on the rails caused severe wear on them in ten years or less. Because initial freight and passenger business was low, many antebellum railroad directors felt that the extra cost associated with dual tracking was not warranted. Single tracking and the relative lack of sidings made bidirectional traffic difficult—a difficulty that became more obvious during the conflict with the North.[28]

The locomotives used on the Charleston & Savannah Railroad were of the standard American design. The smoke box, cylinders, and cowcatcher sat upon a swivel truck of four wheels, and the other parts were carried on four connected drivers. The locomotives were very similar, so much so that they were often used interchangeably for both passenger and freight service. This was not the case, however, on the Charleston & Savannah. Different locomotives were used for passenger, freight, and roadway maintenance. Though they were similar in design, their parts were not always interchangeable. The engines of the Charleston & Savannah differed from each other with regard to cylinder diameter, stroke, and driver diameter. The majority of the engines had a thirteen-inch cylinder diameter and a twenty-two inch stroke. The four oldest engines on the road had a twelve-inch cylinder diameter, and two had a twenty-inch stroke. Half of the engines had drivers with a diameter of sixty inches, and the remainder had either forty-four or fifty-four inch drivers.[29]

Locomotives of the day were distinguished from each other by names rather than numbers. The names of states and geographical features were often used, as were the names of men of local or national importance. On the Charleston & Savannah many of the engines were named for the rivers and streams over which they traveled. The two roadway engines were dubbed *Ashley* and *Stono;* the *Edisto, Combahee,* and *Ashepoo* were freight engines; and two of the passenger engines were the *Coosawhatchie* and the *Isundiga*—the latter being the Cherokee name for the Savannah River. Those named after men of the day included the *Andrew Milne* and the *James Adger,* both freight engines, and the passenger engines *Thomas Rogers* and *Mayor Macbeth* (named after the mayor of Charleston). The remaining engine, the *Southward Ho,* was not put into service until March or April of 1861.

Though smaller and less well equipped, the rolling stock of the 1860s was in many ways similar in function to that of today. The rolling stock was constructed mostly of wood, with metal being used for the axles and wheels. The typical car weighed about four thousand pounds and had a load limit of sixteen thousand pounds. Most locomotives had an operating capacity of approximately three hundred thousand pounds; therefore, most trains of that day had a maximum of fifteen cars. The cars were differentiated according to specific function, but the terminology was still evolving.

Though most locomotives were built in the North, the Charleston & Savannah did have local access to other rolling stock. The railroad obtained cars before 1860 from the shops of Rikers & Lythgoe and later from the factory of Wharton & Petsch, both of which were in Charleston. It was a matter of considerable pride that Southern railroads could attain their cars in the South rather than from what they termed the "Abolitionist" companies of the North. At the beginning of 1861 the Charleston & Savannah had 9 passenger coaches, but 4 of them were under repair. Also traversing the rails between Charleston and Savannah were 3 express and baggage cars, 27 boxcars, 6 stock cars, 1 hoisting car, and 59 platform—or flat—cars. Three of these platform cars were converted into "dumping cars" transporting and dumping loads of gravel, dirt, and sand, among other things. The Charleston & Savannah also had 3 "conductor cars," the precursors of what would come to be known as cabooses. The 108 cars owned by the Charleston & Savannah Railroad were dwarfed in number by the 849 owned by the South Carolina Railroad, but neither could compare with the numbers owned by Northern railroads. The Delaware, Lackawanna & Western owned more than 4,000 cars, and even a more regional railroad, the Michigan Central, owned 2,500 cars.[30]

Locomotives required vast quantities of burned wood for fuel, and railroad companies typically obtained this wood via contracts with local landowners. The directors of the Charleston & Savannah Railroad had used some forethought, and by February of 1861, the road owned more than 6,700 acres of timberland lining the railroad, which they intended to use for fueling their locomotives and making cross ties. Some of the larger tracts the railroad purchased were plantations owned by the Heyward and Price families. Nathaniel Barnwell Heyward sold White Hall Plantation, a 723-acre tract north of the tracks and adjacent to Cuckold's Creek, to the railroad, which secured the mortgage with four $10,000 bonds. William Henry Heyward and Esther Heyward sold right-of-way through their plantation, Pynes, but stipulated that "the Duharra Avenue remain open at all times, and that construction of the railroad should injure the avenue as little as possible." Philip and Emma Price sold Llandovery, a plantation of between 600 and 800 acres, for $12,500, which the railroad was paying in installments.[31]

In addition to the wood used for fuel, the locomotives also required lubricants for efficient operation. Animal oils and tallow were used in large amounts because petroleum products were not yet available. The number of miles produced by burning a cord of wood varied widely from road to road and engine to engine. The engines of the Charleston & Savannah averaged almost 65 miles per cord in 1860, ranging from the older *Ashley*'s 38 per cord to the passenger train *Thomas Rogers*'s 100. On average a gallon of oil would last for 443 miles and a pound of tallow for

109 miles.³² The cost of these materials for the Charleston & Savannah line in 1860 was 3.6 cents per mile traveled. The efficiency of the engines usually depended on the roughness of the terrain over which they traveled and on the loads they were required to pull. Their efficiency decreased as the war effort took its toll and as lubricants became scarce.

The forward-thinking directors of the railroad had provided for the fueling of their engines, but they were very much dependent on the North for many other necessities. Whale oil came from New England, rails from Pennsylvania, railroad spikes from New York, and locomotives from New Jersey. Many Southerners failed to realize their dependence on the North not only for new production but also for the maintenance of the equipment they already had.

There were some places in the South where these items could be produced. Every railroad of significance had its own maintenance shops, which were capable of providing some new production, mostly of rolling stock. The Tredegar Iron Works in Richmond could produce locomotives as could the Nashville Manufacturing Company. The Forest City Foundry in Augusta, Georgia, made railroad-car castings, and the Atlanta Rolling Mill specialized in rolling rails. Even Charleston's Phoenix Iron Works, which had contracted with the Charleston & Savannah Railroad for construction of its turntables and drawbridges, could potentially be called on. Together, though, these companies still did not possess the manufacturing capacity to service the railroad and military needs of the Confederacy.³³

While the South was lacking in its manufacturing ability, one thing it did not lack was cotton. With the invention of the cotton gin, the short-staple cotton grown throughout the South became a worldwide commodity. In 1800 the United States exported $5 million worth of cotton, which accounted for 7 percent of U.S. exports. By 1860 cotton exportation totaled $191 million—an amazing 57 percent. More than three-fourths of the cotton used in the textile industry in England and France was imported from the ports of the American South. The explosion in worldwide demand for cotton led South Carolina senator James Henry Hammond to state in 1858, "Cotton is king." The primary function of many antebellum railroads was to transport cotton to major commercial centers. The Charleston & Savannah was projected to attract King Cotton to some extent to Savannah but primarily to Charleston for factoring and export.³⁴

At the beginning of the war Southerners held the naive belief that cotton would be their salvation. Cotton planters felt that a military victory was only the first step in gaining political independence from the North. Many Southerners believed that the money brought in by an eventually bustling cotton trade would finance the rise of home manufacturing and local industry, which would demonstrate the South's independence. The Confederacy needed recognition as a legitimate sovereignty, and it needed monetary aid from Europe. Consequently Southerners pinned their hopes on cotton. Because England and France were so dependent on it for their textile industries, the Confederate government thought those nations would be forced to recognize the Confederacy and break the Federal blockade. President Davis had a groundswell of public support favoring an embargo on cotton, which many believed would cause a cotton shortage in Europe, forcing England and France into action. What it did was keep the cotton in Dixie and prevent the planters from seeing much profit from their labors.

Cotton factors remained optimistic despite the fact that they had to suspend payments and refuse advances to planters during the secession crisis. They believed that any conflict would be brief and that foreign investors would soon replace their Northern ones. Business should resume on schedule after a short period of adjustment. They managed to sell off the 1860 crop before hostilities escalated and were able to pay their debts to Northern creditors. By this time, though, the planters had another crop in the field, and they were worried about how they would finance it.

Cotton planters were not the only ones in need of cash. Sale of Confederate treasury bonds was beginning to slow, and the Confederate government needed funds to sustain the war effort. The cotton that was being hoarded on Southern plantations could not be used to provide food for soldiers or purchase manufactured goods that had heretofore been imported. Cotton was the foundation of the Southern economy, and without the proceeds from its sale the government would have no basis for its currency. To alleviate this problem the Confederate government proposed a policy whereby the cotton on hand would be used as the basis of a circulating currency.

In May of 1861 the Confederate Congress passed the first of three produce loans, which were designed to provide the government with money. Planters pledged the proceeds from the sale of their agricultural products—such as cotton, rice, tobacco, and wheat—in exchange for 8 percent bonds. Prodigious quantities of cotton and other produce were pledged throughout the South that first summer. Planters in Beaufort, South Carolina, also made liberal subscriptions following an address by Robert W. Barnwell in June, but such displays of patriotism could not change the reality that without a market for their produce, there would be no proceeds. To counter this, Congress amended the law in April of 1862 to allow planters to transfer the produce directly to the government in exchange for 8 percent bonds. Government-owned cotton would provide the Confederacy a much-needed vehicle for purchases abroad and could serve as collateral for foreign loans. Confederate officials hoped that international demand for cotton would compel foreign powers to break the Federal blockade in order to obtain the precious commodity.[35]

Despite initial interest the Confederate government profited very little from produce loans. Although the government was able to acquire more than four hundred thousand bales of the "white gold," much of the cotton in the South remained on the plantations. Many planters chose to ply their cotton in the North or other foreign markets, where they could sell it at better prices. A small fraction found its way through the blockade, thus providing a source of revenue for the Confederacy; however, much of the remainder was eventually captured, stolen, or burned by the

Confederate army to keep it out of enemy hands. In the end the Confederacy realized only about thirty-four million dollars from the loans.

The attempt to deprive Europe of cotton kept the 1861 crop at home long enough for the Union to strengthen its blockade. Since there were no Northern or foreign buyers in the Southern markets and the blockade was preventing further exports, there was no real reason to ship cotton to commercial centers such as Charleston or Savannah. Many planters were directed not to ship any of their cotton to the cities because excessive quantities might create a fire hazard. Some felt that large quantities of cotton in cities would be an enticement for Union invasions of those cities to capture the cotton. The planters and cotton factors were also paying attention to market force, looking toward an early end to the blockade. When it ended, if there were an excessive supply of cotton, prices would plummet, and the planters would lose their profits. The hesitation and uncertainty of the planters and factors during the spring and summer of 1861 made the Union blockade effective even before the Federal navy had enough ships in place to put a choke hold on the South Carolina coast.[36]

Because of cotton's dominance of Southern agricultural efforts, planters had devoted few resources to growing foodstuffs. A year into the war, the food shortage in the South reached a crisis point, and a more concerted effort was made to substitute the production of corn in place of cotton. Starting in 1862, the Confederate Congress urged the voluntary reduction of cotton production, but many state governments went as far as levying punitive taxes on growers whose cotton crops exceeded a specified limit. These measures, in combination with the blockade and the embargo on sales to the North, resulted in a marked drop in cotton production. After growing 4.5 million bales in 1861, Southern growers produced only 1.5 million bales in 1862. This number dwindled to 450,000 bales in 1863 and to 300,000 in 1864. The restriction of cotton played havoc with the Southern railroads, many of whose tracks had been laid to bring the precious bales to market. Until the spring of 1861 many trainloads of cotton had been shipped over Southern rails. In the winter of 1859 more than 400,000 bales had been handled on the Georgia Central Railroad. In 1860 the South Carolina Railroad had delivered 315,000 bales to its Charleston depot. With the outbreak of hostilities and the restriction of cotton exports, railroads relying heavily on cotton revenues saw a 40 to 70 percent decline in their business. Anticipating the Federal blockade, some planters in Tennessee shipped all the cotton they owned southward. With the crisis in Charleston, much of this was diverted to roads supplying Savannah.[37]

Cotton, which in November 1860 cost fifteen cents per hundred pounds to ship, had been driving the rail business.[38] Revenues from transporting cotton had fired its engines, paid its salaries, and maintained its equipment. Now demand for rail shipment was withering. There were other goods to be shipped on the Charleston & Savannah—wool, hides, paper, lard, forage, leather, tobacco, and forestry products —but none could take the place of King Cotton. Charleston and its feeder railroads

Schedule of the Charleston & Savannah Railroad. From the *Charleston Daily Courier*, April 16, 1861

had to wait on the fat profits they had envisioned arriving from Southern cotton plantations.

In its earliest phases of operation the Charleston & Savannah—like most Confederate railroads—had to live or die by its freight business. Revenues from freight transport generally exceeded those from passenger travel. The rate structure was quite informal. A complicated system of tariffs existed, and rates varied depending on the class and type of goods being transported. Interline tariffs were established with very little structure, a situation that was not unique to the South. These rates were often lowered to attract traffic from one center of trade to another. Passenger fares were less volatile and relatively expensive, averaging about 4.5 cents per mile.[39]

The receipts of the Charleston & Savannah Railroad were counter to the trend reported by most of the other Confederate railways in the days before secession. Revenues received from passenger service outpaced those of freight transport by a three-to-two margin. The completion of the road had the misfortune of coinciding with the election of 1860 and growing Southern concern with the impending "revolution." The directors felt that the receipts for through business would increase the overall income of the road. Indeed, in the first month of 1861 alone, through freight business on the line was six times higher than the total for the previous six months combined. Drayton was cautiously optimistic, but because of high interest payments on the debt accrued by the company, he still projected a loss for the year based on the January figures. Whether or not the deficit could be made up by increased passenger and freight business would depend on how the secession crisis was handled. A peaceful resolution would bring about "a new career of commercial prosperity that would give an impetus to our travel and transportation which would in a few years release us from all pecuniary embarrassments."[40]

Clearly the board of directors and stockholders of the Charleston & Savannah were at a crossroads. Owing to the extraordinary expenses of the road's construction, large floating debt, and high bonded indebtedness, the company defaulted on the interest due on their 7 percent bonds in January of 1861. In dire financial straits the road needed an influx of business to have even a slim hope of breaking even for the year.

Many of the railroad's directors were fire-eating secessionists. William F. Colcock and Edmund Rhett were part of a group known as the "Bluffton Boys," who had espoused disunion for many years. Henry Gourdin and his brother Robert had formed a secret society that circulated secessionist pamphlets. The tide in South Carolina was against preserving the Union, and the Charleston & Savannah Railroad was rolling with it.

The directors of the Charleston & Savannah Railroad now faced several harsh political and economic dilemmas. Preserving the South's relationship with the North meant having a supplier of capital equipment and replacement parts for the railroad. War meant the loss of the Northern states as a market for Carolina cotton. With a united country travelers from the North would provide passenger business over the rails, guests for Charleston and Savannah hotels, and customers for local merchants. A military conflict with the North would all but dash the plans for the proposed seaboard route from the Northeast to the Gulf Coast. With the survival of the railroad and the future of South Carolina at stake, the directors had mixed feelings while pressing questions loomed. The South Carolina convention had settled the question of union versus disunion. Many now wondered if there would be war or peace. Riding the emotional wave that inspired secession, most South Carolinians knew the answer. That answer was war.

FOUR

Opening Guns

At the time of South Carolina's secession there remained four Federally occupied installations in and around Charleston harbor: Fort Moultrie on Sullivan's Island, Castle Pinckney, Fort Sumter, and the U.S. Arsenal. Major Robert Anderson, a native Kentuckian who was familiar with Southerners and their doctrine, was transferred to command Fort Moultrie in the latter part of 1860. He realized the fragility of his position and wanted to avoid conflict with the South Carolinians. His orders were to hold possession of the harbor and to defend his position to the last extremity; however, his small force could not effectively occupy all three forts. His superiors advised him to move his men to the fort with the best defensive advantage.

As tensions mounted in and around Charleston, Governor Francis W. Pickens ordered guard boats to patrol the water between Sullivan's Island and Fort Sumter with orders to prevent the Federal occupation of Sumter. State militia leaders were busy on the upper end of Sullivan's Island supervising the construction of batteries. When the Palmetto State finally severed her ties with the Union on December 20, Major Anderson was faced with a difficult decision. Initially leading everyone to believe he would defend Fort Moultrie, he evacuated his entire command from Moultrie to Fort Sumter on December 25–26, along with provisions for four months. According to South Carolinians William Porcher Miles and Lawrence Keitt, this action violated the "gentleman's understanding" they had with President James Buchanan that "their relative military status would remain the same."[1] When the Federal occupation of Sumter was discovered, a fuming Governor Pickens began negotiations with the major.

Pickens wanted the Union commander to transfer his troops back to Fort Moultrie, but he refused. Pickens ordered South Carolina troops—for there did not yet exist a Confederate army—to begin seizing Federal property. The installations fell quickly, as the militia took Castle Pinckney, Fort Moultrie, and finally the Federal arsenal in the city of Charleston. Anderson and his men were surrounded and cut off from munitions and other essential provisions. The only way that the lame duck President Buchanan could resupply Anderson was by water; a fact that was all too obvious to the leaders on both sides. The U.S. military secretly chartered a steamer,

Star of the West, which was loaded with men and supplies to reinforce the garrison at Fort Sumter.²

On January 9, 1861, after the ship crossed the bar of Charleston harbor, a squadron of cadets from the Citadel unabashedly fired into her—the South's first shots against the American flag. After Southern guns at Fort Moultrie joined the fight, the ship was driven off. Governor Pickens then decided not only to discourage but to prevent any more deep-draft military ships from bringing in supplies or reinforcements to the fort. He had four retired ships brought from Savannah to Charleston harbor to be loaded with granite and sunk in the main shipping channels. The only channel that remained unobstructed ran parallel to Sullivan's Island, and a vessel loaded with stones was anchored there to be sunk if necessary. Because of the obstructions in the harbor, many larger merchant vessels were diverted to Savannah, seriously reducing the trade coming into the Port City. The railroad connecting the two cities was now more important than ever before.

In his inaugural address on March 4, 1861, President Lincoln asserted that he was going to do everything in his power to "hold, occupy, and possess" all Federal property lying within Confederate territory. These were strong words, but many Southerners felt the new American president would not be willing to back them up. Indeed, not much happened toward this end in the weeks following his inauguration.³ Meanwhile, South Carolina state troops constructed new batteries in and around Charleston harbor and also strengthened preexisting ones. The governor had his officers ready to lead a strike on the fort if necessary.

In early February, the Convention of Southern States had been convened in Montgomery, Alabama. The South Carolina delegation included two former directors of the Charleston & Savannah Railroad: Christopher G. Memminger and William Porcher Miles. These men advised Governor Pickens that the majority opinion at the convention was against an immediate attack on Fort Sumter. Soon afterward, he was notified that the convention had adopted a constitution for a provisional government and had elected Jefferson Davis president.⁴ The decision to attack Fort Sumter was no longer Pickens's to make. President Davis dispatched General Pierre G. T. Beauregard to inspect the Confederate battlements and make them ready for combat. On March 29, Beauregard received orders instructing him to allow no further communication between Washington and Fort Sumter unless they were first submitted to Beauregard. This served to isolate Major Anderson further from his government.

On that same day President Lincoln informed his secretary of war, Simon Cameron, that he intended to send a fleet to reinforce Fort Sumter. Though his rations were quite low, Major Anderson was instructed to hold out until the reinforcements and provisions arrived.⁵ This surprised Anderson, since rumor had it that Lincoln planned to evacuate Anderson's small force to another location. Even though he was certain the Confederates would fire on the Federal reinforcements, Anderson told his superiors that he intended to stand his post. Hoping to avoid another *Star of the West* debacle, the Federal government officially notified Governor Pickens that they did, indeed, intend to resupply Fort Sumter, and, if they met with resistance, they would reinforce these troops as well. Confederate president Davis was now faced with a dilemma. Would he allow the fort to be resupplied? If not, how could he prevent it? His decision may have come down to his Southern pride. How could the fledgling nation allow the American flag to fly in the midst of one of the most celebrated port cities in the South?

Realizing that further negotiation would gain nothing, President Davis ordered General Beauregard to make a final demand for the surrender of the fort. When Beauregard demanded the evacuation of the fort, Anderson replied that he would be forced to remove his small garrison by April 15 unless he received additional supplies. Knowing full well that Lincoln intended to resupply his troops, Davis felt he was left with no choice but to prevent it by force. Hence, the War Between the States—the American Civil War—began at 4:30 A.M. on April 12, 1861, when Colonel James Chesnut ordered the mortar battery at Fort Johnson to open fire on Fort Sumter. As the shells arced toward the fort, Colonel Chesnut's wife, Mary Boykin Chesnut, watched from a rooftop in the city, as did many other ladies and gentlemen of Charleston. In her now-famous diary, Mary Chesnut wrote of the scene that night: "The women were wild up there on the housetop. Prayers came from the women and imprecations from the men. And then a shell would light up the scene." She expressed surprise the next day that "after all that noise and our tears and prayers, nobody has been hurt," commenting later that it had been "sound and fury signifying nothing—a delusion and a snare."[6]

At his Oak Lawn Plantation, Colonel Ambrosio José Gonzales was awakened by the thundering of the batteries. He threw on his military uniform, strapped on his sword, and hurriedly rode to the Charleston & Savannah Railroad depot at Adams Run, where he boarded a train for the twenty-six mile journey to Charleston. The following day, General Beauregard appointed the Cuban colonel as one of his volunteer aides-de-camp. Over the next four years, Gonzales logged hundreds of miles on the railroad between Charleston and Savannah in his capacity with the South Carolina Siege Train. To him and the many others who pledged themselves in service to state and country, this event would be neither a delusion nor a snare.

The War Between the States was the first military conflict in which railroads played an important role. Essential for transporting troops and supplies and indispensable lines of communication, railroads became the objectives of many military operations. Their benefits were realized early in the Southern war effort, but mostly at the local level. Because the men in power were wed to the theory of states' rights, they allowed local proprietary interests to take precedence in certain enterprises, apparently including the operation of the railroads. Despite their political philosophy, the reality was that the Confederacy needed a centrally coordinated effort to maximize the potential large-scale advantages of the rails.

Although the railroads were of obvious import to the military, the first informal attempt at organizing the Southern rail effort was made by Confederate postmaster general John Henninger Reagan—his motivation being the delivery of the daily mail. Despite the secession and even the initiation of hostilities, the U.S. Postal Service was still delivering mail to the Southern states. Reagan wanted to make provisions for a Southern mail system before the current system was discontinued. Shortly after the bombardment of Fort Sumter, Reagan called for a meeting of Southern railroad officials. When military officials learned of the convention, they also wanted to participate.

Thomas Drayton and the presidents of most of the major secessionist railroads met in Montgomery, Alabama, on April 26. The military interests in attendance availed themselves of the opportunity to exert their influence and quickly made their issues the convention's top priority. The first item on the agenda dealt with regulating passage of troops and supplies. The resolution, endorsed by Leroy P. Walker, the Confederate secretary of war, had three parts. First it suggested that troops be transported at the rate of two cents per mile and that military freight be carried at half the regular local fare; however, appropriate authorization from the quartermaster general or other governmental official would be required. Next it demanded that railroad companies accept payment in Confederate bonds or treasury notes at par if cash were not available. Finally, it gave the quartermaster general the responsibility of designating the class of certificate to be used.[7] Riding a wave of secessionist patriotism, the railroaders approved the plan unanimously.

Also approved at the convention was Reagan's strategy for carrying the mail. His plan called for the railroads to be divided into three classes with reimbursement rendered accordingly. The larger through routes carrying heavy mail and connecting important points would be reimbursed at $150 per mile annually. Completed roads carrying moderately heavy mail loads would be reimbursed at $100 per mile per year, and unfinished or unimportant roads would be reimbursed $50 per mile annually. The Provisional Congress soon enacted the essence of these proposals into law.[8]

Shortly thereafter, another meeting of Southern railroad leaders was called by Samuel Tate, president of the Memphis & Charleston Railroad, and Walter Goodman, president of the Mississippi Central. This convention met in Chattanooga, Tennessee, on June 4 and dealt mostly with nongovernmental matters. The assembly quickly affirmed Reagan's plan for postal routes and agreed to accept Confederate treasury notes at par for all transportation bills. Several overlapping companies arranged for a through passenger schedule between New Orleans and Richmond via Jackson, Grand Junction, Chattanooga, and Bristol. The delegation also attempted to establish classification of nongovernment freight and common tariffs. The last item called for the rail companies to patronize Southern iron mills over those outside the boundaries of Dixie.[9]

During the first three months of the war the South used locomotive power for the first time to concentrate troops. As Union troops led by General Irvin McDowell marched toward Beauregard at Manassas, Virginia, a large force of Confederates under General Joseph Johnston were moved from Piedmont, Virginia, to Manassas

Junction via the Manassas Gap Railroad. Two full brigades, one of which was commanded by General Thomas J. "Stonewall" Jackson, and part of a third were able to reinforce Beauregard at a critical moment in the battle. Even though it was not a perfect operation, the fresh troops eventually turned the day for the Rebels. General Robert Edward Lee also noted the success in moving troops by rail and filed his observations away for later use.

At this time the Charleston & Savannah Railroad was already involved in serving South Carolina's armed forces. Newly organized regiments traveled by rail to camps of instruction situated near the railroad, while others took the cars to Charleston and points north. Locomotives transported carloads of slaves from lowcountry plantations to Charleston to help fortify the harbor and construct new batteries. The military and railroad leaders were pleased by the rapidity with which supplies such as powder, ammunition, and guns were being transported—not only between Charleston and Savannah but also to other areas in Georgia and to Florida. Because of the obstructions in Charleston harbor, the railroad was also crucial for bringing meat, flour, and other staple goods to the citizens of Charleston.

The increase in military traffic—as well as the reduction of shipping into and out of the port of Charleston—had a noticeable effect on the routine operation of the Charleston & Savannah line. Before the conflict began, two-way freight service operated between the two cities on Monday, Wednesday, and Friday. By the first week of July, freight was transported from Charleston to Savannah only on Monday and Thursday, and from Savannah to Charleston on Tuesday and Friday. The reduction in freight services—coupled with what some considered prohibitive tariffs on freight—resulted in what some were calling a "land blockade." Essential goods were not being brought into the lowcountry, and abundant crops of rice were not making it to western markets for sale. As a result commodities such as coffee, sugar, and corn were selling anywhere from 40 to 60 percent higher in Charleston than in New Orleans and Memphis while rice was selling for at least 70 percent lower in Charleston than in other those two centers of trade. The planters of Charleston, Colleton, and Beaufort were making less for their produce and paying considerably more for their household staples. They argued that a lowering of freight rates would actually increase the railroad's business and bring more money into the area. Despite the validity of this argument, the railroad apparently could not justify such a sacrifice during uncertain times. The freight rates did not change.[10]

Passenger trains, which represented most of the railroad's source of revenue at that time, remained on a daily schedule. The average person could still purchase a copy of his or her favorite newspaper from J. M. Quinn, and read it leisurely as the train made its way across the "monotonous rice swamps" and forests of pines, magnolias, and live oaks from Charleston to one of the eleven way stations, or on the entire five hour and thirty minute trip to Savannah. In the first weeks of the war, passengers on the railroad were described by London news correspondent William Russell as "full of politics—the pretty women being the fiercest of the all, . . . [verbalizing] the most unfeminine expressions toward the Yankees."[11] Despite the existence of a state of war, however, some travelers on the railway seemed relatively unconcerned, as is evident in a casual entry in Samuel Burges's diary: "Left on C&SRR boat at 9:15 A.M. Crossed Ashley [River] then by cars to Jacksonboro, then by stage to Walterboro, which place reached about 3:30 P.M. Put up at Blount & Simpson. Found some [men] busy getting up a new company for service. Quite warm."[12]

Thomas Drayton was serving his last months as the head of the Charleston & Savannah road. Having been a military man, he knew the railroad would face hardships if war resulted, but he did not think the Charleston & Savannah would fare any worse than other Southern railroads. Drayton already had a vision of the railroad's ability to concentrate troops at any threatened point along the line. He even prophesied in the company's annual report that if South Carolina's secession resulted in war, "this much neglected railway will be the cheapest and most formidable earthwork that could have been devised to give confidence and security at home and repel invasion from abroad."[13]

He expressed this confidence at a time when the railroad faced a critical financial situation. The company had defaulted on the interest on its 7 percent bonds as of January 1, 1861. Drayton blamed the company's large indebtedness on the extreme expense of completing the road. Even so, the board of directors maintained that the company would not have been in such poor condition "had we not ended our Road on the eve of a great Revolution, and amidst an almost universal stagnation of business." They felt that with the return of peace and a "firm establishment of the Confederate States of America, the road will work out its own deliverance."[14] Drayton had appealed to the Charleston City Council the previous November to lend its aid to the company by endorsing four hundred thousand dollars of the railroad's bonds. He hoped that the city's hypothecation of these bonds would result in their being sold at par. When the city council refused this request, he asked that they "guarantee the payment of $400,000 of its seven percent bonds, at maturity, or grant such other relief as [their] wisdom direct." Again, he was refused. The stockholders were left to their own devices with regard to the company's debt.[15]

Though they postponed the election of directors at their meeting in February 1861, the stockholders did take action on the debt issue. Their initial decision was to issue $525,000 in 7 percent bonds, but the amount was later amended. William Ravenel felt that the company should issue an additional $100,000 and reserve it for future use in case they needed it for completing a permanent bridge over the Savannah River or to pay interest on the company's debt. On the eve of hostilities, April 10, 1861, a committee of stockholders chaired by George A. Trenholm authorized the execution of a third lien, of which Henry D. Lesesne, Francis S. Holmes, and Edward North served as trustees. This "act of generosity and patriotism" allowed the continued operation of the vital link between Charleston and Savannah.[16]

That issue having been settled, the board called a meeting for the following month to elect directors. At the recommendation of the stockholders, the owners of

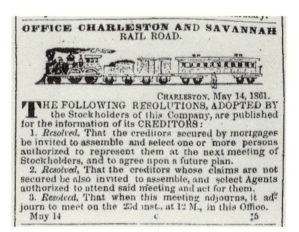

Notice of Charleston & Savannah stockholders' resolutions. From the *Charleston Daily Courier*, May 14, 1861

the 7 percent bonds met to formalize their representation at future meetings. At this meeting a group of bondholders representing about six hundred thousand dollars of those bonds appointed five men to represent their interests before the corporation. They included Isaac S. K. Bennett, Moses Cohen Mordecai, Charles Taylor Mitchell, James H. Taylor, and railroad securities entrepreneur John S. Ryan.

When the stockholders finally elected their new board of directors, there were some new faces on the board. Thomas Drayton was reelected president; however, gone were William F. Colcock, now devoting his attentions to the customhouse, and Thomas M. Wagner, a director since January 1857, who had been commissioned as an officer in the First South Carolina Artillery. Also absent were the names of Lorenzo T. Potter and William B. Hodgson. Replacing these four men on the board were Thomas Ryan, W. F. McMillan, Henry Buist, and William Joy Magrath.[17] If Drayton and the new directors of the company needed an influx of business, they would soon receive it in the form of mail, provisions, munitions, and troops transported for the Confederate government. Unfortunately for the railroad's investors, the next four years involved not only a constant fight for financial solvency but also the periodic threat of an invading Union army.

In addition to Colcock and Wagner, a number of men associated with the Charleston & Savannah Railroad were taking on significant roles in the fledgling Confederate government. One of the railroad's charter directors, Christopher G. Memminger, was named secretary of the treasury. Another former director, William Porcher Miles, was elected to the Confederate Congress. Despite Memminger's and Miles's earlier involvement with Southern railroad management, they were unable to persuade the Congress to legislate any adequate supervision of the loose system of Southern railways.

The military buildup in South Carolina also claimed several men who had been instrumental in establishing, constructing, and operating the railroad. Colonel William E. Martin, a longtime director, formed the First South Carolina Mounted Militia Regiment. Colonel Charles Jones Colcock—an original projector of the

Colonel Charles Jones Colcock, C.S.A., one of the original projectors of the Charleston & Savannah Railroad. Courtesy of the Colcock family

railroad—led a company in this regiment and later commanded an entire cavalry regiment in defense of the railroad. Robert L. Singletary and Henry M. Drane, whose services had been invaluable in completing the railroad, returned to their homes to join military units. Singletary became a captain of the Jeffries Creek Volunteers, Eighth South Carolina Infantry, while Drane served briefly as a commissary and was instrumental in organizing a volunteer corps in Wilmington, North Carolina.[18] The company's treasurer, Edward L. Parker, resigned that December and was replaced by William H. Swinton. Parker and conductor John Strohecker became officers in the Marion Artillery, a unit that participated in the bombardment of Fort Sumter and many later battles and skirmishes in defense of Charleston and the railroad.

Chief engineer and railroad superintendent Edward M. Manigault resigned his position on December 27, and was appointed chief of ordnance for the state. He later became a major in the South Carolina Siege Train. He led his unit in several engagements on James Island and was eventually taken prisoner in the last phases of the war. His replacement was Henry Stevens Haines, a North Carolinian with a wealth of railroad knowledge. Haines was born on Nantucket Island, Massachusetts, in 1836 but lived most of his life with his family in Wilmington, North Carolina. Educated in Massachusetts and North Carolina, he never attended college but was described by his brother as "a student of men and affairs, a first class mathematician and geographer, with a good working knowledge of four languages." His first railroad experience was as an engineer on the Wilmington & Manchester Railroad in

1853. He was next employed in the machine shops and later as a locomotive engineer with the Northeastern Railroad. He was an assistant engineer on the Charleston & Savannah Railroad during the latter portion of its construction. When Manigault resigned, Haines was promoted to superintendent.[19]

It was not long before Thomas Drayton also responded to the call of military service. Drayton had been an officer in the U.S. Army Corps of Engineers and in the South Carolina Militia. His sense of duty to his home state and to his new country was tugging at his conscience. In April 1861 he wrote to his friend and former West Point classmate Jefferson Davis offering his services to the Confederate army. Davis took note of Drayton's offer but decided to wait until all resigning U.S. officers had declined or accepted commissions in the new army.[20] Finally, on September 24, 1861, Davis granted Drayton a commission as brigadier general in the Provisional Army of the Confederate States.[21] Despite his new military post, Drayton kept his position with the railroad until the annual stockholders' meeting in February 1862.

In mid-October 1861 a Federal fleet under Captain Samuel F. Du Pont sailed from Annapolis, Maryland, headed for Southern waters. In addition to their crews the ships also carried U.S. Army troops under General Thomas W. Sherman. After stopping briefly at Hampton Roads, Virginia, they set sail again on October 29 for the South Carolina coast. On November 1 Governor Pickens received word that the fleet intended to invade the state at Port Royal. He requested that additional troops be sent from Georgia and North Carolina to help defend the area.

En route to Port Royal the Federal fleet encountered a strong storm that damaged or sank several ships, the most important of which were several supply ships for the ground troops. The foul weather delayed the operation for a few days, allowing the Confederates to ready their defenses, which proved inadequate despite their preparation. On November 5, with the fleet lying just off the coast, President Davis created the Department of South Carolina, Georgia, and East Florida, and placed it under the command of General Robert E. Lee.

Heading up South Carolina's defensive effort at Port Royal was Brigadier General Thomas F. Drayton. The president of the Charleston & Savannah Railroad commanded the Third Military District of South Carolina and had just moved his headquarters from Beaufort to Hilton Head Island, where his plantation, Fish Hall, lay less than a mile from Fort Walker. Out at sea on the USS *Pocahontas* was his brother, Commander Percival Drayton. Percival Drayton had entered the U.S. Navy as a midshipman at the young age of fifteen and had elected to remain with the Union when his home state seceded.[22] It was one of the first instances in the war that two such high-ranking family members faced one another in battle.

Protecting the harbor at Port Royal were two forts situated nearly three miles across the water from one another. Fort Walker, on the northeastern side of Hilton Head Island, was manned by 622 men under General Drayton and Colonel William C. Heyward. The fort had twenty guns, but only thirteen could be used against a force attacking from the sea. The troops manning these guns included two companies of the German Flying Artillery and Company C of the Ninth (later Eleventh) South Carolina Volunteer Infantry with Major Arthur M. Huger directing the firing. Three companies of the Ninth South Carolina, four companies of the Twelfth South Carolina, and Captain John H. Screven's mounted company provided infantry support. Shortly before the battle, the fort was reinforced by the Fifteenth South Carolina Infantry, Captain John P. W. Read's battery of the Georgia Light Infantry, and a force of Georgia infantry under Captain Thomas J. Berry.[23]

Fort Beauregard lay across the bay from Fort Walker at Bay Point on St. Phillips Island. Its garrison included a force of 640 under the command of Colonel Robert G. M. Dunovant. The men of the Beaufort Artillery and two companies of the Ninth South Carolina were charged with firing the cannon in Fort Beauregard with Captain (later brigadier general) Stephen Elliott directing the firing. Of Fort Beauregard's nineteen guns, only seven could be used against an attacking fleet. Six companies of Colonel Dunovant's Twelfth South Carolina served as the fortification's supporting infantry. An attempt to reinforce Fort Beauregard was made on November 7, but the presence of the enemy ships in the harbor and the remoteness of its position made this an impossibility.[24]

The defensive strategy was made more difficult, if not impossible, by the expanse and depth of Port Royal harbor. Because of the distance between the two forts, guns with the longest range available would be needed. In his planning of the two installations, General Beauregard had intended for ten-inch columbiads and rifled cannon to be put into place. To his misfortune the Confederate government and the state of South Carolina could manage to provide only a total of three columbiads and two rifled cannon for him to divide between both installations.

The weather on November 7 afforded the Federal invaders nearly perfect conditions by which to "vomit forth its iron hail with all the spiteful energy of long-suppressed rage and conscious strength."[25] The Union fleet entered Port Royal Sound at around nine o'clock that morning and directed much of its fire toward Fort Walker. Most of the ships took a flanking position out of reach of many Confederate guns, and at that point, according to Major Huger, "[Fort Walker] was fought simply as a point of honor, for from that moment we were defeated."[26] The enfilading fire from the flanking gunboats could not be answered because the gun on the right flank had been destroyed by a shell. No gun had been mounted on the north flank because the Confederates lacked a carriage for it. The barrage was so severe that most of Fort Walker's guns, including the ten-inch columbiad, became disabled shortly after the battle began. After four hours of intense fighting, with few guns remaining and almost all the powder and ammunition expended, General Drayton ordered the evacuation of Walker.[27]

The fighting at Fort Beauregard was fierce as well, but none of the cannon was dismounted during the engagement. Several men were injured when a caisson exploded and when the rifled twenty-four pound cannon burst, but Captain Elliott ceased firing only when it was obvious that Fort Walker was being evacuated. General Drayton directed his retreat from Hilton Head, and Colonel Dunovant did so from Bay Point, as the entire Confederate command was safely withdrawn to the mainland above the city of Beaufort.

Later that evening Union troops under Brigadier General Horatio G. Wright landed and took control of Fort Walker; however, Captain Du Pont was not certain that the entire garrison at Fort Beauregard had evacuated and refused to put anyone ashore there until the following morning. It was just after sunrise the next day when troops under Brigadier General Isaac Ingalls Stevens raised the Stars and Stripes above the fort.[28]

The easy Union victory at Port Royal had demonstrated to the Confederate leaders their inability to defend the barrier islands and coastal regions with the meager guns that the state and the Confederacy could provide. U.S. Army troops under General Sherman now had a vast base from which to conduct operations along the entire coast of South Carolina and parts of Georgia. The U.S. Navy now had a depot for its blockading force and could more easily implement the military's "Anaconda Plan"—Union general Winfield Scott's proposal to blockade the Atlantic and Gulf coasts, thus squelching Confederate commerce, and to divide the South by controlling the Mississippi River. From an economic standpoint South Carolina had lost the wealth and productivity of the Sea-Island cotton plantations, and its rice plantation were now threatened. The citizenry—and to a certain extent the military—were filled with shock and anxious uncertainty.

While the Federal fleet bombarded Confederate installations at Port Royal, General Robert E. Lee was on board a special train bound for Coosawhatchie on the Charleston & Savannah Railroad. As his train left St. Andrews Station on the western bank of the Ashley River, General Lee was probably already contemplating his defense of the southeastern coast. Since he had cast his lot with the Confederate army, things had not gone well for the general, and President Davis was now sending him into another grim situation. Lee's first assignment had been in western Virginia, where he was sent to coordinate the military effort and quell Union advances in the region. On September 12, 1861, he had unsuccessfully attempted to drive a Federal force from Cheat Mountain, and his October offensive in the Kanawha Valley misfired as well. Critics had dubbed him "Granny Lee" because of a perceived slowness to action and poor decision making in his campaign there.

Despite the criticism President Jefferson Davis appreciated Lee's military expertise and gave him his full support. Davis recalled Lee to Richmond to serve as his military adviser, but his stay there was short lived. Word soon reached Richmond that a Federal fleet was en route to the Southern coast. The Confederate army in South Carolina needed someone capable of uniting several separate commands under a single leader. Only two days before the Federal invasion at Port Royal, Davis gave General Lee command of the Department of South Carolina, Georgia, and East Florida, and immediately dispatched him to Charleston.

On his arrival at Coosawhatchie, Lee was greeted by a scene of sheer chaos. Couriers were hurriedly rushing in and out of camp. Troops were being readied for the unexpected. In the distance was the all-too-familiar echo of artillery fire. He had arrived in time to help pick up the pieces of another troubling defeat. This situation may have been what inspired his remark in a letter to his daughter that the assignment in South Carolina was "another forlorn hope expedition. Worse than in West Virginia. . . . I have much to do in this country."[29]

When General Lee arrived, General Roswell Sabine Ripley was riding to the front to assess the situation. In command of the Confederate troops of that district, Ripley was a transplanted Ohioan with a fiery temperament. Though he tended toward petulance and insubordination, he was a logical and thorough officer. On his return to camp he sat down with Lee to review their grim prospects.

Ripley told Lee that the Federal navy had taken control of Port Royal harbor and that there had been little choice but to order the withdrawal of Confederate troops from those installations. Despite the successful evacuation of the troops, they had in the process left behind most of their tents, clothing, and provisions, as well as all the cannons in Forts Walker and Beauregard. By virtue of their conquest, the enemy controlled the inland waterways separating the Sea Islands from the mainland. The larger ships controlled the deep channels, and shallow-draft gunboats could travel miles inland on the many lowcountry rivers and streams. The Confederate navy lacked the capacity for any resistance to their Northern counterparts, and Lee had only about seven thousand troops—many poorly trained and equipped—at his disposal between the cities of Charleston and Savannah. Stretched between the two cities, on the periphery of the tideland marshes, was the Charleston & Savannah Railroad. If the Yankees took control of the rails, they could prevent the movement of troops and essential supplies between Charleston and Savannah, thus making it easier to mount an offensive against either city.

Lee knew of the effectiveness of the Union navy's guns against the defenses at Port Royal and noted that Federal infantry, covered by the firepower of their gunboats, could be landed "with great celerity against any point and far outnumber any force we can bring against it in the field."[30] He and Ripley decided to initiate a three-pronged defensive effort. The first priority was to shore up the ability of Fort Pulaski, as well as the defenses around Charleston and Savannah, to withstand a heavier bombardment. Second they ordered the obstruction of rivers, streams, and other waterways that could potentially be used for amphibious assaults on interior Confederate positions. The third arm of the plan—the one that directly involved the railroad—was to place the now-scattered Confederate troops at the points most accessible to a Federal advance on the railroad. These troops, which were mostly cavalry, would put up whatever resistance they could—preferably out of range of the Yankee gunboats—until reinforcements could arrive via the railroad. As long as the rails remained intact, troops, guns, and other supplies could be rushed from Savannah or Charleston and concentrate at the threatened point. Drayton had asserted that

General Robert E. Lee, C.S.A. Courtesy of Library of Congress

the railroad would prove to be a "formidable earthwork." General Lee was going to put it to the test.

Until this time the railroad's drawbridge across the Savannah River had remained open except when trains were to travel across it. After the fall of Port Royal Superintendent H. S. Haines wired John L. Roumillat, the company's agent in Savannah, with the order to keep the drawbridge closed at night except during the transit of boats.[31] This move served dual purposes. The rails would remain uninterrupted at night for the efficient movement of supplies and troops into and out of Savannah. The bridge would also impede travel upriver by enemy raiders at night,

Henry Stevens Haines, superintendent of the Charleston & Savannah Railroad, 1861–1866. Courtesy of Lucy Gross and the Haines family

when the monitoring of such movement would not be easy. General Drayton was stationed at Hardeeville and could personally oversee other activities involving the road in that area.

Before the fight at Port Royal, Superintendent Haines had been occupied with maintaining at least local control of the railroad's operations. He was charged with coordinating the movement of government transport trains with that of trains carrying freight, civilian passengers, and mail. He also made collaborative arrangements with other companies to increase the railroad's business. Using the railroad's connection with the Georgia Central at Savannah, he arranged for anyone who traveled to the Southern Commercial Convention in Macon via the Charleston & Savannah to make the return trip free of charge.[32] He also arranged for freight shipped on the Charleston & Savannah to be forwarded via the steamers *St. John*'s and *St. Mary*'s to coastal cities on the Florida-Georgia line and to landings on the St. Johns River.[33] The tightening of the Federal blockade later negated this deal.

The Charleston & Savannah Railroad was having difficulty establishing a commercial identity, and its involvement with the military was only just beginning. In addition to his own rolling stock, Haines now had to oversee the use of cars from the Georgia Central and the South Carolina roads on his tracks. The influx of government travel along the coast had also exposed the most glaring deficiency of the railroad—its lack of a connection with other railroads at Charleston. Because of the inefficiencies encountered there, plans were announced for the connection of the Charleston & Savannah Railroad and the South Carolina Railroad, which would be accomplished by bridging the Ashley at New Bridge and laying track to the South Carolina Railroad's depot at Spring Street. Siting the eastern terminus of the Charleston & Savannah on the west bank of the Ashley, which had originally been a cost-saving measure for the railroad had become a hindrance to the Confederate military. This deficiency would certainly have to be alleviated, but there was also the matter of a Union expeditionary force on the South Carolina barrier islands with which to contend.

General Lee set up his headquarters at Coosawhatchie, a low, marshy hamlet on the coastal plain forty-three miles by rail from Savannah. The village lay at the head of navigation on the Coosawhatchie River and at the junction of the railroad with the stagecoach road from Old Pocotaligo. Coosawhatchie had once been a thriving Indian settlement, but its population had dwindled before the war.[34] Private Joseph W. Turner of Virginia described it as "quite healthy in the winter season but very sickly in the summer."[35] Lee was bothered by the ability of the Federal gunboats to travel up the river within range of his headquarters, and he was pleased that the Union had not followed up its initial naval success with a full-scale invasion. Taking advantage of Federal inactivity, he embarked on numerous trips to inspect the defenses along the coastline and personally direct much of the work.

Lee commanded a patchwork force of infantry, cavalry, and light artillery, which was dotted along the line of the railroad. An observational force stationed at Bluffton under General Drayton consisted of Colonel William C. Heyward's Ninth (later Eleventh) South Carolina Infantry, Colonel William Davie DeSaussure's Fifteenth South Carolina Infantry, and a company of Colonel William E. Martin's cavalry under Lieutenant Colonel (later colonel) Charles Jones Colcock. This detachment, whether by design or coincidence, was led by two men who had a special connection with the land and the railroad upon which they were stationed. Drayton's command soon moved into a forward position covering roads leading to Bluffton and Hardeeville. Colonel Thomas L. Clingman's Twenty-fifth North Carolina Volunteer Regiment guarded the town of Grahamville and the roads leading from Boyd's Landing. Captain James Davis Trezevant's cavalry company lay in observation at the landing to alert the supporting infantry to any force coming ashore. At Huguenin's plantation Colonel Ambrosio J. Gonzales commanded a company of state troops made up of the Palmetto Guard, the Charleston Light Dragoons, the Rutledge Mounted Riflemen, the South Carolina Seige Train, and an eight-inch howitzer battery. The battery was positioned to observe and provide initial resistance at Port Royal Ferry. The Eighth North Carolina Volunteers were placed near Colonel Gonzales to support his battery and Clingman's troop or to defend the road to Eutaw Church.

Dunovant's Twelfth South Carolina Infantry was divided between camps at Hardeeville and Pocotaligo. Colonel James Jones's Fourteenth South Carolina was headquartered at Garden's Corners and was keeping watchful eye on the area between Combahee Ferry and Port Royal Ferry. Colonel Martin's cavalry regiment was spread out at different observational post with the main portion being at Pocotaligo. Additional small groups were stationed at points along the Combahee, Ashepoo, and Paw Paw. Captain Leo D. Walker was supervising the obstruction of the rivers at points below the railroad bridges. Superintendent Haines had dispatched railroad employees to assist with these efforts near the railroad.[36]

Of great concern to General Lee was the lack of troop strength in the area. He was worried that the number of Northern troops seemed to be increasing faster than his were. From his headquarters he lamented that "there are no means of defending the State except with her own troops, and if they do not come forward, and that immediately, I fear her suffering will be greatly aggravated." For some the war had already lost some of its romance. Many South Carolinians apparently preferred serving their home state for an enlistment period of six months to a year rather than enlisting in the regular Confederate Army for the duration of the war. This fact frustrated Lee. When the Confederate blockade-runner *Fingal* arrived in Savannah with a load of guns and powder, Lee was informed that 4,500 of the 9,000 Enfield rifles on board had been allocated for his men. Hoping to use this to his advantage, he traveled to Charleston to meet with Governor Pickens about the troop shortage and the disposition of the guns. Lee offered to issue 2,500 of the Enfield rifles to South Carolina units pledging service for the entire war if the state would also supply arms for the equivalent of two regiments of men willing to serve for the same period. He then returned to the business of readying the defenses of the Southern coast.[37]

On his return to Coosawhatchie, news was somewhat better. He learned that Phillips's Legion of Georgia Volunteers, six infantry regiments, and two field-artillery batteries were being sent down from Virginia.[38] With the addition of the Virginia regiments, and including the garrisons at Charleston and Savannah, Lee now had approximately twenty-five thousand troops under his command. Though their numbers now exceeded those of the enemy, the training of the Confederate troops was slow as was the execution of Lee's defensive plans. Fortunately for the Rebels, the only military operation of consequence in December occurred when the Union navy sank the "Stone Fleet" to obstruct the shipping channels in Charleston harbor on the twentieth, curtailing much of Charleston's commercial and military shipping capability. Lee considered this event, occurring on the anniversary of South Carolina secession, a slap in the face. It also helped to convince him that the Federal objective was not to attack Charleston by way of the harbor entrance, but more to restrict travel in and out. As a result, he stepped up preparations for an attack elsewhere to the south.

Lee's travels were extensive and his pace exhausting. He personally supervised much of the entrenchment and fortification in and around Coosawhatchie and continually tried to speed along the work at all points. Some days he rode his horse, Greenbrier, in excess of thirty miles. Lee had originally noticed Greenbrier when serving in western Virginia. At the time, the colt was called Jeff Davis, and his owner was Major Thomas L. Broun. When the major's brother, Captain Joseph Broun, was transferred to South Carolina with his regiment, he brought the horse, and Lee became reacquainted with the four-year-old "Confederate grey" mount. Lee sought out Captain Broun, who was supervising the fortifications at Pocotaligo and Coosawhatchie, and offered to purchase the horse. At the insistence of his brother, Captain Broun offered Greenbrier as a gift to the general, but he would not accept and, after some negotiation, bought the horse for two hundred dollars. Greenbrier was a fine specimen with excellent endurance. His reputation as a "fine traveler" was the inspiration for changing his name to the well-known Traveler. Consequently South Carolina has its own small place in the story of how General Lee acquired his legendary horse.[39]

At Christmas 1861 the general continued his arduous routine despite the Christmas caroling and hanging of the greens elsewhere in the Confederacy. In a Christmas Day letter to his wife he reflected on the situation locally, in Virginia, and internationally, writing that "the enemy is still quiet and increasing in strength. We grow in size slowly but are working hard." He wrote of his regret that Federal troops now occupied his family residence in Alexandria, and of the loss of his books, furniture, and other possessions. "They cannot take away the remembrance of the spot," he reassured his wife, "and the memories of those that to us rendered it sacred. That will remain to us as long as life will last, and that we can preserve." He also instructed his wife not to place too much hope in military aid from England. He opined correctly that England did not desire a war with the United States despite reports to the contrary. "We must make up our minds to fight our battles and win our independence alone. . . . We require no extraneous aid, if true to ourselves." He also advocated patience and cautioned that it would "be no easy achievement and cannot be accomplished at once."[40] General Robert Edward Lee was the man on whom South Carolina and the Charleston & Savannah Railroad now relied for protection. Though the year 1861 ended quietly for him, 1862 began with a flurry of activity that and gradually grew into a tempest.

General Thomas W. Sherman and his Union expeditionary force on Hilton Head Island and Beaufort had been—much to the surprise and delight of the Confederates—very quiet and not without reason. Though Southerners expected a full-scale invasion, the capture of Port Royal had been primarily to establish a base for the Federal navy's South Atlantic Blockading Squadron. Because of this, Sherman had not been allotted sufficient cavalry or field artillery for inland operations. Also the damage done to a number of supply ships during the voyage to Port Royal hampered their ability to equip infantry adequately for such endeavors. Despite his superior numbers, Sherman believed that he did not have enough men with which to garrison his base at Hilton Head and Beaufort while invading the surrounding countryside.

Sherman did express a desire to take the railroad eventually, but he wanted Washington to send several thousand reinforcements, especially cavalry. From Beaufort he wanted to launch an invasion directed at Bluffton, New River Bridge, and Hardeeville, believing that this would allow him possession of the railroad bridge crossing the Savannah River and prevent reinforcements from reaching Savannah. In a letter to General Lorenzo Thomas, Sherman declared his confidence that an "invasion of this sort would not only give us Savannah, but, if successful and strong enough to follow up the success, would shake the so-called Southern Confederacy to its very foundation." Even though the Yankees landed a force and occupied already evacuated Tybee Island, they did not put together an extensive operation at that time. Sherman's thoughts of damaging the railroad were limited to suggesting that Captain Quincy Adams Gillmore, who later rose the rank of brigadier general, lead a party to burn the Savannah River Bridge. Though ordered, this action was never carried out, and there was no significant movement on the railroad until the New Year.[41]

Brigadier General Isaac Stevens had been growing restless as his Federal troops were busy establishing a base at Hilton Head. He had been collecting flatboats for operations in the shallow channels in the Broad River and its tributaries. He knew of General Thomas Sherman's desire to take the city of Savannah, and he knew how important the city's railroad connections were to its defense. As early as December 9, Stevens had written to General Sherman asking if he should move against the railroad whenever operations were begun against the mainland.[42] This did fit into Sherman's larger plan, but he believed that he lacked the numbers of troops needed to try an ambitious inland expedition at that time.

Stevens knew that every day they waited meant that the Confederates would be able to strengthen their defensive lines further. Indeed, troops were arriving daily from other areas of Dixie, and earthwork batteries were going up quickly under the watchful eye of Generals Lee and Ripley. Knowing it would be some time before operations against Savannah would commence, Sherman was considering a dash against the southern end of the Charleston & Savannah Railroad if he could feel confident of success. He knew his force would have to cross Port Royal Ferry, and establish a position on the mainland. A preparatory assault against Confederate batteries at Port Royal Ferry was planned just after Christmas to keep Whale Branch and the Coosaw open to Federal gunboats, to keep the Rebels off Port Royal Island, and to "punish" the Rebels for firing on the Union steamer *Mayflower*.[43] To lead it Sherman chose the eager Brigadier General Stevens.

At 3:30 A.M. on New Year's Day General Stevens waited with his troops in the predawn darkness at the point of embarkation. The flatboats he had assembled for this sort operation would be put to the test later in the morning. So would his men, almost none of whom—the Seventy-ninth New York Highlanders Regiment excepted—had been under fire in combat. At daybreak the force, consisting of portions of the Highlanders, the Fiftieth Pennsylvania, and the Eighth Michigan, shoved

The Broad River and its tributaries, detail. Courtesy of Image Archives of the Historical Map and Chart Collection, Office of Coast Survey, National Ocean Service, NOAA

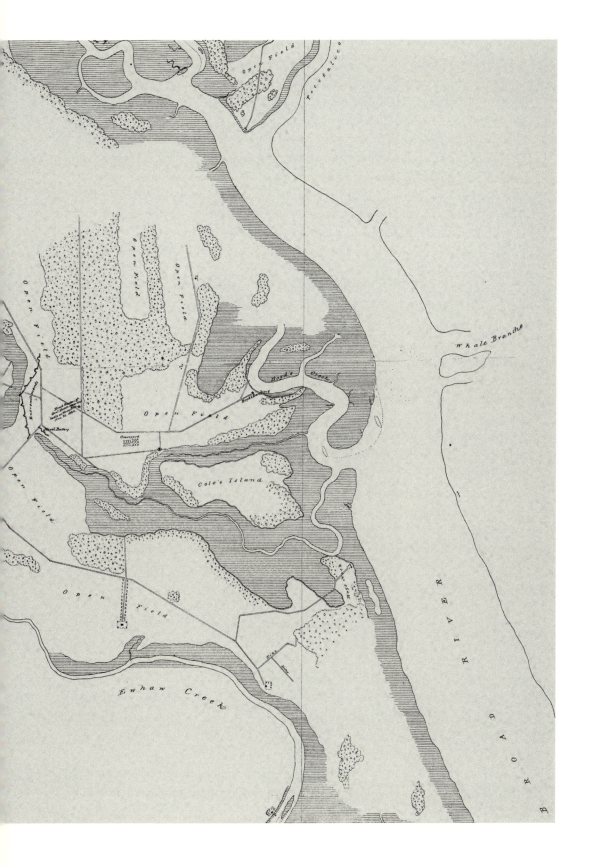

off aboard scores of bateaux. Naval support for their mission was provided by the USS *Ottawa, Pembina, Hale, Seneca, Ellen,* and four gun launches.[44]

The unusual amount of troop activity caught the eyes of Confederate lookouts posted near the bridge and causeway leading to Chisholm's Island. At about 7:00 A.M. the pickets sounded the alarm that Federal troops were landing in force on Chisholm's Island, and within an hour the enemy was advancing toward Kean's Neck Road. General Stevens—with the Highlanders and part of the Fiftieth Pennsylvania—landed near the cotton gin three miles below Adamses' landing. Once he had reassembled his command at the Adamses' house, Stevens began an exhausting three-mile march inland. After leaving the house he sent a portion of his force forward toward the battery and left his artillery in the rear. The gunboats in the river then turned toward the ferry and lobbed shells into the woods and fields in advance of the infantry.

After hearing the firing of the Federal gunboats and receiving word of the enemy landing, Colonel James Jones of the Fourteenth South Carolina Regiment ordered the two guns at Port Royal Ferry removed. A siege howitzer was successfully removed, but a twelve pounder was accidentally overturned in the ditch of the earthwork. Under fire from the enemy's gunboats and threatened by the advance of the Yankee infantry, the Confederates spiked the cannon and quickly vacated the area to avert bloodshed. Colonel Jones then moved his command forward into a line of battle in the woods covering Kean's Neck Road; his left extended toward the river.

Instead of using the main road, General Stevens sent the battle-tested Highlanders forward as skirmishers across fields and along paths shown to them by African Americans who had been picked up along the riverbank. Once the van had passed the Confederate position in the woods, the Rebels lit into them with both musket and cannon. As the gunboats continued to shell the Confederate position, Stevens formed his brigade into a right angle and sent the Eighth Michigan into the woods on the double-quick. The Fiftieth Pennsylvania pushed the skirmishers from the left, and the Forty-eighth New York moved forward from the right, never advancing outside the protective cover of their gunboats. When they had achieved position from which they could not be dislodged, the Highlanders continued their push forward and stormed the fort, finding only the spiked twelve-pound gun in the ditch. Almost simultaneously Federal gunboats entered Whale Branch and opened fire on the battery across from Seabrook Island. Captain Stephen Elliott and detachments of the Highlanders and One-hundredth New York (Roundhead) Regiment stormed the battery but found no guns in place. After destroying the earthwork battery, they crossed back over to Seabrook Island.

Having no opportunity for a flanking maneuver, Colonel Jones withdrew his Confederates to a point out of range of the enemy gunboats. General John C. Pemberton hoped to protect the railroad from a direct assault, but now he knew the enemy's capability to land in force. Deciding to leave Colonel Jones's command in observation at Cunningham's Bluff near Page's Point, Pemberton kept Donelson's brigade and Thornton's Virginia battery in the rear to support Jones's infantry.[45]

Around nightfall shelling from the Union gunboats began to taper off. The Federal troops completely restored the ferry, which made recrossing the river with their ordnance much easier. The entire force of three thousand men had recrossed the river by midnight after nearly dismantling the works at the ferry and destroying nearby buildings. General Stevens was satisfied with the performance of his untested troops. Not only had they behaved well in the field, but they had also proven that a sizable force could be landed any place on the river. The latter point was not lost on General Pemberton and his compatriots, and it neutralized any consolation he may have received from preventing further Federal advance on the railroad that day. Pemberton knew there would be more assaults to come. The only question was when and where.

FIVE

The Most Formidable Earthwork

The action at Port Royal Ferry accentuated to General Lee the fact that he could not afford to expose his men and equipment to engagements against warships or gunboats. His biggest fear was a strong advance on the Charleston & Savannah Railroad. Such an attack at the head of the Broad River would cut the rail connection—and thus the movement of supplies and reinforcements—between the two cities and allow movement on either city with combined ground and naval assaults. Lee hoped the defenses would hold but knew that he would need more arms and personnel for the effort. His defensive works were not yet completed, and he had not received the number of troops he had expected since taking command of the department.

Lee did not feel that his thinly spread units could effectively deter a large-scale enemy landing. He believed his only advantage was to draw the Federals away from the protection of their gunboats to fight on more even terms on ground of his own choosing. He advocated withdrawing his troops from positions vulnerable to naval gunfire, placing them in positions defending the Charleston & Savannah Railroad, which the *Charleston Mercury* already hailed as the "military backbone of [the] tide water districts."[1] Some of Lee's subordinates, in addition to a number of coastal planters, disagreed with his plan for withdrawing the troops in outlying areas, but he remained confident in his overall strategy. He hoped that the local planters would eventually feel the same way.

Lee devised a system of coastal defense based on the theories of Major Joseph G. Totten, who decades earlier had proposed a system of harbor forts that delayed an enemy's advance until local militia could be assembled. Once organized, the militia would move to the area threatened and stem the attack. Lee updated this strategy by adding the element of railroad transport. He planned to keep garrisons at Charleston and Savannah strong enough to meet an initial assault, and—before the enemy could gain the advantage—reinforcements would arrive by rail. Lee continued in his efforts to shore up the harbor defenses and also constructed a number of new works along the Charleston & Savannah Railroad. From this time on, as pointed out by Stephen Wise, the defense of Charleston's harbor works and the railroad were linked together.[2]

Brigadier General Roswell Sabine Ripley, C.S.A. Courtesy of South Carolina Historical Society

After developing a relatively strong series of defenses in Charleston, Lee turned his focus in February 1862 to the defense of Savannah. On February 3 he officially moved his headquarters there from Coosawhatchie so that he could personally supervise the strengthening of Fort Pulaski. Lee entrusted the defense of the Broad River headwaters to General John C. Pemberton near Pocotaligo and General Maxcy Gregg near the Ashepoo. While he fortified Pulaski, Lee instructed Pemberton and General Thomas Drayton to build additional defenses to the north of Savannah. Though Lee was attempting to defend the railroad, he laid the groundwork for an alternative in the event the line was broken. He contacted the mayor of Augusta, Georgia, to connect the Savannah & Augusta Railroad with the South Carolina Railroad to maintain some rail connection between Charleston and Savannah by way of the sandhills. Rumor among some of the Rebel soldiers was that Lee might give up the railroad and concentrate his forces around the population centers at Charleston and Savannah; however, most regimental commanders, including Lieutenant Colonel William Stokes of the Fourth South Carolina Cavalry, knew the importance of the railway to both cities and put little stock into such supposition.[3]

While in Savannah in mid-February, Lee received an order from Secretary of War Judah Benjamin to "withdraw all forces from the islands in your department to the mainland."[4] The administration had agreed with Lee's earlier assessment of Federal naval power. When Lee ordered General Roswell Ripley to abandon the outer Sea Islands and other exposed positions, the evacuation was executed smoothly, but it left many rich cotton plantations near the coast at the mercy of looters, vandals, and Federal raiding parties. Many of the slaves who remained on the plantations fled or were taken across enemy lines into Federal camps. To prevent this, the government advised lowcountry planters to remove their slaves from plantations in the region. While this strategy combated the problem of runaway slaves, it forced planters to abandon that year's rice and cotton crop, which they had just put in the field.

In the spring of 1862, Charles Heyward evacuated his slaves from Combahee to Goodwill Plantation on the Wateree River with surprising ease via the Charleston & Savannah Railroad. Squire Jones, one of Heyward's overseers, recalled that all he had to do was to instruct the head drivers at Rose Hill, Pleasant Hill, Amsterdam, and Lewisburg plantations to "tell all their people on a certain day and a certain hour to assemble at the 'weather house' on the Charleston & Savannah Railroad." Without the slightest commotion the slaves "bundled up their belongings and were at the appointed place long before the arrival of the train. Some carried their chickens with them in baskets covered with cloth, and some wanted to carry a pig or two, but this was not allowed."[5] They then traveled from near White Hall to Charleston, where they transferred to the South Carolina Railroad for the trip to Goodwill.

While Federal raiding parties depleted the plantation resources available to the Confederate troops, the needs of the Confederate army were reducing the effective forces in the lowcountry. The fall of Fort Donelson and Fort Henry in Tennessee and Roanoke Island in North Carolina resulted in the reassignment of several units from Lee's command to those areas. A hopeful, but frustrated, Lee wrote, "I have been doing all I can with our small measures and slow workmen to defend the cities and coast here. Against ordinary numbers, we are pretty strong, but against the hosts our enemies seem to bring everywhere, there is no calculation. But if our men will stand their work, we shall give them trouble yet."[6]

Ripley was concerned that with the depleted force the weak point on the Charleston & Savannah line was near Charleston. He felt that, if the enemy moved a force against Confederate troops near White Point or Simmons Landing, the Federals would be within five to eight miles of the railroad and a short march from Charleston. If the Confederates moved some of their troops from between Ashepoo and Combahee, it would strengthen the Charleston defenses. Though it would make the country between Jacksonboro and Salkehatchie more vulnerable to attack, that area would be mosquito laden in a matter of two months and neither the Confederates nor the Yankees would be able to occupy that area for very long. Ripley advocated relocating troops from that area to near Adams Run in the east and just above Grahamville in the west. If the Federals cut the railroad, the Confederates could maintain communication on the alternate route, and the Union troops would have a long march through unhealthy country to reach either port city.[7]

General Lee sent word to General Nathan G. Evans at Adams Run to have the North Edisto River near Simmons and White Point Landing watched carefully. If the Yankees made a run at him, he was to prevent them from reaching the railroad. If forced to retreat, he should do so in the direction of Charleston so he could be reinforced by Ripley's troops stationed there, and Lee could send Pemberton by train from Pocotaligo to execute a flank movement on the enemy's rear. Though he took Ripley's observations seriously, Lee was also concerned about a possible attack on the western end of the railroad near the Broad River. He told Pemberton to be prepared to move his troops in either direction should the Federals decide to advance in that sector.[8]

At the beginning of 1862 Southern railroads were in crisis. The rapid influx of government business with very little formal regulation was jeopardizing the entire war effort. The military blamed the problems on the railroad carriers, and the railroads blamed the military and other railroad companies. In addition the Federal blockade, lack of industrial capacity, and increased consumption of resources by the military created a shortage of railroad supplies. It was becoming imperative that the railroads of Dixie find a way to cooperate with the government, with the military, and with each other. They also had to find a way to maximize their civilian business despite the increased military usage.

In Charleston the increased transport of military provisions caused a shortage of storage space at the railroad's depots. By the beginning of 1862 the wharf house at the Charleston & Savannah's St. Andrews depot was so clogged with perishable freight that limits had to be placed on its storage. Superintendent H. S. Haines advised factors—and other customers receiving hay, rice straw, and other forage via the railroad—that they would need to claim their shipments within twenty-four hours of their arrival.[9] Furthermore the company would refuse to receive similar freight consigned to those who refused to remove it. Haines also ordered any forage that was unclaimed after three days to be sold at public auction to cover freight and storage expenses. The company wanted to do all in its power to aid the Confederate military, but in these lean times it could scarcely afford to lose freight business.

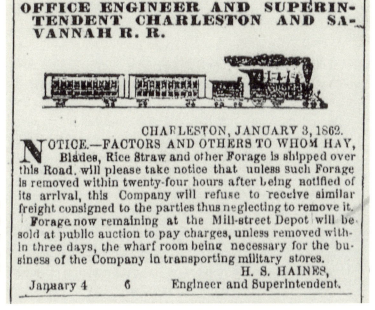

Notice to civilian shippers. From the *Charleston Daily Courier*, January 8, 1862

Railroads throughout the South were facing financial hardships, and the Charleston & Savannah was no different. The Federal blockade was tightening, and the government had called for an embargo on the exportation of cotton. The flow of civilian freight between Charleston and Savannah was reduced by one-third, and the increasing government business on the road was not making up for the loss. The management had to come up with some way to overcome these difficulties.

Often railroads were in competition with the military for resources such as timber and iron. The Confederate navy's ironclad-ship program was a case in point. These ironclads required wood for their superstructures and iron for their armor. The same wood could potentially serve as fuel for locomotives or be built into rolling stock; the same iron could be rerolled into rails, or used for cars, engines, spikes, and untold other railroad supplies. When the call went out for planters along the line to contract with the Navy for their timber, it meant that—not only would the road be in competition for these raw materials—it would also have to transport it to its final destination at the shipyard.[10]

Many Southern railroad companies had no choice but to make cutbacks. The Charleston & Savannah Railroad had defaulted on interest payments before the war started and had not declared a dividend on their stock. Now it was forced to curtail its commercial freight services. The sacrifices necessitated by the war also included reductions in maintenance. Throughout the South there were already significant deficiencies in rails, boiler iron, cylinder castings, wheels, and springs. These could be supplied by only a few Southern manufacturers, who were already encumbered by their military obligations. One wise Charlestonian had noted the South's dependence on the North and foreign countries for its railroad supplies and had called for the formation of a locomotive factory in Charleston.[11] With the already capital-starved area in crisis, this plan never materialized.

Though rolling stock could be built both in Charleston and Savannah, need far outpaced supply. This forced the Charleston & Savannah to borrow rolling stock from the South Carolina Railroad and the Georgia Central. Since the South Carolina Railroad had lent a significant quantity of its rolling stock—much of which the military used for transporting troops and supplies—commercial traffic, even on interior lines, suffered.

Because the South lacked the necessary matériel for maintenance and construction of rolling stock, Southern railroad leaders had held a railroad convention in Richmond on December 19, 1861. The meeting was dominated mostly by the Virginia railroad companies. The delegates were shocked at the magnitude of the problems they faced, but their only idea was to form a committee to devise ways of securing supplies and disseminating information about obtaining supplies.[12] The convention adjourned without much getting accomplished. Each man returned to handle the problems of his own individual railroad. They had tried to do something, but their best effort had resulted only in the formation of a committee.

On January 11, 1862, a special committee was appointed by the Confederate Congress to investigate alleged abuses of railroads by the military. The committee reported that the transportation of troops and supplies by rail and steamship was the responsibility of the quartermaster general and that a paper system to control the alleged abuses already existed. They recommended that the army construct larger depots closer to fields of operation. For the railroads the plan was more radical. The committee asserted that the existing deficiencies were the fault of the fragmentary nature of the rail "system" and the poor distribution of rolling stock. They proposed that the military assume direct control over routes leading through Richmond, Nashville, Memphis, and Atlanta to several armies' headquarters. To a governing body built upon the foundation of states' rights, this proposal was tantamount to heresy. Since it violated their fundamental objection to a strong central government, Congress simply ignored the report and allowed it to die in committee. Yet again the Congress refused to exert any manner of central control over the railroads. The individual roads were left to deal with the military in their vicinity in the best way they could.

Cooperation with the armed forces was not as much an issue with the Charleston & Savannah Railroad as it was with some others. One reason was that the army and the railroad needed one another. Another was that General Drayton and General Lee had prior experience with railroad politics as they applied to the military. General Pemberton did not yet possess such wisdom. In January 1862 General Maxcy Gregg received use of a train for the deployment of troops guarding the bridges and trestles between the Ashepoo and Salkehatchie rivers, and another daily train carried officers visiting the various posts. Gregg and Pemberton wanted to ask the governor to grant them the necessary military authority over the conductors of the railroad in each of those instances.[13]

General Lee knew that military interference with railroad operations had been a frequent complaint of numerous rail carriers, and he tried to avert the potentially inflammatory request. Informing Pemberton that he did not think military authority over the conductors was necessary, Lee expressed the belief that a proper explanation to the railroad's president—which had already been done—would be sufficient for obtaining the desired services.[14] If good relations were maintained, Lee also thought Pemberton could easily get a handcar that he had previously requested from the railroad. Realizing its potential effect on civilian morale, Lee also reminded Pemberton that he did not wish for a guard to be placed on any of the passenger trains. He did not want passenger or mail service to be hindered in the least. He believed that the troops could be posted and relieved by means of the handcar, the freight train, and whatever facilities could be afforded on the mail train without interrupting service.

Notwithstanding the Confederate government's inability to harness the Southern railroads, the Charleston & Savannah road had its own leadership crisis with which to contend. Interest on the company's 7 percent bonds was in arrears, and prospects for their payment remained doubtful. The company's creditors felt they must take action, even if it meant foreclosing on the mortgage and sequestering the railroad's income for payment of the interest to its investors.

With the company's annual meeting only a few weeks away, a group of the bondholders met on January 22, 1862, at the hall of the Elmore Insurance Agency to reassess their position. James H. Taylor was appointed chairman of the meeting and Frederick Richards was made secretary. Other bondholders present at the meeting included Charles Taylor Mitchell, B. D. Lazarus, John S. Riggs, the Reverend J. B. Kendrick, A. B. Dunkin, James McCall, E. M. Seabrook, H. P. Gibbs, Richard Yeadon, Moses Cohen Mordecai, C. D. Carr, and James Butler Campbell.[15]

The railroad's board of directors had met previously with representatives of the creditors and expressed a desire to "meet the views of the bondholders." The directors hoped to avoid the bondholders' proposed call for a foreclosure, especially in view of the critical social, political, and economic conditions in the state. Because of its initial capital investment in the railroad, the Charleston City Council commanded a strong voting position in the company. Mayor Macbeth pledged his willingness, as a representative of the city government, to offer seats on the board of directors to at least six of the bondholders. This would give them more of a voice in the day-to-day running of the railroad and a better vision of the company's physical and financial condition. Richard Yeadon proposed a resolution insisting on the nomination of at least six directors by the bondholders, and Campbell proposed that three people be appointed to meet with representatives of the city council to nominate the board. Both resolutions carried, and the group nominated Campbell, Yeadon, and Lazarus to represent them.

These men met with a special committee of the city council and submitted their final report to the bondholders on February 6. After the meeting was called to order, Campbell and his special committee submitted their nominees for the railroad board, all being from among the ranks of the bondholders. The nominees were M. C. Mordecai, Frederick Richards, C. T. Mitchell, William C. Bee, James H. Taylor, and Theodore D. Wagner. Those representing the city council then introduced the names of their nominees: Mayor Charles Macbeth, George Walton Williams, Henry Gourdin, Otis Mills, Richard W. Bradley, and Henry Buist. Without much wrangling these names were readily approved by both parties. With that done Henry Ravenel made a motion that if the stockholders failed to elect the approved slate of directors, the bondholders should take possession of the company through foreclosure.

Just days before the meeting the city council had made a change that threatened to pit the bondholders against the City of Charleston for control of the railroad. Without prior warning council excluded one of the bondholders' nominees and one of their own and had chosen their replacements without the consent of the bondholders.[16] This angered many stockholders and bondholders, who felt that the city

William Joy Magrath, president of the Charleston & Savannah Railroad, February 1862–October 1862, photograph by W. B. Austin, Vandyke Studio, Charleston. Courtesy of South Caroliniana Library, University of South Carolina, Columbia

council was abusing its power. One could argue that the city had earned that right with its substantial investment in the railroad, but the council had made a good faith pact with the bondholders. Capital investment did not, in the eyes of most, justify betrayal.

As the date of the company's annual meeting approached, there was an air of uncertainty about what would become of the railroad's management. President Thomas Drayton's presence had been nonexistent because of his duties with the military. The bondholders and the Charleston City Council agreed that his successor should be William Joy Magrath, but a group of stockholders were trying to garner support for the nomination of Edward W. Marshall.[17] This disagreement, coupled with the bondholders' power play, promised to make the meeting interesting. Who would lead the road for the remainder of the war? The answer lay in the hands of the men appointed the previous summer to represent the bondholders should they decide to foreclose on the mortgage. Three of them—Mordecai, Mitchell, and Taylor—had been nominated to be directors, and a fourth, John Ryan, had saved the road from failure once before.

There had been much negotiation and quarreling before the meeting, but when the stockholders met on February 12, the proceedings were apparently quite civil. With the endorsement of both the bondholders and city council, William Joy Magrath was elected president to succeed General Drayton. Magrath was a quiet and even-tempered gentleman. A native of Charleston and a younger brother of Judge

Andrew Gordon Magrath, William Magrath was mostly self-taught and had acquired a keen knack for business while working with his father, a Bay Street merchant. He had been a director of several railroad companies, and many considered him one of the best-informed railroad men in the community.[18] The board of directors felt that this austere fellow was just the person to lead the Charleston & Savannah Railroad through this difficult period. Unlike its previous president, who had been splitting his duties between the Charleston & Savannah Railroad and the Confederate defenses around Hardeeville, Magrath, the board hoped, would be more focused on the railroad. The new board of directors differed slightly from those previously nominated. Nine of the twelve nominees of the joint committee were elected. Instead of Richards, Williams, and Mills, however, the stockholders selected Thomas Ryan, John H. Steinmeyer, and the man who had overseen the construction of a major portion of the road, Captain Robert L. Singletary.[19]

Despite their previous threats the bondholders did not foreclose. Perhaps it was because they had succeeded in placing five of their six nominees on the board. Maybe they considered such a move at such a critical time to be detrimental to the long-term prospects of the road. Was the exclusion of two men from a company's board worth an ownership struggle? For whatever reason, the answer was "no." The company proceeded with a new president and a balance of power split between the owners of the 7 percent bonds and the Charleston City Council. The city government's stranglehold on the management of the company was loosened and the threat of foreclosure gone. The railroad's management could now turn its attention to increasing commercial business and confronting the ever-present danger of a military assault on the road.

A common theme among Southern railroads at that time was deficiency—of capital equipment, of revenue, of supplies. The most immediate deficiency Magrath had to tackle was that of the railroad's lack of rolling stock. Because there was little central control of the distribution of rolling stock, individual roads had to comb the countryside to fulfill their needs. Shortly after becoming president, Magrath met with State Attorney General Isaac W. Hayne and informed him that the unpredictable demands of the military were exacerbating the already limited travel over his railroad. In order to conduct the business of the railroad and continue to fulfill the needs of the military, he would need more rolling stock. At Hayne's recommendation the South Carolina Executive Council approved a loan of twenty-five thousand dollars to the railroad. This loan was granted on the condition that any state expenses owed to the railroad for the transport of troops, provisions, or other property would be considered payment by the company for part of the loan. The council also stipulated that any cars or other property purchased with the money would be mortgaged to the state until the loan was paid off.[20]

In early March, Magrath traveled to Augusta, Georgia, in search of railcars. He had been authorized to purchase twenty cars for the sum of $12,500, but he could not find any sellers. Railroads throughout the Southeast were experiencing the same problems as the Charleston & Savannah and either could not or were unwilling to spare any of their rolling stock. Magrath telegraphed railroad superintendents in Macon and Chattanooga, but he had no luck there either. He made a conscious decision to stay away from Savannah in his quest, even though this seemed the logical place to ask. The previous summer, the Central of Georgia had loaned the Charleston & Savannah nearly two dozen cars but had done so reluctantly.[21] Acknowledging their prior hesitation, he was not optimistic that they would provide the needed rolling stock. He was concerned that such a refusal would create resentment between the two cities.

After failing to procure cars in Atlanta, Magrath looked closer to home. Drawing on his connections with the South Carolina Railroad, he was able to secure some cars from them. There were so few available, however, that he also had to enlist the services of Charleston carmakers Wharton & Petsch to alter cars that the company already owned. This was just a stopgap measure, but it was better than nothing.

With additional money at his disposal, Magrath set out to rectify another inadequacy of the railroad. Until the war, traffic on Southern railroads had not justified the construction of many sidings, also called "turnouts." Construction of additional turnouts would aid the removal of provisions, livestock, and other freight from the cars without obstructing the main line. These improvements carried an $8,200 price tag, and were completely funded by the state's loan.[22] After taking steps to improve the efficiency of the road and increase its carrying capacity, Magrath turned to another problem he had inherited—the lack of a bridge across the Ashley River.

In November 1861 plans had been announced for a connection between the Charleston & Savannah and the South Carolina Railroad's Spring Street depot via New Bridge. This important connection was nowhere near being finished when General Lee was reassigned to Richmond in spring 1862. Lee considered the bridge vital, and hinted that the Confederate government might lend financial assistance toward its completion. Thus far, that aid had not materialized. When General John C. Pemberton took command, he reiterated the essentiality of the project, calling it not only "a matter of great public interest" but also "emphatically a military necessity."[23]

Magrath saw the connection's practical importance from both a military and commercial standpoint. An Ashley River bridge was needed, but the cash-poor railroad could not undertake the project alone. Magrath wrote to Hayne that if the South Carolina Executive Council was not willing to pay for the project, it should at least find a less expensive way to allow the passage of railcars into Charleston. The inefficient passage across the Ashley and lack of rolling stock soon became more noticeable, as the enemy was beginning to stir.

On March 4, 1862, General Lee was ordered back to Richmond. A few days later Major General John C. Pemberton officially took command of the Department of South Carolina, Georgia, and East Florida. Though Lee did not see the effort around Savannah to its completion, the works he had helped to establish in Charleston,

Savannah, and along the railroad line were the foundation of an effective defensive line that were tested many times in the coming years.

General Pemberton had been assigned to South Carolina as a part of General Lee's staff, and had been recently promoted from brigadier general. After Lee's departure, Pemberton set up his headquarters near the railroad at Pocotaligo. Though he assumed command of a well-planned defensive scheme, he had to deal with intrusive politicians, stubborn colleagues, and a significant reduction in troop strength. Lacking Lee's gentility, the Pennsylvania-born Pemberton had a strained relationship with Governor Francis Pickens. In several communications to Pemberton after he left, Lee urged him to "maintain harmony between the State and Confederate authorities" by improving communication with state officials.[24] Another thing Lee had done before he left was impress on Pemberton the importance of defending the cities of Charleston and Savannah and, to that end, maintaining the line of the Charleston & Savannah Railroad.

On March 19–20, 1862, a Federal force under Lieutenant Colonel John H. Jackson executed a reconnaissance in force on the May River between Bull Island, Savage Island, and the mainland, with a force consisting of the Third New Hampshire and a detachment of the Third Rhode Island Regiment. They left Seabrook Island on fifteen flatboats, but a late-afternoon squall and impending darkness temporarily thwarted their plans. They put their boats ashore to protect their fieldpiece and waited for the weather to break. The next day, the mini-armada resumed its mission, monitored closely by mounted pickets on Hunting Island and the mainland. The Federal troops occasionally exchanged fire with the vedettes, but no one on either side was hit. As the boats moved slowly along in the river, Confederate pickets sent word of their movement back to General Drayton's headquarters.

After maneuvering between the islands and mainland for most of the morning, the Federals landed separate forces at Kirk's and Baynard's plantations, capturing four Rebel pickets in the process. On landing Jackson shelled the woods around the perimeter of Baynard's plantation to disperse the pickets and sent out skirmishers to flush out any additional resistance. After learning of a relatively large Rebel force on the mainland, he held his ground and sent for an additional fieldpiece.

Jackson's hesitation gave General Drayton time to move troops into a forward position that would allow him to cut off the enemy if he overextended himself. Because of an overestimation of the Union troops by Major John B. Willcoxon, who commanded the pickets, General Pemberton initiated a series of defensive maneuvers based on the belief that the Federal force was part of a larger movement against the railroad. General Daniel S. Donelson was ordered to move several companies of the Twelfth South Carolina by rail to Grahamville, and the rest of his command marched toward Buckingham to oppose the Federal advance. General Maxcy Gregg sent Colonel James Jones's Fourteenth South Carolina Infantry by train to Coosawhatchie and then to the earthworks near Bee's Creek. This well-orchestrated defensive movement met with no resistance; the advance on the railroad never materialized. Having placed his second artillery piece, Jackson advanced on the village of Bluffton, and after a brief reconnaissance the Federals retired to their boats. Having achieved their objective, to feel out the Confederates' defensive plan, the Federals steamed back toward their base on Hilton Head with no losses.[25]

This Union operation had elicited a good demonstration of Pemberton's overall defensive strategy. Cavalrymen were stationed as pickets at critical observation points to skirmish with the enemy while word of its advance was relayed up the chain of command and eventually telegraphed to district headquarters. Units in closest proximity could be moved forward to meet the attack while troops from nearby posts could be moved by train to fortify interior positions defending the railroad. This dry run had demonstrated the responsiveness of the Confederate troops to a perceived threat and highlighted the importance of accurate intelligence in preventing exaggerated responses to such small reconnaissance missions. On a small scale the mission had made evident the symbiotic relationship between the Confederate military and the Charleston & Savannah Railroad. The railroad needed the military to keep its rails from being cut by the invaders; the military had to have the rapid-deployment capability and communications provided by the railroad.

Alarmed over the increased activity in the region, President Magrath asked General Pemberton how the railroad would be impacted by additional military traffic. Pemberton tried to reassure Magrath that he did not anticipate any interruption of the business of the railroad in the ensuing months. He did not foresee a great increase in military or other government transport that summer, but he did allow for the possibility of emergencies arising. Of course this would necessitate the forwarding of troops from either Charleston or Savannah to support points of attack along the line. Since he would need to travel to Savannah more frequently and might be needed at other intermediate points along the railroad, he thought he might need a special train for his personal use in those instances.[26] Otherwise, he felt business should continue as usual.

The morning of May 29, 1862, began routinely for Colonel (later brigadier general) William Stephen Walker. A native of Pittsburgh, Pennsylvania, he was raised by his uncle, Senator Robert J. Walker of Mississippi. William Walker began his military career as a midshipman in the U.S. Navy. After serving in the Mexican War (1846–48), he was commanding a Federal gunboat at the time of the secession. He soon resigned his commission in the navy and entered the Confederate army, serving in the office of the inspector general and as an infantry officer. By May 1862 he had risen to the rank of colonel and had been assigned to command the Third Military District of South Carolina.

On May 29 Colonel Walker had been away inspecting his troops and was riding the train back to headquarters. A lot had happened in the previous two months to complicate the defense of his section of the lowcountry. After the Confederates had spent months buttressing the walls of Fort Pulaski, the Federals had bombarded it into submission in April. The Union army now controlled entry into and exit from

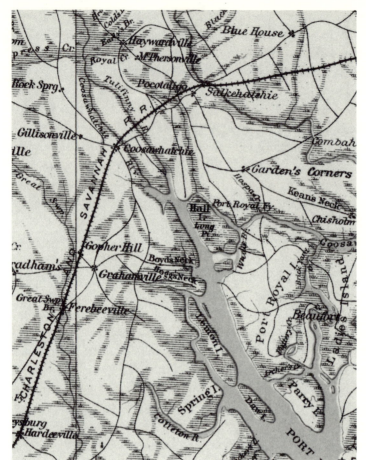

Troop movements during the First Battle of Pocotaligo, May 29, 1862. From *Atlas to Accompany the Official Records of the Union and Confederate Armies*, plate 91:4, detail; courtesy of Rare Books and Special Collections, Thomas Cooper Library, University of South Carolina

the Savannah River. On May 13, 1862, a black man, Robert Smalls, and a crew of fellow slaves stole the steamship *Planter*, smuggled family members aboard, piloted it out of Charleston harbor, and handed the ship over to the Federal blockading squadron. More important than the loss of the *Planter* was the intelligence that Smalls would surely pass along to the Federal commanders about the abandonment of Coles and Battery islands at the mouth of the Stono River.

About a month earlier, during the last week of April, Pemberton had moved his headquarters from Pocotaligo to Charleston, and a week later, Walker had been named commander of the Third District, encompassing the area between the Ashepoo and Savannah rivers. With only a shadow of the force originally intended to man the defensive works, Walker was faced with defending an expansive section of rice fields, rivers, and plantations that lay in front of the vital rails connecting Charleston and Savannah. The depletion of his ranks had begun in April 1862, when Pemberton had received orders from Richmond to send six regiments from his department to General Beauregard in Tennessee. From the Beaufort District, Pemberton sent the Eighth and Sixteenth Tennessee regiments under General Donelson.

Gradually, as other troops were dispatched to Northern Virginia, North Carolina, and Tennessee, Pemberton and Walker's capacity to resist a Federal assault was diminishing. As important as the Charleston & Savannah Railroad was to the Southeast, the fact was that some other areas took precedence in the grand scheme of the Confederacy.

Pemberton also had to contend with the Union occupation of Edisto Island and the threat of an attack on the railroad from that location. With the cities of Charleston and Savannah more vulnerable to attack, Pemberton had pulled back nearly all troops stationed on the railroad between the Ashley and Okatie rivers except for the cavalry already along the line. Earlier in the year Pemberton and General Thomas Drayton had pressed for an alternative railroad connection between Charleston and Savannah, should the enemy cut the direct line. Of some consolation to Walker was the fact that General Drayton still had a relatively strong contingent of infantry and light artillery in his Fourth District and could send reinforcements by train in the event of an attack at Pocotaligo or Coosawhatchie. As Walker returned to his headquarters on the afternoon of May 29, his train slowed to an unscheduled stop. A gray-clad trooper rushed into his car with a telegram. The Yankees were landing near Port Royal Ferry, and were beginning a march toward the railroad.

Union general Isaac Stevens was intent on reaching and cutting the Charleston & Savannah Railroad. With adequate intelligence data and what he considered sufficient troops, he decided to make a quick hit on the railroad at Pocotaligo. He entrusted the operation to Colonel Benjamin C. Christ and the Fiftieth Pennsylvania Volunteers, which were supported by detachments of the Seventy-ninth New York, the Eighth Michigan, Rockwell's battery of Connecticut artillery, and the First Massachusetts Cavalry.

The Federal force began crossing Port Royal Ferry aboard flatboats at 3:00 A.M. on May 29. The crossing was easy enough for the infantry, but the cavalry and artillery had a more difficult time. There was also a critical delay caused by the need to feed and water the animals that transported the artillery's fieldpieces. Once across the infantry began marching at 5:00 A.M. The cavalry followed about an hour later, but the artillery did not get underway until 8:00 A.M.[27]

About two miles from the ferry, the Federals drove in the Rebel pickets, some of whom dashed off to alert headquarters. These mounted pickets continued to harass the Yankees along their march with hit-and-run fire. The Federals marched through Garden's Corners and up Sheldon Church Road to Castle Hill Plantation, easily deflecting these mini-assaults. At around 10:30 A.M., they approached Old Pocotaligo.

At about 9:00 that morning, word of the Union march reached camp. Orders were given for several companies of the First South Carolina Battalion of Cavalry, the Rutledge Mounted Riflemen, and a company of cavalry under Captain D. Blake Heyward to get booted and saddled. Major Joseph H. Morgan of the First Battalion was the ranking officer in the field, and the Rutledge Mounted Riflemen were commanded by Captain William L. Trenholm, a brother of Charleston & Savannah

Brigadier General Isaac Ingalls Stevens, U.S.A. Courtesy of Library of Congress

Railroad investor George A. Trenholm. These three commands assembled at Pocotaligo Station and began a ride to meet the enemy.[28] General Pemberton telegraphed General Drayton to send a regiment at once from Hardeeville to Pocotaligo, but warned the men to be cautious as they approached. Drayton was to communicate with Colonel Walker as to the disposition of the troops before the train left Hardeeville. Colonel William Stokes was on his way to Pocotaligo from Grahamville with the main body of the Fourth South Carolina Cavalry. At the tiny village of Ravenel a train was made ready for the Seventeenth South Carolina Regiment, which was sent on its way.[29]

On the approach to Old Pocotaligo was a causeway over Screven's Canal flanked on either side by marsh and rice fields. Less than one hundred yards from the causeway, the Confederates had dismantled a small bridge over which the advancing Union troops had to cross, leaving only the six-inch stringers. After his men disassembled the bridge, Trenholm deployed his mounted riflemen among the live oaks in the woods adjacent to the road. Morgan dismounted his cavalry and placed them along the left bank of Screven's Canal. Heyward's company was held in reserve and was poised to charge if called upon.[30]

As the vanguard of the Fiftieth Pennsylvania appeared, the Union soldiers on the point noticed the dismantled bridge and began scanning the area for the enemy. As they came within forty yards of the causeway, some of the excited secessionists opened fire. The Yankees randomly returned fire, and were quickly positioned by Colonel Christ—five companies to his left and two companies to the right of the canal. The Confederates under Morgan and Trenholm kept up a slow and deliberate fire on the Federals as they exposed themselves. They returned fire more frequently, but inflicted little damage because of the protected positions held by their Southern counterparts.

As this prolonged firing went on, Colonel Christ was expecting the artillery to arrive at any time. After two hours of fighting, however, he knew he must act quickly—without the artillery—if they were to reach the railroad before sundown. Captain Charles Parker volunteered to lead his company across the string pieces of the torn-up bridge. With cover from the troops on the Union left, Parker led his men across the six-inch boards and into a ditch to the enemy's right. As he exhorted his troops to move forward, Parker was fatally wounded, but his company pressed on. Eventually three hundred men under Lieutenant Colonel Thomas S. Brenholtz were able to cross over and flank a detachment of the Confederate Rutledge Mounted Riflemen under Lieutenant R. M. Skinner. Having been cut off from the causeway, the flanked Confederates sought cover in a ditch on their left.[31]

After the Federal flanking maneuver, the Rebels quickly retreated into the woods and fell back to a position three-fourths of a mile to the rear toward the railroad. The temporarily jubilant Northerners immediately relaid the bridge, allowing the remainder of their force to cross, and began to chase the retreating enemy. As the Confederates retreated, the small force came upon an ammunition train and reinforcements from Hardeeville under Colonel Walker, arriving on the field after a hurried train ride. The Federal troops were already exhausted from their morning march, three hours of skirmishing, and chasing the Confederates through the swamps and forests. The Federal pursuit fell apart after penetrating as far as Thomas Elliott's plantation. Milton Maxcy Leverett observed that many of the Union soldiers stopped to rest at a spring. "They could have taken the railroad then," he wrote, "but they seemed to have been afraid, they were there at least two hours before the cars came up with reinforcements."[32]

As Colonel Walker entered the field, he sprang into action. He had orders from General Pemberton to burn the buildings at Pocotaligo Station if the troops were driven back as far as the railroad, but he had no intention of giving up easily. The

Brigadier General William Stephen Walker, C.S.A., (above). Courtesy of South Carolina Historical Society

Brigadier General Stephen Elliott, C.S.A., photograph by George S. Cooke. Courtesty of South Caroliniana Library, University of South Carolina, Columbia

resupplied Confederate troops and their reinforcements were deployed in a ditch crossing the road, while some of Trenholm's command were left in observation. The first Confederate reinforcements to arrive were three pieces of the Beaufort Volunteer Artillery under Captain Stephen Elliott, which Walker placed on high ground commanding the road. Next came the train of Captain Allen C. Izard's company of the Eleventh, which arrived from Hardeeville at the same time as Captain B. F. Wyman's company of that regiment, which had marched from McPhersonville.[33] Walker placed these troops as skirmishers along a canal perpendicular to the road. Thus far, the overall defensive plan was working. The initial resistance allowed time for reinforcements to arrive by rail, and the strengthened force was now in a stout defensive position in front of the railroad.

By the time Colonel Christ could consolidate his scattered Yankee force at Old Pocotaligo, three more hours had elapsed. Hearing reports of Confederate artillery and infantry arriving by train and cavalry pushing up from Grahamville, Colonel Christ reassessed his situation. His men were physically tired, were relatively low on ammunition, and were facing a Confederate presence that was growing by the hour and now held a more favorable defensive position. Heeding General Stevens's directive to avoid a larger engagement, he made the decision to burn the wagon bridge over the Pocotaligo River and retreat back toward Port Royal Ferry. To the Confederate soldiers it appeared that, as soon as the Federals saw the approaching train, they "commenced to retreat and couldn't be caught up with."³⁴

Hearing that the Federals were retiring, Colonel Walker sent the main body of his force after them. They chased the Yankees back down the Stony Point road with the Eleventh South Carolina leading the way to prevent an ambush. The cavalry soon picked up the chase, but because the Federals had burned Pocotaligo Bridge, they were too late to cut them off.³⁵ General Stevens had heard of the Confederate reinforcements and sent some of the reserve troops up to Garden's Corners in support of the retreat. The chase ended as darkness descended on Garden's Corners, when the Union rearguard fired into the pursuing cavalry, killing one of the mounted Confederates and severely injuring another. The reinforcements sent by Stevens afforded a somewhat more peaceful crossing, and the entire force was safely across the ferry by 3:00 A.M. on May 30.³⁶

Bolstered by the arrival of Phillips's Legion of Georgia Volunteers, Colonel Walker's troops set out early the next morning for Port Royal Ferry. Arriving there, they found that the enemy had recrossed during the night. Frustrated at his inability to do more damage to the invaders, Walker ordered Captain Elliott's artillery to throw a few shells at the Federal pickets posted at the ferry house across the way.³⁷ The fight over, both sides sat back to count their losses. The Union had lost two killed, nine wounded, and one prisoner. The Confederates counted two dead, six wounded, and two prisoners.³⁸ Though they had come to within a mile or two of the railroad, the assailants were no longer on the mainland. The rails were safe—at least for the present.

General Stevens considered the operation another successful reconnaissance mission. It confirmed his initial suspicion that the Rebels were not concentrated in force on the railroad. The tardiness of the artillery, which had not shown up at the causeway until the end of the skirmish, had thwarted the day's effort. Had it arrived with the main force, they could have easily dislodged the Confederate resistance and reached the railroad at Pocotaligo Station. From there, asserted General Stevens, they could surely have disrupted the railway from Salkehatchie all the way to Coosawhatchie.³⁹

Yet the Confederate command had put up an effective defensive effort. Even though the reinforcements from Ravenel and Grahamville had arrived too late to engage the enemy, Walker's men had successfully defended the railroad against superior numbers. The skirmish had also backed up the earlier assertion that the farther the Yankees could be drawn away from the protection of their gunboats, the easier it would be to stop an assault on the railroad. The Southern troops appeared to have executed their strategy effectively; however, they were fortunate that the Federal force had not been larger and that they had been unable to bring up their artillery sooner.

After this close call General Pemberton immediately contacted the head of the Charleston & Savannah Railroad. He highlighted to President Magrath the importance of maintaining direct communication between the cities of Charleston and Savannah and stressed that he did not have adequate troops to station at all possible points of attack. He knew that the army and the railroad needed one another for survival and proposed an arrangement he hoped would be mutually acceptable. He requested that the company keep sufficient rolling stock ready to move at a moment's notice, but asked that Magrath charge the government the cost of "running expenses only." In exchange he would keep a regiment of infantry and a detachment of artillery, in addition to the cavalry already stationed along the line, at a healthy point close to the road, such as McPhersonville. He wanted the rolling stock to be kept near Salkehatchie Station to use the turntable there for movement in either direction.[40] If these proposals were acceptable, Pemberton wanted the initial transport of those units to McPhersonville to be made free of charge. After careful thought, Magrath assented to the proposal. He had a railroad to run, but he also knew that the military was essential to that railroad's survival.

SIX

Secessionville to Pocotaligo

General Lee had instructed General Pemberton to protect the city of Charleston at all costs. Since New Orleans had already fallen to the enemy, Lee felt that the loss of Charleston would cut the South off from the rest of the world. Lee and Pemberton believed that Union generals David Hunter and Henry W. Benham were likely to leave the railroad alone, given the unhealthiness of the area and the difficulty of operating in the marshy ground adjacent to it. The troops stationed along the interior lines of the road were likely to suffer more from disease than from Yankee minié balls. For this reason Lee advised Pemberton to withdraw defenders from these areas to bolster the defenses of Charleston and Savannah. On March 27, 1862, Pemberton made the decision to abandon Coles Island at the mouth of the Stono River. He knew that his troops would not be able to defend the island successfully and feared the loss of the men and guns located there. With the help of Colonel Ambrosio J. Gonzales, Pemberton chose to erect a system of interconnected earthworks on James Island, which was a stronger defensive position, and placed his men farther out of the range of Union gunboats. His decision was unpopular in South Carolina, primarily because it gave the Federal navy complete control of the waters of the Stono River from its mouth up to Fort Pemberton.

Shortly after the failed dash on the railroad at Pocotaligo, General Hunter loaded nearly all the Federal troops stationed at Port Royal and Hilton Head onto transports and set sail for Charleston. Having been provided information by former slave Robert Smalls, Hunter knew that the Confederates had abandoned Coles and Battery islands and that the mouth of the Stono River was virtually unprotected. On June 2, 1862, troops under General Horatio G. Wright, who had been occupying Edisto Island, were ferried across to Seabrook Island, marched across Johns Island to Legareville, and were then transported to James Island for the anticipated assault on Charleston. During the march hundreds of soldiers—mostly New Englanders—fell victim to the hundred-degree heat and nearly 100 percent humidity as well as the mosquitoes and sand fleas that were ever present on the Carolina barrier islands at that time of year. That night they were subjected to torrential rains and a menacing display of lightning. General Isaac Stevens's men left Beaufort aboard transports on June 1 and went ashore the following day on James Island under fire from the

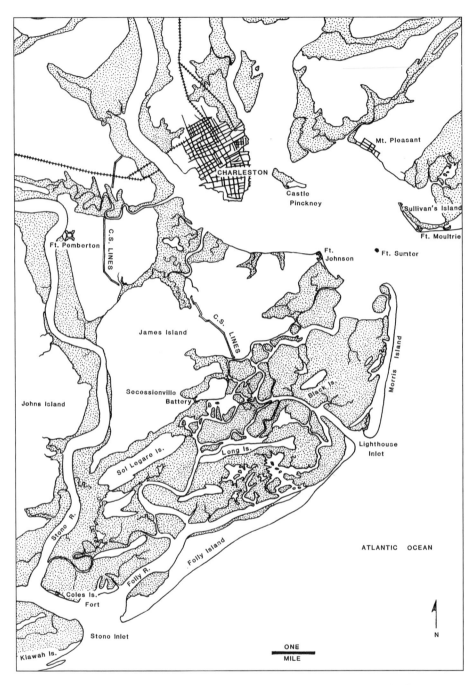

Confederate defenses at Charleston, spring 1862. From *Gate of Hell: Campaign for Charleston Harbor, 1863,* by Stephen R. Wise, © 1994 University of South Carolina

Confederate infantry stationed there.[1] Supporting fire from Union gunboats held the Confederates at bay, allowing the Federal troops to push forward across a causeway to Sol Legare Island, where they set up camp. Though they did not suffer through the same grueling march as General Wright's command, Stevens's men were battered by that night's electrical storm and got little sleep. The soldiers who continued their march the next morning were in no condition to fight.

Confederate colonel Johnson Hagood was surprised that General Nathan G. "Shanks" Evans had not hit the enemy at a time when they were so vulnerable. Such a move should have easily driven them from Johns Island. On June 3 General Pemberton telegraphed Secretary of War George Randolph in Richmond, notifying him that an enemy force was advancing toward the railroad from Seabrook Island. To counter the Union movement Pemberton ordered General Thomas Drayton to send the Eleventh South Carolina from Hardeeville to Adams Run to support General Evans's right flank. Drayton advised General Alexander R. Lawton to arrange rail transportation for two regiments to be sent from Savannah. When General Evans reported the enemy advancing on his position with cavalry, infantry, and artillery, Pemberton ordered an additional regiment up from Hardeeville and one from Savannah, one of which he intended to hold at St. Andrews depot to provide assistance against the Union advance. He called back all officers and enlisted men who were on leave and put the cavalry stationed around Grahamville on alert.[2]

Later that day, elements of the Twenty-eighth Massachusetts, One-hundredth Pennsylvania, and Seventy-ninth New York skirmished with the Twenty-fourth South Carolina Regiment on Sol Legare Island. Cannon from Union gunboats quashed an attack by the Carolinians, allowing the Federals to continue their push toward the port city. Another heavy skirmish took place at Grimball's Landing on the western side of James Island on June 10. General Pemberton had ordered several units forward to establish a battery and drive away Union gunboats in the area. They met a well-entrenched enemy supported by an unopposed naval presence. A Confederate force consisting of the First South Carolina, the Fourth Louisiana, and the Forty-seventh Georgia charged the Union lines but were repulsed as their efforts to contain the Union advance failed again.

With the enemy's massive commitment of resources, Pemberton felt more comfortable taking troops away from the interior lines of the railroad. He ordered General Evans to move two of his regiments to Charleston to shore up the city's defenses. Pemberton also informed W. J. Magrath that he expected an attack via Johns and James islands and requested that Magrath have several trains ready at a moment's notice. He also asked the railroad to halt its passenger and freight service on June 9 because the trains would likely be required for the transport of troops or stores.[3]

The original Federal plan to make a quick strike on Charleston was no longer a possibility, owing largely to Mother Nature. Benham had assumed that the divisions under Stevens and Wright would arrive on James Island simultaneously, but he was wrong. He also thought he would have a superior force that could move on the city with the support of Union gunboats. Because of the distance required, however, Captain Percival Drayton did not feel his guns could damage the Confederate positions as much as would be required for the planned operation. General Hunter was aware of these developments and told Benham not to make any major assault on Pemberton's defenses unless he was strongly reinforced or unless he received a direct order from Hunter to that effect. Hunter returned to Hilton Head to attend to other matters.

Tower Battery sat on a neck of land bordering Big Folly Creek to the south and the tidal shallows of Simpson Creek to the north. It was built in the shape of an *M* so that attackers would be funneled into its center and subjected to enfilading fire. There was a dry ditch along the western edge of a cotton field neighboring the fortification. To the west of that lay a second cotton field, thus creating nearly seven hundred acres of open field across which the enemy would have to march.[4] When the Federals began construction of an artillery battery on Sol Legare Island on June 13, Rebel artillerists in the Tower Battery kept up a steady fire on them. Since the Confederates were constructing extensive works in the area and were strengthening the fortifications of the Tower Battery, General Benham felt it imperative that he conduct what he termed a "reconnaissance in force," with the goal of capturing and holding the tower.

Inside the fortification Colonel Thomas Gresham Lamar commanded five hundred Confederate soldiers who were dead tired on the morning of June 16, 1862. They had worked almost ceaselessly for several days strengthening the battery and trading shots with the Union artillery. The Confederates were so tired that Colonel Lamar allowed them a couple of hours sleep before daylight. Lamar expected an attack and felt fairly certain that its focus would be Secessionville. He hoped that, with many of his troops enjoying some restorative sleep, his pickets would give ample warning of any enemy movement.[5]

In the Federal camp troops were preparing for battle even before their Southern counterparts began their slumber. Though he objected to the attack on Secessionville, General Isaac Stevens had his Second Division ready for battle at 3:30 that morning. With General Wright's First Division marching on his left flank in support, Stevens's division began its silent march with the Eighth Michigan in the lead, followed successively by the Seventh Connecticut and the Twenty-eighth Massachusetts. As the Eighth Michigan marched across the cotton fields with fixed bayonets, they were as quiet as predators stalking prey. Their prey was entrenched and waiting for them.

The first warning from Confederate pickets reached Colonel Lamar at about 4:00 A.M. When the vanguard of the Federal advance was within range, the Rebel artillerists released a devastating blast of canister and grapeshot into their ranks splitting the line in two. Pressing on, some of the Union troops reached the ditch at the base of the earthwork and fought their way up the embankment. Those who

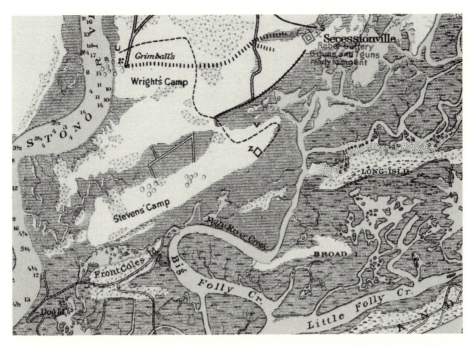

The Battle of Secessionville, June 16, 1862. From *Atlas to Accompany the Official Records of the Union and Confederate Armies,* plate 23:7, detail; courtesy of Rare Books and Special Collections, Thomas Cooper Library, University of South Carolina

eventually reached the parapet "poured a murderous fire" down into the Rebel position before dashing down the embankment and into the battery. Fierce fighting followed, and—just when it appeared the Confederate line might not hold—the Ninth South Carolina (Pee Dee) Battalion arrived to save the day. These fresh troops, commanded by Lieutenant Colonel A. D. Smith, beat back the Eighth Michigan, which withdrew into the hedgerows in front of the battery. After re-forming his men, Stevens twice more charged the Confederate works across the open field, and twice more they were driven back. After again gaining a brief foothold, they retreated into the hedgerows with heavy losses.[6]

For days the Yankees—with the protection provided by their gunboats—had been able to move at will on the islands on Charleston's periphery. In fact Pemberton's troops had not been able to halt the Union advance because of the gunboats that controlled the Stono River. General Benham had hoped that the fleet could put pressure on the defenders of Tower Battery, but they were ultimately of little use. As in the lower part of the state, the Union army could land in force, but the farther they could be drawn from their gunboats the more vulnerable they became. The Tower Battery was just out of accurate range of the Federal gunboats, and their errant shelling endangered friend and foe alike.[7]

During the heavy fighting in front of the works, portions of the Third New Hampshire and Third Rhode Island regiments moved down the adjacent peninsula and fired on the battery from the rear. They could not get close enough to the fort because of the boggy marsh bordering Simpson Creek and were never as much of an interior threat as those in front of the battery. Colonel Johnson Hagood dispatched the Charleston Battalion and the Fourth Louisiana Regiment, who dispensed with these invaders.

The terrain and the narrowness of the strip of land leading to the front of the Confederate works made it impossible for the Yankees to concentrate large numbers of troops on the flanks of the battery. Stevens's and Wright's brigades were stymied in the hedgerows, and the Confederates had turned back the flanking maneuver. General Benham knew that additional frontal assaults would result in the needless sacrifice of his men and ordered a general retreat at 9:00 A.M.[8] The Eighth Michigan, which had led two charges across wide-open ground, had lost one-third of their regiment that day. Other units suffered similarly. Because of their ill-advised attack and decisive defeat General Hunter arrested General Benham, who was subsequently transferred out of the department.

If the Union force had won the Battle of Secessionville, it would have resulted in the flanking of the Confederate defensive line on James Island. Had they followed up their victory by moving across James Island and capturing Fort Johnson, Charleston would have been ripe for the picking. If Charleston fell, its railroads and waterways would have given the Union an excellent staging area from which to move on other areas. A Confederate loss at this juncture might even have caused a diversion of troops to South Carolina from Virginia, depleting the defenses around Richmond. It could potentially have changed the course of the war in its early stages, possibly shortening it.[9] The battle left the Union dazed and discouraged; the Confederates were exhausted but elated with their success.

Hunter replaced Benham with General Horatio G. Wright, who was ordered not to advance on Charleston. Expecting an attack from Pemberton, Wright strengthened his defenses around Grimball's plantation, but such a strike never came. The Union activity on James Island, thus, consisted of defensive maneuvers and a relatively meaningless engagement at Simmons Bluff. With little being accomplished General Hunter ordered Wright to "abandon James Island and go 'to some more healthy location.'"[10]

Under the cover of Union gunboats Wright successfully evacuated his entire command and dispersed them to the three primary points of Federal concentration —Hilton Head, Beaufort, and North Edisto—by June 9. Though defeated and temporarily demoralized, the Union army could count several accomplishments since November 1861. Union troops were firmly embedded on Hilton Head Island and had established a large depot for supporting the South Atlantic Blockading Squadron. Partly because of their naval superiority and partly because of General Lee's strategy, the agricultural wealth of the South Carolina Sea Islands was decimated. Federal gunboats also controlled the Carolina coastline and its inland waterways, enabling them to land substantial numbers of troops with which to threaten the Charleston & Savannah Railroad.

For its part the South had repulsed the Union at Secessionville. They had thwarted a major expedition targeting the city of Charleston and controlled a line of defense encompassing the railway between Charleston and Savannah. There was much to celebrate, but no time for over-confidence. With the threat to Charleston gone for a time, the troops sent there by General Hugh Mercer in Savannah were recalled to bolster the defenses of that city. Things in Charleston seemed to be somewhat closer to normal.

There were problems elsewhere in the Confederacy, however, and these crises could not be ignored. Federal troops were again threatening Northern Virginia, and more troops were needed there. Since it was the "sickly season" on the South Carolina coast, President Davis hoped that General Pemberton could spare up to half of his effective force. On July 13 Pemberton was ordered to dispatch such a large force that he feared his coastal defenses would be debilitated. Two days later, two brigades left South Carolina for Richmond. Brigadier General Nathan G. Evans commanded one, consisting of the Seventeenth, Twenty-third, and Eighteenth South Carolina; Holcombe's Legion, under Colonel Peter F. Stevens; Captain Walter Leake's battery of the Virginia Light Artillery; and Captain Robert Boyce's battery of the South Carolina Light Artillery. The other brigade—consisting of Phillips's Legion from Georgia, the Fifty-first Georgia Infantry, the Fiftieth Georgia Infantry, the Fifteenth South Carolina; and the Third Battalion South Carolina Infantry—was commanded by Brigadier General Thomas Fenwick Drayton.[11] As Drayton boarded the steamer at the railroad's St. Andrews wharf, he was leaving his family and his beloved home state, as well as the railroad that he had helped build and defend. He did not know it, but he would not be reassigned to defend South Carolina, or the Charleston & Savannah Railroad, for the remainder of the war.

For W. J. Magrath and the Executive Council, Secessionville was a wake-up call. Had the Union army broken through, it would have had a clear path to the railroad and the city of Charleston. Magrath feared that if any portion of the railroad fell into the enemy's hands, all the company's rolling stock—valued at an estimated five hundred thousand dollars—would be lost. Magrath would lose his company, and the Confederate army's plan for concentrating and deploying troops would be shattered.

The importance of connecting the Charleston & Savannah Railroad with the Northeastern Railroad and the South Carolina Railroad was never more evident than now. Generals Lee, Ripley, and Pemberton had endorsed the construction of the bridge across the Ashley River, and its merit was becoming increasingly apparent. Because of the national shortage of materials, rolling stock was becoming as important to the state governments as to the individual railroads. A bridge connecting the Charleston & Savannah line with the other two roads would make it easier to salvage the cars should the enemy take control of the rails. The governor and Executive Council had authorized State Attorney General Isaac W. Hayne to take all the necessary measures to complete this project—provided it could be built for less than thirty-five thousand dollars.[12] Hayne believed that it could, and appointed

Magrath, Northeastern Railroad president Alfred F. Ravenel, and South Carolina Railroad superintendent H. T. Peake to a commission in charge of the enterprise.

In May, Magrath had placed the work under the direction of Charleston & Savannah Railroad vice president Bentley D. Hasell. A skilled engineer, Hasell had joined the company after stints as chief engineer at the Memphis & Ohio Railroad and the New Orleans, Jackson & Great Northern Railroad. He possessed the necessary expertise, but the lack of resources at his disposal often frustrated him. With the intensification of military activity in the area, he often found himself in competition with the military for materials, machinery, and labor. In late May, General Pemberton had asked Magrath to "furnish every facility of railroad transportation available" for hauling timber needed for the obstructions in Charleston harbor. The railroad even had to lend the military its ferryboats because of the dilapidated condition of the government-owned steamers. In June, General Pemberton threatened to suspend all work on the bridge unless Governor Pickens sent more laborers to the coast as he had requested.[13]

By August work was progressing more rapidly but the bridge was still less than half finished. Hasell reported that just under half of the planned 152 piers had been driven; 54 of those were capped; and only three hundred feet of longitudinal braces had been placed on the caps. None of the stringers had been placed, but at least he had the necessary timber in his possession. His work on the piers had been delayed, yet again, because he had to loan one of his two pile drivers to the Confederate government for use in the harbor.[14] Despite his progress, Hasell doubted the project would be finished before the enemy renewed its efforts against the South Carolina lowcountry in the fall. As the defenses around Charleston harbor were being strengthened, the Ashley River continued to daunt the railroad and military officials who eagerly desired a bridge across it.

Later that month, the focus shifted to what the railroads were charging for government freight and troop transportation. On a national scale the rate schedules of the various Southern railroad companies had no more structure than the unfinished Ashley River Bridge. The previous January, the Greenville & Columbia Railroad had raised its fare for military personnel to three cents per mile and charged the same rate for government freight as for commercial freight. This set a precedent among many of the other Confederate roads, whose profits were being choked by the escalating operating costs. By the end of the summer, President R. R. Cuyler of the Central Railroad of Georgia decided to take action. He called for a convention of the presidents and superintendents of all railroads in the Confederacy to consider the current state of the railroads, rolling stock, and the rising costs of railroad operation. He also intended to discuss the regulation of fares and freight in proportion to the changing economic conditions in the Confederate States.[15]

W. J. Magrath and Bentley Hasell traveled to Columbia, South Carolina, to represent the Charleston & Savannah at the convention, held at the Columbia City Hall on September 4, 1862. The group unanimously resolved that they would give their best efforts to transport troops and government property. They called on President Davis to issue an order that officers of the government should not interfere with the loading or running of the trains because prior interference had been detrimental to the business of many railroads. The Transportation Committee recommended that passenger fare for discharged, furloughed, and active-duty soldiers should be two cents per mile on main lines and three cents per mile on branch lines. Commissioned officers would be charged full fare. Freight tariffs would be a little more complex; the committee's recommendation ranged from sixty-five cents per mile per pound for first class to fifteen dollars per carload for fourth class. The rates would increase by 50 percent if the freight were carried on passenger trains. The proposed rates would go into effect on October 1.[16]

Railroads in the South were beginning to encounter serious shortages of iron for maintaining their rails. At the beginning of the war, there was only one mill for rolling rails in the entire Confederacy, and the Confederate navy was now consuming a large quantity of iron. The Committee on Supplies recommended that two mills be built for the rolling of railroad iron; one in the coal region of Alabama and the other in the Deep River area of North Carolina. Rail companies in Virginia, North Carolina, and South Carolina would pay for the North Carolina plant, and the companies in the remaining states would fund the Alabama plant. The committee also submitted that the companies should be able to obtain supplies impressed by the government at cost if they were proven necessities.[17]

The only other issues addressed by the convention involved minor scheduling matters. One of these was particularly pleasing to Magrath and Hasell. Because of the difficulties and delays in settling disbursements between connecting lines, it was recommended that through tickets should not be sold beyond Charleston, Savannah, or Augusta in either direction.[18] This would not guarantee increased revenue for their company, but it would put them at the terminus for many through-ticket holders traveling the coastline route in both directions.

The convention approved the resolutions and sent them to Quartermaster General Abraham Myers in Richmond. He was heartened by the rates proposed—largely unchanged from those approved at the Chattanooga convention a year earlier—and gladly certified them. Collectively they had come up with rates they felt were quite reasonable, but they still had to do something to satisfy their stockholders. The Charleston & Savannah, like many other companies, had to pass along some of the burden to private citizens. With their concessions to the Confederate government as a likely rationalization, they soon announced an increased charge for transporting livestock.[19]

While Magrath and Hasell worked to protect the profits of the railroad, the Confederate armed forces were doing their best to protect the railroad after the loss of so many men from the region. After the departure of Evans's and Drayton's brigades, Colonel Johnson Hagood was assigned command of the Second Military District and was promoted to brigadier general. Hagood's command guarded the

Charleston & Savannah Railroad in addition to protecting the local planting interests and serving as an advanced guard for the city of Charleston. His district encompassed the area south of Charleston from Rantowles to the Ashepoo River, and his headquarters was at Adams Run, twenty-five miles from the city. His troops were of mixed arms, and ranged in number anywhere from one thousand to four thousand because of the constant shifting of troops into and out of the district as emergencies required. The region was considered "fatally malarious" in the summer months, and the enemy made use of its many rivers, streams, and creeks for many small raiding incursions that proved a constant nuisance.[20] For the defense of Charleston the assignment was far more important than glamorous.

Since the majority of the area was a network of boggy swamps and tidal creeks, Hagood dismissed the idea that the Federals would attempt any kind of advance on Charleston or the railroad at any point other than between Pon Pon and Rantowles Creek. Hagood planned three successive lines of defense similar to Colonel William S. Walker's defensive strategy around Pocotaligo. The first was south of the Willtown and Rantowles road and was designed to resist just long enough to ascertain the strength and objective of any invading force. The second line was behind Caw Caw Swamp and was designed to delay the advance long enough for reinforcements from the Third Military District to arrive. The third line of defense was behind the Ashley River, and guarded the line of the South Carolina and Northeastern railroads in the event the enemy penetrated further inland. The region from the Edisto River westward to the Ashepoo was defended by Company B of the Sixth South Carolina Cavalry, which picketed Bennett's Point on Bear Island. This small force was reinforced by Captain George H. Walter's Light Battery of the Washington Artillery and sometimes supported by a detachment of infantry if available.[21]

Most of the camps scattered through the Second District were in the midst of mosquito-infested rice swamps. Camp Lee was just across the Ashley River from Charleston, and was often the first camp encountered by soldiers stationed in and around the city. Camp Croft on Johns Island was the base camp for picket duty at Rockville. Camp Pillow, which was established in an open field on Curtis Plantation, was level and good for drilling troops. Some camps, such as Camp Simons, were split into two separate divisions. Camp Simons Number 1 was at Rantowles Station on the Charleston & Savannah Railroad, and Camp Simons Number 2 was at the village of Ravenel, also positioned on the railroad.[22] The camp at Ravenel lay fifty miles from Pocotaligo, a two to three hour trip by rail, depending on the urgency of the trip.

Camp life for these soldiers was varied during the second summer of the war. Many of the harsh realities experienced by troops in Tennessee and Virginia had not yet come to the Carolina coast. Officers often were able to have their families come for extended visits, staying temporarily in the abandoned summer houses of the planters around Adams Run. With the railroad nearby, care packages from home were easily accessible for those whose families could afford to send them. Many of the green recruits had little military experience, and they spent their days drilling and learning military procedures. General Hagood established schools of instruction in each regiment and personally inspected his men. He also organized boards for examining the competence of his officers. There was an air of gravity in the camps; however, the soldiers were able to enjoy a significant amount of free time.[23]

For many younger soldiers there was much fun and adventure. Days were spent fishing the waterways, crabbing in tidal creeks, and hunting game. Some of the troops also enjoyed taking nature walks, playing ball, or taking time out for a boat ride. At night around the campfire, there was hardly ever a shortage of fiddlers. One of the old hands would saw out a tune while others sang or danced reels until lights out.[24] Once tattoo had sounded, they would retire to their tents for unsettled sleep and swatting the biting, malaria-carrying pests that infested the lowland camps.

Military work around Adams Run and the railroad did not amount to much at this point in the conflict, entailing little more than drilling, parading, and mastering weaponry. At other times troops busied themselves by making furniture for the camps and foraging for food on neighboring plantations and islands. Those stationed around Adams Run were ordered to build a road across the marsh and a bridge across the Stono River at Church Flats to the mainland. There was, however, a great stir in camp if the troops received orders to reinforce a particular area when danger threatened. On command they boarded a train, or marched to the threatened area, only to find false alarms more often than real danger.

Such an incident occurred on July 10, 1862, when two gunboats fired into the woods near Port Royal Ferry. Pickets sounded the alarm that the Yankees had again crossed over at the ferry. Passengers on the afternoon train to Savannah circulated the rumor that the Federals were marching inland toward the railroad. A company of the Eleventh South Carolina Infantry boarded a train at Hardeeville and proceeded to Pocotaligo, where the troops rested on their arms until officers ascertained exactly what was happening. After several hours Confederate scouts determined that no Federal troops were on the mainland, and the men of the Eleventh reboarded the railcars and returned to Hardeeville late that evening.[25]

Another aspect of "army work" combining excitement, boredom, and danger was picketing. Picket duty was a long, lonely, and often monotonous detail. When ordered out to picket, a soldier rustled up a week's rations, grabbed a blanket, shouldered his rifle, and marched to his post. Most of these stations were on potential enemy landing sites, and they were to watch for Yankee ships or boats in their vicinity. They often ended up watching the waves lapping on shore, with the monotony broken by the occasional gull flying overhead or the blowing of a porpoise offshore.[26] When not standing guard, pickets would go out in search of something to eat.

Under General Hagood's command these posts were manned by mounted sentinels, or vedettes, who were posted at every landing along the defensive line. A strong infantry guard was posted at a church about a mile from Adams Run, and the main body of troops bivouacked at Adams Run for strategic and sanitary purposes.

The pickets' horses were never unbridled or unsaddled, and only half could be fed and watered at a time. The soldiers had quick access to their guns as they were always on the highest alert. Only half the men were allowed to sleep in the daytime, and all remained on alert during the night. Fires were not allowed when it might be possible for the enemy to spot them. The commanding picket officer was notified at the movement, or even suggestion of movement, of the enemy. Any advance was reported to the superintendent of pickets, Major John Jenkins of the Third South Carolina Cavalry and the vedettes would fall back, keeping the enemy in sight and putting up any resistance possible.[27]

Maintaining the defense of the city and the railroad also meant keeping the soldiers in the field. Poor hygiene in many camps had resulted in outbreaks of measles, mumps, and diarrheal illnesses. These, along with insect-borne diseases, resulted in many sick furloughs and deaths. Sanitation, nutrition, and battling the excessive heat were at certain times larger problems than defending their land from the Yankee invaders. In this vein the district's chief surgeon, Dr. J. F. M. Geddings, introduced sanitary guidelines for the sickly season. Trash and sewage were handled appropriately, bedding was kept dry to prevent disease, and soldiers received preventative therapy for malaria. Officers inspected food to protect against serving the troops unripe or rotting fruit and vegetables. Dr. Geddings instituted frequent roll calls to prevent straggling and forbade such military duties as drilling and parading during the heat of the day. A shelter was built for the camp sentinels to prevent overexposure to the sun.[28] As a result of the efforts of General Hagood and Dr. Geddings, the efficiency of the force in the Second District was never seriously impaired during the sickly season.[29]

There were few events of any military interest in the district between July and October other than the occasional exchange of fire between pickets. The men were becoming restless in their daily routine and hoped for an opportunity to fight the Yankees. When the Federals on Hilton Head began an advance on the Third Military District on October 22, 1862, Colonel William S. Walker sent General Hagood an urgent dispatch calling for reinforcements. General Hagood immediately assembled the Seventh South Carolina Battalion at the railroad depot, stopped and emptied a passing train, and sent the troops to Walker's aid at Pocotaligo.

In late October, Major General Ormsby MacKnight Mitchel became the next Union general to attempt to cut the railroad. After taking command of the Department of the South, he addressed his troops with a fiery speech, telling them that they "had been altogether too long inactive, that the capture of Port Royal was intended to be but the beginning of operations that should end in the capture of Charleston and Savannah; that now we should prepare ourselves to move upon the enemy's works."[30] General Mitchel's ambition was fueled by successful operations against Jacksonville, Florida, and the Bluffton Saltworks in South Carolina. His goal was to cut the railroad, thus disrupting communication between the two cities and making each vulnerable to a sudden strike.

Major General Ormsby MacKnight Mitchel, U.S.A.
Courtesy of South Carolina Historical Society

As the steamy summer almost imperceptibly transitioned into autumn, Confederate colonel William S. Walker kept a watchful eye to the south of his McPhersonville headquarters. He knew his district well and knew that it would not be long before the Yankees made another attempt on the railroad. Colonel Walker believed that this move would come somewhere between Grahamville and the Combahee River bridge because it was the most remote section of the railway. He lamented that the Confederate high command's policy of abandoning their defensive posture—favoring one primarily of observation—would make many of the district's earthworks obsolete. They could be used in hampering the forward progress of an assault, but the "final stand" should be closer to the railroad to allow time for the arrival of reinforcements. Rebel pickets had noted increased activity the south of the Coosaw River, and on October 14 General Pierre G. T. Beauregard, who had recently replaced General Pemberton as head of the Department of South Carolina, Georgia, and East Florida, received intelligence of an upcoming Federal attack. He immediately called on General Hagood at Adams Run to have a thousand troops ready to move at a moment's notice, requested that General Mercer in Savannah be prepared to send two thousand, and readied two thousand of his own troops to be sent from Charleston in case of an emergency. Beauregard suggested several plans for defending likely points of Federal attack; however, none of these contingencies addressed landings on Mackay's Point, Gregorie's Neck, or the Coosawhatchie River above Bee's Creek.[31]

From his base at Port Royal, General Mitchel planned an expedition up the Broad River, hoping to land a substantial force and destroy as much of the railroad as he could in one day. He had sent scouts and spies to all the key points on the Charleston & Savannah Railroad between the Savannah and Salkehatchie rivers and decided that he should make a move on Pocotaligo. The units sent on this assignment included the Forty-seventh, Fifty-fifth, and Seventy-sixth Pennsylvania; the Third and Fourth New Hampshire; the Sixth and Seventh Connecticut; the Forty-eighth New

York; the First Massachusetts Cavalry; the Third Rhode Island Artillery; and a detachment of the First U.S. Artillery. General Mitchel was originally to command the mission himself, but he became incapacitated by yellow fever. He informed General John M. Brannan that he would be in command only a couple of hours before the troops set sail.[32]

They left Hilton Head between 11:00 P.M. and midnight on October 21, with a thick fog rising from the river. Poor visibility made communication between the lead ships and those in the rear quite difficult. For the voyage they had procured the services of contraband slaves who had piloted these rivers and streams for their former masters on many occasions. They steamed ahead through the mist for their objective, Mackay's Point at the mouth of the Pocotaligo River. Mackay's Point was at the southern tip of Mackay's Neck, formed by the Pocotaligo and Tullifinny rivers. The Federals had chosen this point because it could be protected from flank attack by gunboats in either river. From there it was about a seven- or eight-mile march to the railroad. Brannan had sent elements of the Seventh Connecticut that night to capture the pickets near the landing area, but they were only partially successful. General Brannan, aboard the *Ben De Ford*, arrived at Mackay's Point at about 4:30 A.M. on the October 22, well ahead of the other boats, which were delayed about three hours when of one of the large transports ran aground. This delay gave the Rebels an opportunity to begin defensive action and to telegraph Charleston and Savannah for reinforcements. When the main body of the Union force arrived at the landing site, most of the troops unloaded and prepared for the ensuing march. General Brannan also sent a detachment consisting of the Forty-eighth New York, the Third Rhode Island Artillery, and a regiment of the New York Engineers further up the Coosawhatchie River.[33]

Sometime between 8:30 and 9:00 a.m. Colonel Walker was notified at his McPhersonville headquarters of the Federal landing party at Mackay's Point. Almost simultaneously, he received a second dispatch that four additional boats were ascending the Coosawhatchie River toward the village of Coosawhatchie as well. He ordered the majority of his troops toward Old Pocotaligo and sent the LaFayette Artillery under Lieutenant L. F. LeBleux and a section of the Beaufort Artillery under Lieutenant (later captain) Henry M. "Hal" Stuart to Coosawhatchie to protect the town and the railroad's trestle. Walker also sent telegraphs requesting reinforcements from General Beauregard in Charleston, General Hagood at Adams Run, and General Mercer in Savannah.[34] Hagood immediately put the Seventh South Carolina Battalion on a train bound for Pocotaligo, and ordered several companies of the Twenty-sixth South Carolina on the march from Church Flats to Rantowles Station to board a later train. General Beauregard sent word back to Walker to "hold [his] ground, and fight every bridge." He also told him to have someone watch the railroad so that the troops could be taken off the trains and sent where needed.[35]

Knowing that Captain B. F. Wyman's company of the Eleventh South Carolina was stationed near Coosawhatchie, Colonel Walker sent them to support the artillery

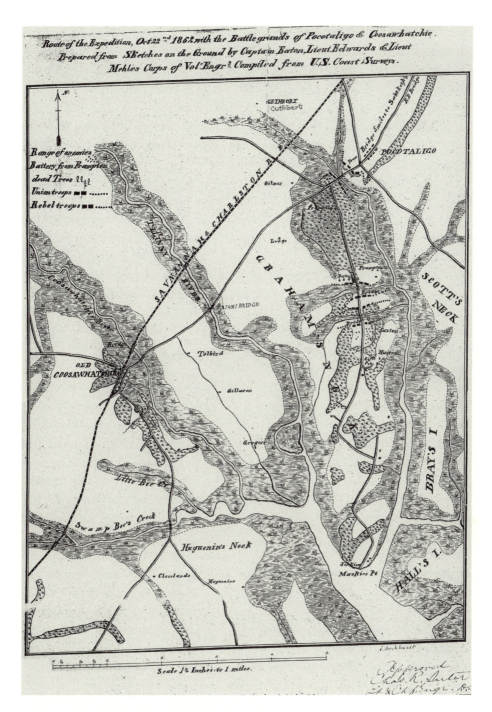

Route of the Second Battle of Pocotaligo, October 21–22, 1862.
Courtesy of South Carolina Historical Society

in the area. He also ordered Colonel Allen C. Izard's company of the Eleventh from McPhersonville to Pocotaligo, and sent five companies of the Eleventh South Carolina Infantry stationed near Hardeeville to reinforce Wyman. Companies C, D, and K, were nearest to the depot, and climbed aboard the first available train to Coosawhatchie. Companies E and G were in Bluffton and had to march fifteen miles back to Hardeeville and take a later train.[36] Troops were on their way from the direction of both Charleston and Savannah. Walker hoped they would arrive in time.

General Brannan's troops formed on Mackay's Neck and began their march through the fields and deserted farm communities toward Pocotaligo. They made slow progress owing to the frequent stops to reconnoiter, and they encountered their first resistance from a battery at Caston's plantation (also referred to as George Parsons Elliott's plantation).[37] Moving to meet Brannan's advance, Walker sent an advance guard made up of a two-gun section of the Beaufort Volunteer Artillery supported by Captain Joseph Blythe Allston's First South Carolina Sharpshooters and two companies of cavalry under Major Joseph H. Morgan, to skirmish while he positioned his main force adjacent to a salt marsh near Dr. Thomas Woodward Hutson's home. The Federal force had marched a little more than five miles and was approaching the Rebel position when Captain Stephen Elliott's artillery opened fire on them. Quickly, Colonel Tilghman H. Good deployed the Forty-seventh Pennsylvania and the Fourth New Hampshire and brought up the artillery. As soon as they were fired on, the small Confederate force withdrew up Mackay's Point Road and dismantled a bridge behind them. Pressing the fight, Chatfield's brigade chased the retreating Rebels up the road toward Caston's plantation, where Elliott deployed his artillery on either side of Mackay's Point Road. As the pursuing Federals came into view, Elliott's guns and Allston's sharpshooters opened fire and—after another brief encounter the Confederates again withdrew—taking apart another bridge along the way.

These delaying actions had given Walker enough time to assemble his force in a strong position near Frampton's plantation, a little over a mile to the north. The earthworks here were part of the defensive line planned by Generals Robert E. Lee and Roswell Ripley nearly a year before, using the natural topography to funnel invading troops into concentrated boluses, making them easier targets. The Confederates were entrenched in the woods with a deep swamp in front, impassable except for a narrow causeway on which the bridge had been destroyed. Along Walker's front and flank was an impenetrable thicket intersected by a deep ditch, passable only by a narrow road. The Beaufort Volunteer Artillery took a position on and near the road commanding the causeway, and the Nelson Artillery was posted in an open field, screened from the enemy by a grove of trees. The Rebel artillery shelled the woods with grapeshot, shell, and canister, and their infantry poured a heavy fire into the pursuing Federals. With ammunition for their artillery running low and the Yankees unable to mount a flank attack, Brannan decided on a risky move. He sent

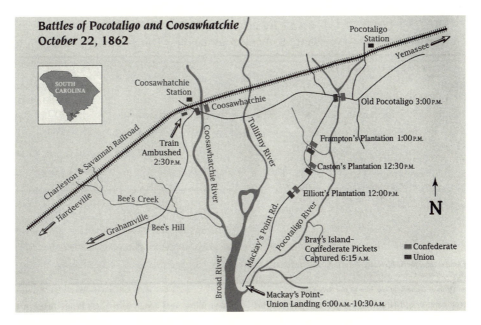

Map by Linda Hathaway. Courtesy of South Carolina Historical Society

the First U.S. Artillery to the causeway and advanced the brigade forward through the thicket to the edge of the swamp. The Forty-seventh Pennsylvania formed a line of battle and advanced against the Confederate position, but the overwhelming fire of the Confederates from across the marsh prevented them from crossing. As Brannan continued to press his men forward, the Confederate artillery lost so many men and horses that several of their guns became disabled. After holding this line for about forty-five minutes, Walker began to fear that the Union forces would soon cross the creek and hit his unprotected flanks. He called for his men to pull back to a position two-and-a-half miles in his rear near Old Pocotaligo.[38]

The Confederates retreated into the woods, and the Federals stormed after them. With Izard's company of the Eleventh South Carolina covering the retreat, Walker led his men back over Pocotaligo Bridge, where many of the soldiers took cover among the houses and trees in the small village. The handful of artillerists who were left hurriedly manned their guns in a position that commanded the causeway and bridge. As Izard's company raced back across the bridge, they dismantled it behind them. The Federals reached the point where Coosawhatchie Road passed through a swamp to Pocotaligo Bridge and received a blast from Elliott's field battery. In their front the Yankees faced a narrow, fifteen-foot deep river, the opposite bank of which was high and steep. The river stemmed the tide of the onrushing Northerners, and they again spread out in Walker's front with the Seventy-sixth Pennsylvania occupying the extreme Union right.[39]

Throughout the day Brannan's troops frequently heard train whistles above the din of battle, each indicating the possibility of reinforcements arriving from camps along the line. Rumors penetrated the Union ranks that a sixteen- to thirty-car train was spotted entering Pocotaligo with fresh troops. Though this story may or may not have been true, the Confederates would have used whatever means necessary to discourage their enemy. During the fight Captain Elliott withdrew one gun of the Beaufort Artillery and placed it three hundred yards on the Confederate right. Since he executed the maneuver out of the enemy's sight, the shelling from a new and unexpected position gave the appearance of a reinforcing unit. At about the same time Walker sent forward the Charleston Light Dragoons, whom he had kept a half mile to the rear in reserve. As they advanced, Walker instructed them to shout in order to convince the Federals that reinforcements had arrived at Pocotaligo. As they joined the fight, they rattled the air—and the nerves of their enemy—with a Rebel Yell, which was echoed by those already engaged. When Lieutenant Colonel Patrick H. Nelson's Seventh South Carolina Infantry Battalion finally arrived on the train from Adams Run, Walker directed half of the force to the front to relieve the exhausted troops and sent the other half to protect the right flank.[40]

With the Federal troops stalled in front of them and the Confederates not making any movement to cross back over the river, two companies of the Seventh Battalion under Captain J. H. Brooks made a rush on the Union left flank. The Seventy-sixth Pennsylvania under Colonel Dewitt C. Strawbridge extended their line and hit the Seventh with a volley, causing the Rebels to fall back. Two more Federal companies came up with muskets ablaze to bolster the Union left. They kept up their fire until many of the Federals depleted their ammunition. Losing daylight and dangerously low on ammunition, Brannan ordered a general retreat back toward Mackay's Point at about 6:00 P.M. As they retreated, Walker sent small detachments of the Rutledge Mounted Riflemen and Captain Manning J. Kirk's Partisan Rangers to chase the Yankees, but their chase was impeded by the destruction of bridges by the Federal rearguard.[41]

Farther up the Coosawhatchie, the Federal diversionary force had also encountered stiff resistance. Soon after daylight Colonel William B. Barton proceeded up the Coosawhatchie River with orders from General Brannan to reach the Charleston & Savannah Railroad, to destroy as much of it as possible, and to destroy its bridge over the Coosawhatchie River. Along with his Forty-eighth New York, he also had under his command the Third Rhode Island Artillery and a detachment of the New York Engineers Corps, who carried implements necessary for tearing up the railroad. They sailed aboard the *Planter*, the side-wheeled steamer that Robert Smalls had run out of Charleston harbor mere months before, with the support of several gunboats. When they reached a point approximately two miles from Coosawhatchie, the *Planter* ran aground at the nadir of low tide, preventing any further progress upriver. The Union troops disembarked at this point on very swampy ground, which made the landing quite difficult. After trudging through the muck and struggling to get a borrowed twelve-pound boat howitzer to solid ground, they finally reached the main road.[42]

At his headquarters in Grahamville, Third South Carolina Cavalry commander, and one-time railroad projector, Colonel Charles Jones Colcock was battling a prolonged bout of fever. In his stead, he put Lieutenant Colonel Thomas H. Johnson in command of his troopers and sent them along with two companies of Major Joseph Abney's First Battalion South Carolina Sharpshooters to monitor the Federal movements near Coosawhatchie. As they approached Bee's Creek Hill, he received information, later determined to be erroneous, that a portion of the Union force was landing at Seabrook Island. Fearing an attack at Grahamville, Johnson divided his force, leaving three of his companies at that point, and proceeded toward town. Two miles from Coosawhatchie, Johnson learned that the Yankees had landed and were moving toward Bee's Creek. He dismounted his men and formed a skirmish line, but the Federals did an about-face, drove in a small band of pickets, and began their march toward town and the railroad. Johnson had his men remount and dash around to the left to cut them off before they reached their objective.[43]

As the Northerners reached a point about a mile from town, they heard a locomotive whistle, creating some excitement among the ranks. A contraband slave who was serving as their guide informed Colonel Barton that it was the "dirt train," which usually passed by at that time of day. In fact it was the dirt train, but, instead of dirt or sand, it carried Companies C, D, and K of the Eleventh South Carolina Infantry under Major J. J. Harrison. Company D was also known as the Whippy Swamp Guards, a former militia unit formed by cousins and neighbors in what is now the area near Crocketville, Miley, and Earhardt, South Carolina. These soldiers were heading to Pocotaligo to reinforce Captain B. F. Wyman, when they came upon Captain A. R. Chisholm's company of the First South Carolina Sharpshooters marching along the track approaching Coosawhatchie. Stopping briefly, the train picked up the appreciative marksmen and was soon chugging toward the depot. Most were completely unaware of the danger that lay ahead.[44]

The public road at this point ran parallel to the railroad tracks. The Union advance scouts reported that the train was stopped, giving Barton time to have his infantry and howitzer deployed a few hundred yards from the railroad in ambush. Most of the Confederate troops were sitting in the open on the train's six platform cars, with the officers in the two boxcars. As the train came out of the trees and into view, Barton's men directed "a heavy fire upon it with grape, canister, and musketry."[45] Many men fell wounded with the first volley and a number jumped from the cars and headed for cover. Killed in the ambush were Major J. J. Harrison of the Eleventh South Carolina, Captain Grafton Geddes Ruth of the Whippy Swamp Guards, and the train's fireman. The driver of the engine was also severely wounded. Noticing the slowing of the train, the conductor, J. H. Buckhalter, jumped into the cab and drove the train across Coosawhatchie Bridge.[46]

The eager New York Engineers began tearing up the rails with clawbars, cut the telegraph wire, and built fires for burning cross ties and bending rails. They dislodged several rails and cut the telegraph in several places as Colonel Barton took

The Forty-eighth New York Volunteer Regiment attacking a Confederate train, engraving after a sketch by W. T. Crane. From *Frank Leslie's Magazine,* December 6, 1862; courtesy of South Carolina Historical Society

the main force and pushed toward the railroad bridge. As the Yankees approached, the troops who rode the imperiled train into Coosawhatchie unloaded and formed ranks on the far side of the river alongside the artillery that protected the bridge. Lieutenant L. F. LeBleux's LaFayette Artillery and Lieutenant Hal Stuart's Beaufort battery, entrenched at the approach to the trestle, lobbed several shells at the oncoming Blue Coats, damaging the town's church in the process. At this time Lieutenant Colonel Johnson's cavalry neared, further strengthening the Confederate presence. Facing superior numbers and fearing that further delay would endanger his entire force, Colonel Barton recalled his troops, including Captain Samuel C. Eaton's engineers, and retreated back to the cover of their gunboats. The cavalry gave chase, but they were hindered by the destruction of bridges by the retreating Federals. They dismounted and advanced toward the *Planter,* firing a few volleys into the boats, but fire from Federal gunboats kept them at bay. As soon as the boats were floating again, they steamed back to Mackay's Point to join General Brannan.[47]

About two hours after the fateful train attack, another train, carrying the Thirty-second and Forty-seventh Georgia regiments, approached Coosawhatchie. The engineer stopped after being signaled to do so by an African American man who had jumped from the train that had been ambushed. The man informed the driver of what had happened and told him that the Yankees had torn up the rails just ahead. The troops got off the cars and were sent forward as skirmishers, but they found that the enemy had already left. With the area secured the Georgians set to work repairing the track and telegraph lines. They replaced the rails and cross ties in quick order, and the second train was finally able to reach its intended destination. After arriving in Pocotaligo, Buckhalter wired railroad vice president Bentley D. Hasell of his safe passage: "The enemy has been defeated, both at Pocotaligo and Coosawhatchie. The Yankees tore up four bars of iron at Coosawhatchie, and cut down some telegraph poles and the wire. I have mended both track and wire. The enemy's gunboats now lie at anchor below Coosawhatchie."[48]

The brave conductor's telegraph to Hasell certainly summarized the day, but it in no way told the full story of the furious activity that had taken place. The Union force returned to their base at Hilton Head the following day. The only rewards for their efforts consisted of a few swords and rifles, several prisoners, and the colors of the Whippy Swamp Guards, captured when Barton's troops ambushed the train. The Federals suffered 337 dead and wounded, and the Rebels reported 163 killed, wounded, and missing.[49] Colonel Walker had wisely positioned his slim forces for the Federal attack, and General Mitchel's vaunted "third expedition" had failed. He never got another chance to cut the railroad, for he succumbed to yellow fever not long after the battle. Things remained quiet along the Charleston & Savannah's rails for the remainder of the year.

SEVEN

The Business of War

After the Battle of Pocotaligo, the Charleston & Savannah Railroad found itself not only recovering from the first major assault on their line, but also looking for new leadership. Just two days before the battle, John Caldwell, president of the South Carolina Railroad, had resigned his position. In its search for a new president the rival company called on William Joy Magrath, president of the Charleston & Savannah and formerly vice president of the South Carolina Railroad. On October 22, 1862, as the Charleston & Savannah Railroad was under Union fire, Magrath accepted the offer to become the next chief executive of the South Carolina Railroad. During his eight-month tenure as president of the Charleston & Savannah, he had seen the railroad through some trying times. He had kept the road in business, but difficulties remained. The demands of the military were putting a strain on the company, and the Ashley River Bridge remained unfinished. To handle these and other weighty matters, the Charleston & Savannah's directors announced on October 31, the elevation of vice president Bentley D. Hasell to president. Assuming direction of the railroad, Hasell knew that he must continue to strive for Magrath's goal of safe and efficient travel while maintaining a good relationship with local military leaders.[1]

The road's change of command occurred just after a string of incidents that raised doubts about the safety of rail travel between Charleston and Savannah. For many railroads in the Confederacy the war brought an increase in accidents, especially on those roads subjected to heavy troop movements. Though a direct cause may not have been identified, factors such as a lack of sidings, variability of schedules, emergencies, and distractions caused by soldiers certainly must have contributed. The Charleston & Savannah had been relatively free of accidents during the first year of the conflict and had been able to avoid the sort of high-profile accidents that plagued other roads, such as the South Carolina and the Wilmington & Manchester. This had changed in September, when the railway experienced two accidents within four days of one another.

On September 24, 1862, a Savannah-bound passenger train had derailed near Pocotaligo, but none of the passengers was injured nor was any freight permanently damaged. This had not been the case three days later, when Private Thomas Jefferson

General Pierre G. T. Beauregard, C.S.A., photograph by Quinby & Co. Courtesy of South Caroliniana Library, University of South Carolina, Columbia

Graham of the Ninth South Carolina Battalion was returning after a sick furlough to a unit encamped near the railroad. For reasons that were unclear, the locomotive overtook him as he walked along the tracks, and he was mortally injured. He was taken to the depot, where he died a short time later.[2] These events, added to the Federal army's attack on the train at Coosawhatchie in October, had the potential to shake people's confidence in the Charleston & Savannah. Hasell knew that his engineers had performed well until now and that he should be able to remedy any lapses in their operational vigilance. He hoped that the public's faith in the "Boys in Gray" protecting the rails would instill the same confidence about traveling on his railroad.

In late November, General Beauregard received intelligence that the fleet of recently promoted Admiral Samuel Du Pont had left Hilton Head carrying a force large enough for an invasion. He notified General Ripley that he should be prepared to "meet him on all fronts" and that he would command all troops transferred into his district.³ Beauregard ordered the preparation of enough railroad cars to transport two regiments at a time over both the Charleston & Savannah and the Northeastern railroads. Movable troops received three days' rations and ammunition and were advised to be ready to travel at short notice. Instead of moving on Georgetown, as Beauregard suspected, the fleet proved to be headed for the North Carolina coast, where they supported a Federal expedition to Goldsboro, North Carolina, led by General John Gray Foster.

Once Beauregard was confident of the enemy's objective, he had the Forty-sixth Georgia; the Sixteenth, Twenty-fourth, and Twenty-fifth South Carolina; Lieutenant Colonel Patrick H. Nelson's battalion; and several sections of artillery in motion by December 15. With the reduction of the Union naval presence at Hilton Head, Beauregard felt more comfortable in sending troops from Savannah to reinforce the Confederate force in North Carolina. Harrison's and Wilson's brigades boarded the cars in Savannah and sped off in the direction of Charleston. After disembarking at the St. Andrews depot, they marched across New Bridge to the Northeastern Railroad depot and boarded a train bound for Florence on December 16. They duplicated the scene in Florence and in Wilmington, finally arriving in the field two days later. Their trip was plagued by overloaded trains, worn-out locomotives, and inadequate wood and water. General William Henry Chase Whiting lamented the inefficiency of the Wilmington & Manchester Railroad as well as the lack of a system for running the trains.⁴ Although they taxed the ability of the railroads between Charleston and Wilmington, the Carolina and Georgia regiments arrived in time to help turn back the Federal expedition, Once Foster's force had withdrawn to New Bern and the threat had subsided, Whiting transferred Beauregard's troops back to Charleston and Savannah.

Back in Charleston, Bentley Hasell found himself relieved because of the reduced threat of a military action on the railroad, but he was frustrated by the source of the railroad's defense. Like many Southern railroad executives, Hasell enjoyed the military's protection of his railroad, but he resented the control that military leaders and the government attempted to exert over the railroads. When Beauregard demanded cars for the transportation of reinforcements or ordered entire trains to remain in reserve in Charleston or Savannah, the railroad complied mostly out of necessity. As he prepared to attend an upcoming railroad convention in Augusta, this matters and others weighed heavily on his mind.

The army's failure to reciprocate the company's cooperation aggravated the tenuous relationship between the railroad and the military. Soldiers often abused the rolling stock, breaking fixtures and knocking out boards to improve ventilation on hot days. Many cars were used not only for storage of the soldiers' baggage but also for sleeping quarters. After the Battle of Pocotaligo, Beauregard ordered the Charleston & Savannah to send an entire train to Savannah to be held in reserve at the Central of Georgia depot for transporting troops back to Charleston. Beauregard also instructed President R. R. Cuyler of the Georgia Central to have a train in reserve at the Charleston & Savannah's Ashley River depot for moving troops in the direction of Savannah. Such demands rankled the proprietary railroad chiefs and further depleted the roads' already low shipping capability.[5]

The Charleston & Savannah Railroad was so occupied with military business that, when called on by the citizens of Charleston, it was unable to join the city's other major rail companies in transporting firewood into the city. Firewood was in short supply, and was being consumed in large quantities by the military, the railroads, the salt works, and blockade-runners. Railroads such as the Charleston & Savannah were so occupied with the transportation of troops, supplies, and government freight that they were unable to carry much firewood into Charleston. As temperatures dropped in late November, the shortage of wood neared the point of crisis. Families did without the much-needed fuel for heat and cooking. Though unable to use its own rails, the Charleston & Savannah Railroad was ultimately able to participate in the effort by leasing a locomotive and fifteen cars to the Northeastern Railroad expressly for the shipment of wood. In so doing, the Charleston & Savannah Railroad, along with other cooperating railroads serving the city, prevented the disruption of services between Charleston and Florence.[6]

Impressment of locomotives and rolling stock often disrupted the company's freight, passenger, and mail service. Even when the trains were able to pass, the military frequently interfered with the mail service by detaining the postal agents. Not only that, the postmaster's office refused payment to the railroad for the days on

Charleston & Savannah notices to passengers crossing the Ashley River. From the *Charleston Daily Courier*, September 23, 1862

which the mail was detained.[7] Certainly, these were minor inconveniences in the grand scheme of things—and Hasell was as patriotic as the next Confederate—but he had a railroad to run. The trains provided the troops with food, clothing, letters from home, and reinforcements during times of emergency. In return Hasell thought the railroad deserved better treatment than it got; however, to him anything was preferable to government control of the railroads.

Aside from interfering with rail travel, the military also contributed to the disruption of the company's steamer service at the Ashley River terminus. General Pemberton had requisitioned the company's steamers to help carry materials and supplies for the defenses in and around Charleston harbor. The wear and tear on the ferryboats was heavy, and a shortage of labor and spare parts hindered their upkeep. Unable to find replacement boats, the company had to discontinue its steamer service to and from the city. Passengers and freight ended their trips at St. Andrews depot and required transportation to and from Charleston across New Bridge by carriage. Opportunistic drivers began charging outlandish tolls for driving passengers to and from the depot. Because of the inconvenience, Hasell decided that the railroad would pay tolls for passengers to and from the trains as long as they provided appropriate traveling credentials for themselves and their drivers.[8] The cost of lost opportunity in not having the steamers was mounting. The directors' decision to save money during construction by not building a rail bridge over the Ashley was now costing the company in terms of efficiency and real dollars. Not only that, but the lack of a permanent railroad bridge across the Ashley River was a significant impediment to Confederate government transportation. The solution was clear—the Ashley River Bridge must be completed.

Hasell had been in charge of the bridge's construction since May. Progress was slowed by shortages of labor, supplies, and equipment. The strains of the war effort continued to take its toll on the South Carolina and Georgia coastal plains—especially with regard to railroad iron. Iron was in such short supply at this point that Lieutenant George Mercer wrote: "Iron is particularly needed. Agents have been sent over the whole country to purchase or collect it in any form. . . . The large cylinders intended for the Charleston & Savannah Railroad bridge have been melted into shot and shell. . . . Whole railroads are being torn up, so that the iron may be used for gunboats." Mercer felt that the situation represented not weakness but a "spirit of patriotism and self-sacrifice, a moral power that is impregnable."[9]

In fact, very few new rails were produced in Dixie after 1861. The rolling mills proposed at the Columbia railroad convention had not yet materialized, and the railroads were in competition with the military for any iron that could be found. When the South Carolina Railroad offered a quantity of well-worn rails taken from some of its sidings, Hasell secured them for use on the Ashley River Bridge. This was not a matter of frugality; these rails were all that were available at the time. He also requested 100 to 110 tons of rails from the Confederate government for the project.[10] Piecing together whatever materials he could obtain and using whatever

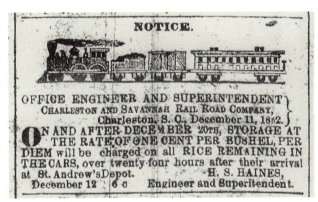

Charleston & Savannah notice of rice-storage rates. From the *Charleston Daily Courier*, December 12, 1862

labor he could hire, Hasell's men completed the bridge by mid-December 1862 except for its draws and approaches.

Gaps such as the one at the Ashley River haunted the railroads of the Confederacy. Deficiencies in rolling stock and a lack of coordination between adjacent railroads also caused congestion along the lines connecting the Mississippi to Richmond. At the time the Charleston & Savannah owned only forty-four boxcars, nine of which were not in running order. Making matters worse, twenty-eight cars that had been loaned them by other roads had been returned at the lenders' request. Of its forty-nine platform cars anywhere from twenty to thirty were in use during the week hauling government timber for harbor obstructions, gunboats, fortifications, and bridges.[11]

Because of the scarcity of rolling stock, the railroad had nearly five hundred thousand bushels of rice that it was unable to transport. Every day the bridge remained unfinished risked the loss of cargo because of weather damage or even capture by the enemy. To help speed turnover of boxcars, the company instituted a policy of assessing a daily fee for every bushel of rice left unclaimed for more than twenty-four hours. The shortage of cars also affected the military. While the railroad was relied on for shifting troops to points of attack, Hasell estimated his company's troop-carrying capacity—using the full complement of its rolling stock—to be only about 4,325 men without arms, ammunition, cannon, and horses. He did not feel that his railroad could fulfill all that the government expected from it without an additional fifty boxcars. If the government could offer some assistance, Hasell informed Colonel William M. Wadley on December 24, 1862, that the company "would be willing to purchase at a fair price for their future business, one half of that number."[12]

When the South Carolina Executive Council had authorized them to purchase twenty cars nine months earlier, the Charleston & Savannah had been able to cajole only ten cars out of their neighbors. With his former colleague and ally Magrath now at the helm of the South Carolina Railroad, Hasell saw an opportunity to alleviate this problem. He was presented the prospect of buying ten boxcars from another company, but again the Charleston & Savannah needed the assistance of the state government to complete the transaction. At the time Hasell became chief executive, he inherited cash reserves of only $30,000 with which to operate. Passenger and freight business were on the rise, but during that year the company had already spent more than that amount toward its connection with the South Carolina and Northeastern railroads and more than $8,000 on sidings. Not wanting the chance to slip by, Hasell petitioned the Executive Council for $6,500 to purchase the rolling stock at the same time he requested the state's $35,000 apportionment for the bridge.[13]

On December 3, 1862, President Davis had commissioned William M. Wadley as a colonel and appointed him assistant adjutant general of the Confederate Army. A former railroad executive from Georgia, he was charged with the supervision of government railroad transportation throughout the entire Confederate States of America. He knew firsthand about railroad operations and the potential benefits the railroads afforded the armed forces. He also knew of the Southern railroads' problems: the lack of coordination between neighboring roads, shortages of rolling stock, and interference by the military. He wished to promote interline cooperation and alleviate some of the deficiencies so that the railroads could be used to their fullest potential. Achieving this goal would require some central coordination, and he did not want the strained relations between the railroad companies and the government to undermine such an indispensable part of the Southern armamentarium.

Colonel Wadley's first undertaking in his new role was to invite the presidents and superintendents of all the Confederate railroads to a convention in Augusta, Georgia. Bentley Hasell represented the Charleston & Savannah at the meeting, which convened on December 15, 1862, at Augusta's Masonic Hall. According to the Richmond newspapers, the purpose for the convention was to "devise some facilities for the transmission of freights which, for lack of a system, have accumulated on the lines, greatly to the inconvenience of the Government and the public."[14] Wadley agreed with that line of thinking and had in his mind a smooth system of transportation for a nation at war, which included a plan for through train schedules and a real system of freight car interchange.

His plan called for more cooperation between adjoining railroads and centered on his office serving as a coordinating center for all Southern railways. He intended for the superintendents of all the companies to serve as liaisons with his office, submitting regular reports that would aid his efforts to organize the tangle of rails into an efficient system of transportation. He wanted the companies to allow their cars to travel on neighboring roads under the watchful eyes of the superintendents, who would see to it that the rolling stock was maintained and returned in a timely fashion. Locomotives transferred to other lines would remain under the control of the driver of the lending company but would be subject to the rules of the borrowing company and the Assistant Quartermaster's Office. Colonel Wadley would dictate the terms of prompt return and maintenance of the rolling stock, as well as the compensation to the companies. In his proposal he also addressed the incessant difficulties caused by gaps between adjacent railroads. He offered government aid in the construction of connections between railroads provided they were deemed necessary. When these breaks in the rails necessitated the transfer of troops and freight, Wadley planned to have the process supervised by agents appointed by the Confederate Quartermaster Department.[15]

In exchange for permitting more government control of railroad operations, Wadley envisioned more government assistance to the companies. The Confederate government would underwrite the expenses incurred while returning to the various railroads all rolling stock and equipment that had been assigned by government officials to other roads. They would also provide the companies a substantial amount of cash toward the purchase of supplies, labor, and equipment. If a military commander needed the services of a local railway, the request would go through Wadley's office, and the company would receive compensation for any delays encountered.[16] Colonel Wadley had laid the groundwork for a coherent, workable plan for Southern railroad operation, but the common denominator of each proposal was government involvement. To most of the forty-one railroads represented, that was unacceptable.

After Wadley briefly outlined his plan, the chairman appointed a committee to study it and quickly moved on to matters that the railroads considered higher priority. Though Wadley had wanted to avoid a heated debate on the subject, the railroaders next addressed the issue of tariffs—specifically tariffs for government business. The committee on tariffs introduced a passenger fare increase of one half cent per mile and increases through the whole schedule of government freight. They also recommended restrictions on special services offered to the government. These proposals were rapidly adopted, but Wadley's plan was debated and rejected. The railroad men at the convention felt that having a pool of rolling stock could potentially hurt the interests of some individual railroads.[17] They were also leery of a central government's intervention into their business. Perhaps feeling some regret at slapping away a hand offered, the delegates later passed a resolution expressing an "earnest desire to cooperate" with Wadley in the execution of government transportation. This rang hollow to the ears of the assistant quartermaster.

After the convention adjourned, Wadley wrote to the various railroad presidents asking them to use his office as a coordinator and requesting that their superintendents submit weekly reports on local conditions. In his report to the government he recommended that the government not accept the tariff proposed at the convention, but he was not entirely without empathy for those managing the railroads under such tough circumstances. Remaining true to form, many railroad managers resisted or rejected any central government control. After receiving Wadley's postconvention communiqué, Hasell informed him of his company's position. "I beg leave to state," he replied, "that although the officers of this road will take pleasure in cooperating with you to the fullest extent, yet it is not intended thereby to place the control of the road in other hands than where the charter and the voice of the stockholders have signified such management should be held—namely in the President and Board of Directors."[18] The message was clear. The Charleston & Savannah would cooperate but only on its own terms. This philosophy, so prevalent in the Civil War South, was exactly what was causing many of the problems Wadley tried to address. This truth would unfortunately be realized only when it was too late.

Though his ideas had been rejected at the convention, Wadley remained sympathetic toward the railroad companies. Having been a railroad man himself, he had experienced firsthand the difficulties of operating a road. The root of at least some of the opposition, he felt, lay in the manner in which military officials disregarded the private property of the railroad companies. Rolling stock was often ordered from one road to another without any provision for its return or for its condition on return. Impressment of cars was also common and interfered with the ordinary routine of railroad employees. Such practices had been carried out frequently by Generals Pemberton and Beauregard on the Charleston & Savannah road. If this continued, Wadley feared railroad managers would lose interest in conducting a business in which they were vested with little responsibility or control.

In a letter to Adjutant General Samuel Cooper, Wadley called attention to the deterioration of rolling stock and machinery, as well as the scarcity of men available to operate the railroads and repair the machinery. All but the wealthiest rail companies were in dire need of supplies. Many of the more recently established roads could only afford the most ordinary of repairs, even at the outset of the war. With the extra stress placed on the roads by military and government business—and with not enough civilian business to sustain adequate earnings—things were looking grim. Wadley suggested that the government aid the railroads by allowing iron foundries and rolling mills that had been engaged only in government business to furnish the needed materials for their operation and maintenance. He also advocated exempting railroad employees from conscription and assigning previously enlisted men to work on the railroads. In his opinion these difficulties must be addressed or the roads would soon be unable to meet the demands placed on them by the Confederate government.[19] He could recommend all he wanted, but he knew the president and legislature would have the final say on whether to tender the necessary aid or let the railroads flounder. It was also up to the railroads to accept the aid as well as some measure of centralized control. Wadley was trying to bring the two sides together, but the proprietary sentiment of the day was making his job quite difficult.

Far removed from Richmond, the defenders of the Charleston & Savannah Railroad were fighting an entirely different war than their brethren in other theaters. As the Christmas holidays approached in 1862, the Army of Tennessee readied for battle near Murfreesboro, and Lee's Army of Northern Virginia jubilantly celebrated its victory over General Ambrose E. Burnside at Fredericksburg. The soldiers encamped in the Carolina lowcountry were growing weary of monotonous camp life. Many of these men were between forty and fifty years of age, serving ninety-day tours in the reserves. They missed their families and friends and fretted over the businesses they left behind. Diseases such as measles, smallpox, and pneumonia took a larger toll than the Union army, and with the holidays approaching thoughts of home kept the men from performing their various duties with the same verve as the younger soldiers in the regular army. The magnetic pull of home was often so great that some deserted. Weekends often provided a much-needed physical, emotional, and spiritual respite that helped the soldiers' flagging morale.

In return for their service and sacrifice, they received a mere eleven dollars per month. It was an insignificant amount in contrast to the money their businesses were losing back home. Yet most remained loyal to the Confederate cause, galvanized by the desire to keep the Federals from invading their beloved homeland. Sometimes a Southerner was able to pay a substitute thirty dollars a month to serve in his place, but many felt like Private D. Daniel Louis of the Eleventh South Carolina Reserves, who stated, "I refuse to take one, it is not just."[20] Private Louis and the other reservists had much in common with Charleston & Savannah Railroad president Bentley Hasell. They were trying to keep businesses afloat while providing valuable service to the Confederate military. They were bringing in considerably less profit, and were being asked to make substantial sacrifices for the Confederacy.

The troops stationed along the tracks in places such as Camp Allen and Camp Pocotaligo could not brag of battlefield exploits like those of their counterparts in Virginia and Tennessee, but their service to their state was no less important. In December 1862 the men of the Eleventh South Carolina dutifully guarded the western end of the railroad. Companies C, D, and K went into winter camp at Camp Elzey and Camp Jones near Hardeeville, and Colonel F. Hay Gantt detailed Company G to Camp Heyward on the Savannah River to picket the trestle of the Charleston & Savannah Railroad. Company H remained at Coosawhatchie to guard the railroad and picket the rivers in that sector, while Company I did the same at Pocotaligo. Guard duty was taxing on the soldiers, and in an effort to soothe frayed nerves regimental surgeons often ordered barrels of whiskey for distribution to the troops during inclement weather. With so few men to guard such a wide area and with little human interaction other than visits from the officer of the day, it was lonesome duty.[21]

Through it all the railroad continued to serve as a lifeline that kept the soldiers in touch with the civilian world. The trains carried the cornbread and molasses cakes that livened up a tired camp menu, which often included beef and rice for breakfast, lunch, and supper. They brought visitors from home to cheer the depressed or carried troops into town for a concert or play. They also carried letters to and from home, allowing men to keep up with childhood milestones, family finances, and gossip about town.

During the first months of the war, the mail train left Charleston at 10:15 A.M. and arrived in Savannah at 5:00 P.M. every day except Sunday.[22] That was, however, during the time before military activity and insufficient maintenance aggravated the inefficiencies already existent in the Southern railroad system. Trains frequently missed connections, significantly hindering mail delivery as public disenchantment mounted. The Confederate Post Office Department under Postmaster General John H. Reagan came under fire, often for incidents that were beyond its control. Railroad executives found it difficult to balance their profit motive with the needs and demands of the military, the Post Office Department, and a demanding public.

From the beginning, relations between Confederate railroads and the Post Office Department had never been particularly happy. Despite initial appearances, many railroad executives disliked the classification of the roads—especially those designated as second and third class. Furthermore railroad superintendents frequently had to rework entire schedules to accommodate postal regulations. Railroad officials began to feel as if Reagan had drawn them into a hard bargain at the Montgomery convention of April 1861.[23] Conversely Reagan regarded the railroad leaders as spoiled and unpatriotic, going as far as saying that the railroads "as a general thing, [were] doing a better business than they ever did on account of the war, while all other interests are suffering."[24] Believing the railroads were trying to avoid accountability and government control by refusing to execute their contracts with the post office, Reagan decided to withhold payment beginning on June 30, 1862, to those roads refusing to contract with the department.[25]

When payments to the Charleston & Savannah Railroad were delayed in the spring of 1862, the railroad's treasurer, William H. Swinton, was surprised to find that Thomas Drayton had never signed and executed a contract with the Post Office Department prior to his resignation. Another misunderstanding that Swinton discovered—partially the fault of Charleston postmaster Alfred Huger—was the company's failure to provide seven-day mail service after November 1861. All was forgiven, however, when Swinton returned a signed contract along with figures detailing the company's services rendered. Payment resumed, and the railroad put its seven-day service into effect soon thereafter.[26]

By the end of 1863 the majority of Southern railroads had contracted with the Post Office Department, but there was little evidence that service had improved. Disruption of schedules occurred more often as the speed of locomotives diminished and the frequency of breakdowns and accidents increased. As enemy activity escalated along the Carolina coast, military monopolization of the Charleston & Savannah's rolling stock created still more delays. The transport of troops and munitions often took precedence over the timely delivery of mail to Savannah, and on these occasions the postmaster general was less forgiving. He refused payment for delays on October 22 and December 4, 1862, because it was "the only means of protecting [the Post Office Department's] interests against the interference of those over whom it [could] exercise no control."[27]

Whether fairly or unfairly, when the mail was delayed, the public's criticism focused squarely on the postal service. Savannahians blamed many of the delays on the fact that the mail was not separated before arriving at the point where the railroads diverged to Charleston and Augusta. Even though Charleston was on the most direct route to Savannah, the Richmond mail arrived there as often from Augusta as it did from Charleston, and even then, it was often a day late. On September 8, 1862, the *Charleston Daily Courier* pointed out that trains on the Northeastern Railroad often arrived with barely enough time to transfer the mail to the Savannah train, but there was no time to open and sort the bags—a requirement at the time. Either more time would have to be allotted in the schedule for Charleston, or a bag for Savannah separated further up the line.[28] As dysfunctional as the system was, though, it had to suffice. Even the express companies were subject to the same annoyances; only they did not carry the stigma of being government agencies.

If Bentley Hasell and the directors of the Charleston & Savannah Railroad were leery of their national government, they did not always feel the same way toward that of their state. The Charleston & Savannah Railroad was second only to the South Carolina Railroad in the amount of its stock that was held by the state of South Carolina.[29] To complete the bridge across the Ashley River, the Charleston & Savannah had to petition the state, but the players in Columbia were now different. Immediately after the Augusta railroad convention of December 1862, Milledge L. Bonham had been elected the new governor of South Carolina. Bonham appointed a new Executive Committee, minus Attorney General Isaac Hayne, and Hasell worried that his railroad would suffer because of the change of administration.

The railroad's fiscal year had just ended, and Treasurer William Swinton was anxious to clear the books and accounts. In early January 1863 Hasell penned an urgent letter to Governor Bonham calling attention to the request he had made to Hayne the previous month. He rationalized that since he fully expected the cost for the bridge to equal the amount allocated by the Executive Council, the state should go ahead and reimburse the company the full amount. After studying the issue, Bonham approved the reimbursement. Since it would benefit the Confederate government as much as—if not more than—the state, he subsequently decided to petition the Confederacy to refund the state treasury the amount of the bridge.[30]

With the funding secure Hasell turned his attention to gaining the cooperation of the Charleston Bridge Company. The company was building what they termed a "temporary structure" on their company's western causeway, and they held the upper hand. Their insular president, Alex H. Brown, felt his company commanded ownership of the bridge since the state, not the railroad, had provided the funds for its completion. After some haggling, Hasell was able to secure from Brown a tentative agreement for a "commutation of the tolls" for transporting of passengers and nongovernment freight across the Ashley to the connection with the Northeastern and South Carolina railroads.[31] This would be the last major accomplishment for Bentley Hasell as president of the Charleston & Savannah Railroad. Thomas Drayton had built the railroad. W. J. Magrath had guided it through its first direct military threat. Hasell had seen the company through what seemed to be the interminable construction of the Ashley River Bridge; however, he was not the head of the company when the first train passed over that bridge. In February the railroad's directors elected a new president, and in April Hasell became president of the fledgling Shelby & Broad River Railroad.

EIGHT

Singletary's Inheritance

Captain Robert Legare Singletary had been one of the "white knights" of the Charleston & Savannah Railroad. A native of Marion District, South Carolina, he was married to the former Sarah Jane Evans, sister of Brigadier General Nathan G. Evans. Singletary and his associate Henry Drane had finished construction on the road only two years after they had completed similar work in Florida on the Fernandina & Cedar Keys Railroad. When South Carolina seceded, Singletary returned to Marion and enlisted as a captain in the Eighth South Carolina Infantry. Later that year, he resigned that position with his unit and served as the superintendent of labor for the Charleston defenses. Captain Singletary, as he was still known, was elected to the board of directors of the railroad in March 1862, and when the Charleston City Council met in mid-February, they tapped him to replace Bentley Hasell as president of the company. With this endorsement stockholders formally elected him at the company's annual meeting on February 18, 1863, and Hugh R. Banks filled his spot on the board.[1]

From the outset the railroad's new president had no shortage of unresolved issues with which to contend. He was concerned about how his railroad would fit into the grand scheme of the Confederacy, but he had to solve two more important local problems that had plagued the company since it began operation in November of 1860—bridging the Ashley River and paying the delinquent interest owed to the company's creditors. In addition to ensuring the security of the railroad from military attack, he realized the importance of confronting the company's indebtedness—every bit as much a threat to the company's long-term prospects as the Federal military. After assuming office, he searched for ways to increase revenues and improve net earnings. Fortunately for the company Singletary was unlike Magrath and Hasell in one important way. Like Thomas Drayton before them, he had invested an enormous amount of toil in the road, and he knew every turn, slope, and crossing on the line. He also had a heavy financial stake in the road because Drayton had paid him mostly in company bonds rather than cash. No one wished to see the company succeed more than he.

The cost of maintaining a railroad had become almost crippling. The prices of fuel, lubricants, chairs, spikes, provisions to feed employees, and other materials—as

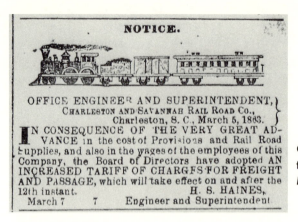

Charleston & Savannah notice of fare increases. From the *Charleston Daily Courier,* March 7, 1863

well as employees' wages—were advancing at an alarming rate in a time of shortage. The company had not paid the first dividend to its investors since it began full operations in November 1860. Though it was an unpopular decision at a difficult time, Singletary and the directors decided it would be necessary to increase both freight and passenger fees starting in March 1863. Seeing an opportunity in the relative military quietude that spring, the company also decided to add a second daily passenger train to the schedule in April. They hoped that the prospect of making a round trip between Charleston and Savannah in less than twenty-four hours would attract more business.[2]

Another way Singletary found to obtain currency was by releasing an individual from his bond pledge, placing the bonds on the open market, and selling them for a sum greater than the face value of said bonds. He would then apply the surplus to the payment of delinquent interest on those bonds. Singletary and William Swinton had done this with notes of the Bank of South Carolina and another individual creditor, and the company had realized an eighteen thousand dollar surplus. One of the directors, William C. Bee, thought this might be setting a dangerous precedent, especially during uncertain economic times. He urged caution, pointing out that, should the market price of the bonds fall below their present value, the directors would be liable to the creditors for the amount of the depreciation.[3]

Encouraged by the company's net earnings in February, Singletary continued his efforts to reduce the company's debt and increase revenues. He executed the sale of 250 shares of Farmers and Merchants Bank stock that had been pledged to secure debts to three banks in Savannah. With the proceeds he paid the debts secured by the stock and netted a surplus of more than $5,000, part of which he used to clear a $1,600 debt owed to the Central Railroad and Banking Company of Georgia since 1860. He also hoped to do the same thing with the stock the railroad owned in the Kings Mountain Railroad. In addition to selling stock Singletary also proposed that the company begin to sell off certain of the company's nonessential real-estate holdings, provided their sale would be sufficient to cover their cost plus interest. These properties included a large portion of White Hall Plantation, valued at $20,000; Hayne Hall Plantation; and several other smaller tracts in the city.[4]

Not stopping there, Singletary continued to search for ways to increase the railroad's cash holdings. He and Treasurer W. H. Swinton pored over the company's books and were surprised to find that the Confederate Treasury Department owed them a substantial amount of money. When the government originally contracted with the railroads at the Montgomery convention, delegates had mutually agreed that payment would be in Confederate treasury bonds. As early as the summer of 1861 Thomas Drayton had noticed that the government had an annoying habit of delaying the issue of the bonds. Not only that, Drayton noticed, instead of bearing the date the payments were due, the bonds bore the date of their issue. Viewing this practice as a means of reducing government expenditure at the expense of the railroad company, the company immediately contacted the Quartermaster Department. Assistant Adjutant General Abraham Myers told Captain (later major) Hutson Lee to reassure Drayton that future bonds would bear the date on which the accounts were to come due and reassured him that the government had no ill intent toward the railroad.

Future bonds did, in fact, bear the due date of the accounts, but they also bore the date of issue, and that date was the one the treasury used to make its interest payments. This apparently went undiscovered until the audit, and Singletary immediately wrote to the secretary of the treasury to demand back interest owed in the amount of $4,443. Transportation rates on the railroad were already a bargain for the government, which paid only half the customary charge for private business. This was a financial hindrance to the company, and Swinton was adamant that they should not have to suffer the further insult of the government's refusal of this claim. After careful review, the treasury made good on Myers's past promise, and ordered Major Hutson Lee to settle the account for the amount claimed.[5]

The sale of stocks, bonds, and real estate—added to the increased receipts from higher volume and higher rates—had an immediate impact. Bolstered by a three thousand dollar per month lease payment from the Northeastern Railroad and improved future earnings projections, the company under Singletary was operating at a solid surplus. Soon Singletary was able to declare the company's first ever dividend. On April 30, 1863, the company notified its creditors that they would pay all interest due on their 6 percent state-guaranty bonds as well as on the 7 percent second-lien bonds. This announcement was cheered in Savannah, Charleston, and in the hamlets along the line, where many of the railroad's investors lived. An article in the *Charleston Daily Courier* hailed the company as it "[took] its place, ere long, among dividend paying railroads." "Honor," the editorialist offered, "to the enterprising men who built this important commercial and military Road, and who may be said to have redeemed it from failure or destruction, and who have raised it to its present high and palmy state of prosperity."[6]

Not long after Singletary assumed the helm his headaches began with regard to the Ashley River Bridge. Singletary had written to Alex Brown of the Charleston Bridge Company, anticipating the completion of the project and expressing a desire to receive and disembark passengers at the Spring Street depot. Singletary also asked about a mutual arrangement that would allow railroad employees to cross the bridge to and from their workshops west of the Ashley. On March 14, Brown sent back an uncompromising reply. He felt that Singletary had misunderstood the terms and conditions on which the bridge had been constructed. Since it was paid for with state funds and was built on the Charleston Bridge Company's private property, Brown balked at the idea of his company's land being used for another's private enterprise. His company, Brown asserted, "will not sanction any such use of their private property as you indicate in your note." The railroad could use the bridge only for military necessities, and, he added, "any other attempted appropriation will be enjoined as an interference alike with the chartered rights and property of the Bridge Company and inconsistent with the understanding had on the subject before the work began." He concluded by tersely stating that the "experimental contract" between him and Hasell was "not entirely satisfactory."[7] Brown also said that he would either modify or annul the contract soon and would notify him of the final outcome.

By the first week of April 1863 the bridge was finally completed. It had taken more than sixteen months to finish; the construction process had outlasted three of the railroad's presidents. The company's engineers had originally felt such a project to be ill-advised; however, it had proved a necessity both to the company and to the Confederacy. The directors of the railroad had been excited about the bridge's construction, but they were now having to face not being able to carry freight and passengers across it for anyone other than the government. Since the company had been unable to construct a turntable on the eastern bank of the Ashley, engineers had to deal with the nuisance of backing their trains in one direction or the other to cross the bridge. Nonetheless the road was finally linked to the South Carolina and Northeastern railroads at the Spring Street depot; only the mercenary attitude of the Charleston Bridge Company dampened their enthusiasm. Still the event was noteworthy, and Captain Henry Drane, now the superintendent of the Wilmington & Manchester Railroad, had returned to Charleston to celebrate the connection with Captain Singletary. On April 7, 1863, the first train rumbled across the bridge and into the city. Not a word about this short, but important, ride was included in the city's major newspapers, for on this very day the Federal navy attacked Charleston harbor.[8]

At the beginning of 1863 there were an estimated 14,500 troops of all arms stationed along the Carolina coast, more than half of these in forts or along the direct approaches to Charleston. In the Third District, the recently promoted brigadier general William S. Walker commanded approximately 3,000 troops—more than half that number consisting of cavalry—compared to a year before when General Lee commanded 11,000 in the same district. At the same time the Union army began concentrating troops at Port Royal. General Walker predicted that the Yankees would attempt to hit the railroad and then move on Charleston by land. The arrival of several Union monitors off the coast under Admiral Samuel F. Du Pont further alarmed the Confederate command.

Such a buildup by the Union army and navy concerned General Beauregard, who requested and received reinforcements from Wilmington. These troops arrived in Charleston and Savannah by rail in early February. As more troops accumulated at the Federal base at Port Royal, Beauregard continued to strengthen the Third District. The Charleston & Savannah Railroad was instructed to keep at Pocotaligo a special train capable of carrying one thousand men, and additional naval support was requested for the protection of the Savannah River trestle. Beauregard ordered Colonel Clement Stevens and seven companies of the Twenty-fourth South Carolina aboard the cars at Pocotaligo and to transfer them to James Island. Batteries were constructed and forts were strengthened. The Confederates watched and waited, but nothing happened. The capture of the Federal gunboat *Isaac P. Smith* in the Stono River was the only event of military significance in the South Carolina lowcountry that winter, but despite the relative quiescence of the period the Confederate command knew that the Federals' work on the coast was not finished. There was sure to be an attack. When it occurred, the railroad must be at the ready.[9]

By the first week of April, Beauregard had been able to expand the available force to more than 22,500 troops along the coast, with 12,850 positioned in and around Charleston. In January U.S. Secretary of the Navy Gideon Welles had wired Admiral Du Pont that he was sending four monitors and the flagship *New Ironsides* for an assault on the port of Charleston. To assist his mission the army provided General David Hunter with 10,000 troops, but made it clear to Du Pont that the operation would rest solely on the success of the naval force.[10] As the ships lay anchored just off the bar a few days before the attack, General Beauregard was making preparations. The entire channel had been marked with buoys to give the Confederate gunners more exact ranges for their targets. Between Fort Sumter and Fort Moultrie, obstructions and strategically placed torpedoes awaited the passage of enemy ships. The first line of defense included Fort Sumter, Fort Moultrie, and Battery Gregg on Morris Island. The secondary defenses consisted of Fort Ripley, Castle Pinckney, and the batteries on James Island. If Union vessels were able to penetrate these, they would then encounter the guns mounted at the Battery on the Charleston peninsula.

On April 7, 1863, the day on which the first train crossed the new Ashley River Bridge, the Union navy began its attack. Led by the monitor *Weehawken*, the squadron entered the main ship channel at about 3:00 P.M., and by day's end Confederate artillerists had completely or partially disabled five of the eight Union ships engaged. Receiving the worst damage of the day was the double-turreted monitor *Keokuk*, whose hull was so badly damaged that her captain disengaged and steered her to a point adjacent to Morris Island, where she dropped anchor and later sank. The battle continued for nearly two-and-a-half hours, when a frustrated Admiral Du Pont signaled for a general retreat. Secretary Welles's plan for storming the gates of Charleston had to wait until another day. In his report to General Hunter, Du Pont wrote that he had "tried to take the bull by the horns, but he was too much for us." After the battle he decided that Charleston could not be taken if assaulted solely by a naval force. The discouraging news of Du Pont's defeat was not received well in the North. Secretary Welles was angry that the admiral had abandoned the harbor after one brief attempt and that he felt it unwise to attempt another, similar venture. In fewer than two months Welles relieved Admiral Du Pont from his command of the South Atlantic Blockading Squadron. It seemed logical that future attempts by the Union to take Charleston would, at least in some way, target the railroad.[11]

Two weeks after the attack on Fort Sumter, General Beauregard was relieved of his command of the Department of South Carolina, Georgia, and East Florida and sent to assume command of the defenses around Richmond, Virginia. Union general Joseph Hooker and the Army of the Potomac crossed the Rappahannock River and were soundly defeated by General Robert E. Lee's Army of Northern Virginia at Chancellorsville in May. Despite the Confederate military success in Virginia during the spring of 1863, railroads throughout the Confederate States of America—like General Pemberton's troops in and around Vicksburg—were struggling. The Confederate dollar was worth only about $.25 in Federal money, and the prices of supplies had increased anywhere from four to twenty times depending on the equipment desired. The amount of rolling stock was grossly insufficient for the task at hand. To address these and other issues Colonel William Wadley of the Quartermaster Department called yet another meeting of railroad representatives in Richmond on April 20, 1863. Wadley was quite active at this convention and was able to impress upon the carriers their importance to the government and vice versa.

The delegates approved resolutions that formally recognized the new Railroad Bureau directed by Wadley and several assistants, but the railroad men made it clear that there would be no direct governmental control over their companies' operations. The new bureau would supervise the extension of public aid for railroad supplies such as rails, lubricants, spikes, and rolling stock, even if it were achieved by obtaining them from less strategically important railroads. Other recommendations included the importation of skilled labor from European countries, use or erection of storage facilities along railways to facilitate the prompt unloading of cars, more strict prohibition of military interference with the running of trains, and diversion of as much traffic as possible to inland waterways. The resolutions passed at this convention were more far-reaching than any others, but the delegates feared they would have little impact because the Confederate government was unwilling to enact any effective regulatory legislation relative to the railroads.[12]

Finally, a week after the convention adjourned, the Confederate legislature produced the government's first railroad law. It gave the executive power to require any carrier to devote its facilities to support the army except what would be required to run one passenger train per day. It required railroads to adhere to through schedules set by the government and ordained the wide distribution of locomotives and other rolling stock. It provided for the impressment of those roads that refused to cooperate, and it authorized the taking of rails, shop machinery, or other portable property from the premises of one company to assure the proper maintenance of another. The only hitch was that it placed the entire supervision of the carriers in the hands of the quartermaster general and made no mention of the Railroad Bureau or Wadley. The very day that they passed the new legislation, the Confederate Congress decided to remove Wadley from his position in the Quartermaster Department. His replacement, named later that June, was Captain Frederick W. Sims, a native of Washington, Georgia, and a former transportation agent on the Georgia Central Railroad, who in the course of his service was promoted to major and then lieutenant colonel.[13]

At the time the Charleston & Savannah Railroad was riding a tide of financial prosperity that many railroads in the South were enjoying despite their other difficulties. President Singletary had used a combination of creative financing, stock sales, expansion of services, and other modes of increasing cash flow to free his company from the grasp of bankruptcy. Other companies had simply increased their rates for private customers. While public reaction to the rate increases was not favorable, the carriers contended that they were in line with those for other goods and services.[14] The economic conditions that allowed the Charleston & Savannah to pay all the delinquent interest on their bonds allowed other companies to realize good profits and declare exceptional dividends. The owners of the railroad securities were probably quite pleased with the return on their investment until they took their money to town and actually tried to buy something with it.

After the defeat of the Federal navy at Charleston, newspapers throughout South Carolina and the Confederacy hailed the defenses of Charleston harbor. With the electricity of victory still in the air, Captain Singletary soon decided on a plan that, he hoped, would similarly excite investors in the Charleston & Savannah Railroad. At the company's monthly board meeting in May, he recommended that the company immediately pay the interest due on the mortgage secured by the company's bondholders on April 10 (third-lien bonds). He also proposed that the company notify its unsecured creditors of their right to receive the third-lien bonds as payment for the debts due them as soon as the government paid the balance it owed the company. If they would accept the bonds as payment, creditors would receive the interest in arrears on those bonds for the previous two years. This policy would take effect immediately.[15]

Having just declared the company's first dividends a couple of months before, Singletary felt that the timing was right for his plan. "I know of nothing," he stated, "more likely to aid this than to do that which is clearly our duty to do, namely—to pay the interest in arrears upon such bonds as have already been issued." At that point the company had a surplus on hand and cash assets of $125,000 available, exclusive of the actual cash balance already appropriated for payment of the interest owed on the first and second lien. With little military activity along the railroad and the confidence of the citizenry on the rise, the company's net income for May was expected to be at least $25,000, and the income for June was estimated at $20,000, giving a projected surplus of $170,000. Singletary had calculated the ensuing year's interest at $100,300, leaving a clear surplus of $69,700, excluding future earnings. If the railroad cleared at least $15,000 per month through June 1864, its surplus could potentially exceed $250,000.[16]

Singletary emphasized the fact that, as of May, unsecured creditors had been slow to accept the third-lien bonds issued. The prospect of receiving two years worth of back interest might entice the creditors to accept the bonds as payment of their debt. Even if they did, resulting in payment of back interest and interest through June 1864, the company would still have a surplus of nearly $140,000. The company would not need the surplus in the event of a disaster or for permanent improvements to the road, unless the company's net earnings fell to less than $11,500 per month. In his opinion, the increased business anticipated upon the completion of the Port Royal Railroad would make this virtually inconceivable.

It should be noted that Singletary was probably one of the company's largest individual creditors by virtue of his and Henry Drane's work on completing its construction. Singletary asserted that the "prospects for payment rest on the faithful and successful administration of the road." His interests were also those of the other creditors, and the interests of the stockholders were "identical and inseparable." Singletary had a vested interest in the company's paying its creditors, but he had to keep the company financially solvent. Singletary felt that it was just for unsecured creditors to require payment of their debts from a debtor who has the means of relieving that debt, and the directors agreed. After some discussion Mayor Charles Macbeth made a motion that the treasurer be authorized to pay the interest on third-lien bonds already sold or taken by creditors in actual payment of their claims at par. Swinton assured the directors that the sum of the company's unsecured debt did not exceed $525,000, and they adopted the resolution.[17]

On May 27, 1863, the *Charleston Daily Courier* announced that the holders of the third-lien bonds would receive the interest due them at the beginning of March upon presentation of their coupons at the Bank of Charleston. Singletary's position as both the chief executive and a major creditor of the company had likely factored into his business strategies, and so far they appeared to be rather effective. Singletary's proposals interested the bondholders, who had accepted $297,000 of the third-lien bonds by the end of the year. Those most accepting of the offer were creditors who were owed for work done or materials furnished. Many of the largest creditors, however, were unwilling to agree to the terms. Singletary's proposal stayed on the table through November, by which time the company's accumulation of cash allowed it to satisfy its debts on those accounts with cash.[18]

With their newfound successes the directors voted to reward the company's officers with a salary increase—a welcome raise because of the decreased purchasing power of Confederate currency. In a surprising show of good faith Singletary refused his raise, while approving those of Superintendent Haines and Secretary-Treasurer

Swinton. There were other surprises and disappointments for Singletary and his subordinates in the coming months. The Port Royal Railroad remained unfinished until after the war, and did not send the anticipated freight to Charleston or Savannah. Though the state auditor implied to Singletary that the state might be willing to assist the Charleston & Savannah in purchasing the Ashley River Bridge property from the Charleston Bridge Company, this aid did not materialize.[19] The second passenger train, which began its service in April, had to be discontinued in July because of the interruption of travel through Charleston.[20] Despite such setbacks, Singletary determinedly faced very uncertain times.

NINE

Unbroken Lines

During the summer of 1863 the railroad and the Confederate army encountered a new sort of soldier. Men who had once been slaves—who had helped to harvest the corn and cotton, grade the railroad beds, and maintain plantation houses—were now in Federal uniform. On June 2, 1863, Union colonel James Montgomery—with the aid of the well-known black abolitionist Harriet Tubman—led a black regiment on a raid up the Combahee River. The Second South Carolina Volunteers—later known as the Thirty-fourth U.S. Colored Troops—landed early that morning at Field's Point, not far from Combahee Ferry. There was some confusion among the Confederates picketing the ferry, and after a delay of two hours several mounted Rebels sped off to Green Pond to sound the alarm.

It was 9:00 A.M. when General William S. Walker received the telegram notifying him that the Federals had landed between two hundred and three hundred men at Field's Point and that a gunboat was destroying a pontoon bridge at Combahee Ferry. He ordered his entire command to consolidate at Pocotaligo Station and quickly put three companies of the Eleventh South Carolina and Captain William L. Trenholm's Rutledge Mounted Riflemen aboard his special train bound for Green Pond. Walker also sent small detachments of cavalry and artillery to Combahee Ferry and Salkehatchie Bridge, but once all his troops had arrived and deployed, the raid was over. The Federal raiders burned four residences in the area, including that of Colonel William C. Heyward, and six mills. They also made off with about seven hundred slaves, who appeared—to the Southerners observing the proceedings—to have had some prearranged plan with their captors.

The Union's employment of freed and contraband African Americans as soldiers disturbed many South Carolinians. They saw the black soldiers charging the Confederate defensive works and navigating the state's inland waterways as more than just an insult to the Southerners' pride. They could also be viewed as symbols of the crumbling foundation of the Southern economy and the Southern way of life. Although the railroad had a brief introduction to African American soldiers in the summer of 1863, it was not their last meeting with them.[1]

Without the use of slave labor the Charleston & Savannah Railroad probably would not have been completed, and the same could be said for most Southern

African American soldiers in battle, engraving after a sketch by Thomas Nast. From *Harper's Weekly*, March 14, 1863; courtesy of Library of Congress

railroads of that era. In the days shortly after the company's charter, there were many who doubted the ability of blacks to assist in its construction. Thomas Drayton, however, had persisted in his desire to use them to build his railroad. He had argued that an African American could do more work per day than a white laborer and at one-third the cost. As late as 1863 the Charleston & Savannah owned only three slaves valued at $2,700, preferring to hire their labor from neighboring planters rather than purchase slaves outright. Thus slaves owned by the Heywards, the Lowndeses, the Elliotts, and the Colcocks laid the foundation for the railroad that was now so fervently defended by the mixed band of Confederate regulars and South Carolina militia.

In 1856, while the *Charleston Daily Courier* had been congratulating Drayton for his "application and adaptation of our own distinctive species of labor,"[2] the Reverend Thomas Wentworth Higginson had been busily writing abolitionist tracts and quietly supporting John Brown and his followers. Born in 1823, Higginson lived in Boston, Massachusetts, where his father was the bursar of Harvard College. After receiving his degree from Harvard, the younger Higginson became an ordained Unitarian minister. A hard-core abolitionist and women's rights advocate, he had retired from his ministry by 1860 to devote all his time to writing. After Southern secession he taught himself about military strategy, fortifications, and principles of attack and defense, and soon thereafter received a commission in the Fifty-seventh Massachusetts Regiment.

After the Union victory at Port Royal and the seizure of the Sea-Island cotton plantations, many Northerners considered the myriad slaves in the area to be free. A large number of fugitive slaves braved the malarial swamps and Rebel patrols to make their way to Beaufort, attracted by the idea of freedom. The freed and contraband slaves were fed, employed doing odd jobs, and given a basic education. In 1862 General David Hunter, before his removal as commander of the Department of the South, had recruited a band of freed and fugitive slaves to serve in the military, but military leaders in Washington never recognized the regiment. The unit was later re-formed after a significant number of troops left Hilton Head to reinforce General George McClellan on the Potomac. Federal commanders authorized General Rufus Saxton to raise five thousand African American troops in the Department of the South, and the First South Carolina Volunteer Regiment became the first regiment of freed slaves mustered into service in the U.S. Army. To command them General Saxton chose Colonel Thomas Wentworth Higginson.

In June of 1863 Brigadier General Quincy A. Gillmore took over command of the Department of the South. Gillmore, who was known best for his role at Fort Pulaski, was ready to move on the city of Charleston. He devised a plan by which his command would take the lower end of Morris Island, reduce Battery Wagner, and bombard Fort Sumter into submission. In so doing he hoped to open the way for the Federal fleet to reach Charleston.

Major General Quincy Adams Gillmore, U.S.A. Courtesy of South Carolina Historical Society

Back on Hilton Head Island, Colonel Higginson was ready to show what his regiment could do. He had not been enduring the heat, humidity, and sand flies of the South Carolina lowcountry for nothing. In training his enthusiastic new charges, he had seen their pride in their newfound liberty and wanted to give them the chance to fight for their people's freedom. His men had participated in raids in Georgia and Florida; now Higginson sought to involve them in a larger-scale operation. A raid up the South Edisto (or Pon Pon) River had been planned and approved by General Hunter, but was put on hold at the time of his dismissal. Colonel Higginson petitioned General Gillmore, who decided to add the raid to his overall plan as a diversionary tactic.[3] To take Morris Island the Yankees planned three separate maneuvers: the primary attack on Morris Island across Lighthouse Inlet, a demonstration on James Island to persuade Beauregard to divert in that direction, and Higginson's raid up the Pon Pon to cut the Charleston & Savannah Railroad and prevent reinforcements from reaching the city from the west.

The raiding party left Hilton Head on the afternoon of July 9, 1863. Two hundred fifty men of the First South Carolina Volunteers and a section of the First Connecticut Artillery sailed aboard three vessels toward their destination just southwest of Charleston. The steamer *John Adams* was an armed ferryboat carrying two Parrott guns (a twenty pounder and a ten pounder) and two howitzers. The transport *Enoch Dean* sported a ten-pound Parrott and a small howitzer. The tugboat *Governor Milton* carried two twelve-pound Armstrong guns. As they slowly ascended the Pon Pon, Higginson's senses were acutely aware of his surroundings, "the dark and silent banks, the rippling water, the wail of the reed-birds, the anxious watch, the breathless listening, the veiled lights, the whispered orders."[4]

The Federals had chosen the Pon Pon River for this raid partially because they knew that its large rice plantations were still being actively worked. It was also a river by which they could easily reach one of the many railroad bridges on the Charleston & Savannah line. As Higginson's small convoy steamed silently toward their objective, they found travel somewhat more difficult than anticipated. In addition to the increasingly dense fog rising from the river, Higginson complained about "the vexation of an insufficient pilotage, for our Negro guide knew only the upper river, and as it proved, not even that."[5] At around 4:00 A.M. on July 10, they anchored twenty miles upriver at Willtown Bluff, near Morris Plantation.

Since Colonel Montgomery's raid up the Combahee River the previous month, the vigilance of Rebel pickets along inland waterways had increased. The fog, while causing navigational problems for Higginson, actually worked to his advantage, making his troops virtually invisible to the Confederate vedettes. Captain Frederick C. Schulz's battery of the Chesnut Artillery was assigned picket duty at Willtown Bluff that evening. Willtown Bluff was the key to ascending the Pon Pon, and heavy wood pilings obstructed the river there. Captain Schulz commanded a small detachment—only three guns—supported only by a handful of cavalry since a company of infantry had been sent to reinforce the Charleston defenses the night before.

The camp was just beginning to stir when Rebel guards spotted the three Federal vessels.

As the first light began to appear in the east, Higginson signaled to Captain (later lieutenant colonel) Charles T. Trowbridge to begin removing the obstructions. Seeing a boat near the obstructions, Lieutenant Thomas G. White gave the order to fire on the small steamer. Assessing the situation from aboard the *Enoch Dean*, Higginson examined his position relative to the battery and noted Trowbridge's slow progress removing the pilings. He then turned toward the riverbank, squinting to get a clearer image of what first looked to him like black dots appearing from the rice fields. These dots turned out to be scores of slaves, mostly women and children, who were crowding the banks with the expectation of being carried off to freedom by the Yankee raiders.

After the first shot, the Confederates were able to lob two more shells at the Federal force before a friction primer lodged in one of their guns and a cannonball was forced into another gun without a cartridge. Colonel Higginson and Lieutenant James B. West landed with a smaller force and began to move on the Rebel position. Lieutenant White observed the Federal column advancing along a side road leading to the rear of his position. Having no infantry support except for seven vedettes and only one effective gun, White feared the capture of his entire command. He ordered a hasty withdrawal, leaving behind clothing, equipment, and their breakfast fires burning.

Higginson and West's small force easily took possession of the bluff, allowing Trowbridge's men to take up the obstructions without the fear of Southern shot raining down on them. The task was still more difficult than anticipated. Having mistakenly assumed that the pilings would be decayed, the Federals' plan had not allowed for the three hours it eventually took to remove the obstructions. Because of this delay, the tide was too low for the boats to travel safely, and they had to wait until noon to proceed toward the railroad trestle.[6]

On hearing of the Federal landing, Colonel Hugh Kerr Aiken ordered fifty men of the Sixth South Carolina Cavalry under Lieutenant Colonel L. P. Miller to support the force at Willtown Bluff. The remainder of his command followed under the command of Major Thomas Ferguson. By the time Colonel Aiken arrived on the scene, White had already withdrawn his position. Aiken quickly deployed his entire force as skirmishers and advanced on Higginson's men at Willtown, while also trying to drive off the blacks remaining on the bank of the river. Skirmishes between Aiken's troopers and the Yankees remaining at the bluff occurred intermittently the rest of the day with little result.

When the tide was finally high enough, Higginson gave the order for the steamers to proceed. At their first effort to ascend the river, the *John Adams* ran aground and was unable to dislodge. The *Milton* and the *Dean* had better luck, and pressed on without the *Adams*. A little upstream the *Dean* again came under fire from Schulz's battery. Lieutenant White re-formed his men at a point above Barnwell's plantation and began to shell the invaders. The heavier and more operational guns of the Federals, however, persuaded White to retire after a brief exchange.[7]

The day had been extremely hot, and the higher the Federals ascended on the river, the more tortuous the channel became. The *Dean*, apparently riding a little low, had frequently run aground but was able to continue despite some minor engine problems. At a point about two miles below the Charleston & Savannah railroad bridge crossing the Edisto, however, the *Dean* became lodged more firmly in the muck. Not wanting to lose any more time or tide, Higginson signaled Major J. D. Strong aboard the *Milton* to go forward and complete their mission. As the ships approached the bridge, the soldiers had their combustibles ready, poised to burn the bridge just ahead. The men of the First South Carolina were preparing to attack a railroad that many of them had no doubt helped to build a few short years before, but it did not happen that day.

Advancing to a point opposite Dr. Joseph Edward Glover's plantation about a quarter mile from the bridge, the *Milton* encountered a battery of Walter's Light Battery, a section of the Washington Artillery commanded by Lieutenant Samuel Gilman Horsey. With the *Dean* able to offer very little assistance, the *Milton* came under brief, but furious, fire from the battery. The shells passed through both thinly made boats and exploded on the other side. Most of the injuries were from the shells or the splintering of wood. The former slaves in the First South Carolina manned the guns alongside the men of the First Connecticut, but their efforts proved vain. The accurate Confederate gunners soon thwarted the mission by disabling the *Milton*'s engine. With the *Milton* dead in the water—able only to float with the tide and current and nearly thirty miles from the mouth of the river—Major Strong turned his efforts toward navigating safely back downriver.[8]

Just downstream of the activity, Colonel Aiken had ordered a section of the Marion Artillery under Lieutenant Robert Murdock to reinforce Lieutenant White's force at Gibbes Plantation, where the combined force engaged the *Dean* and *Milton* as they descended the river. As they approached the original point of conflict, the *Milton* again ran aground. By this time the *Adams* could travel freely and the *Dean* had come back to Willtown Bluff to pick up the infantry occupying the works and any additional contraband they could hold. The *Adams* traveled upstream to aid the *Milton*, but the combination of the Confederate shot and falling tide prevented her from freeing the grounded craft. With the situation appearing unsalvageable, Strong ordered the vessel abandoned and pushed her guns overboard. When Higginson learned of this decision, it was too late to do anything but order the ship set afire to avoid its capture.[9]

Around sunset Colonel Aiken ordered a final advance on Willtown Bluff, but his men found the position abandoned and the enemy moving downriver out of range of their guns. Aiken dispatched a section of the Sixth South Carolina Cavalry with part of the Chesnut Light Artillery to a point below Morris's mill, but their random firing had little effect on the retreating Federals. The raiding party headed back to

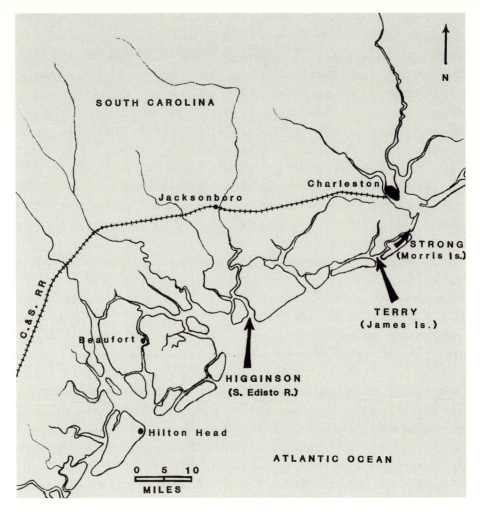

Union assaults planned for July 9, 1863. From *Gate of Hell: Campaign for Charleston Harbor, 1863*, by Stephen R. Wise, © 1994 University of South Carolina

headquarters at Hilton Head with six bales of good-quality cotton, two prisoners, and about two hundred contraband slaves formerly owned by the Morris and Manigault families. They had also burned Colonel Morris's mill and a number of barns containing significant quantities of rice.[10] To an abolitionist such as Higginson, the day certainly was not a total failure. From a military standpoint, however, the same could not be said. The Confederates had driven off the Yankee raiders and had captured two brass rifled guns from the wreckage of the *Governor Milton*. More important, they had protected the railroad and preserved the main conduit for reinforcements sent to Charleston.

General Gillmore's other diversionary maneuver began on July 9 and 10, when General Alfred Terry landed on James Island with a force of 3,800. General Beauregard had received word of a possible Federal attack somewhere in the area after a reconnaissance detachment discovered a number of Yankee barges collected in the creeks approaching Lighthouse Inlet. Beauregard's dilemma was that he did not know exactly where the main assault would occur, although he suspected Morris Island. General Terry's landing was a serious threat to Charleston, but with limited numbers of troops in the area Beauregard could ill afford to jeopardize his other positions. He had 2,900 men of all arms in the vicinity and decided to place them in defensive positions to protect the approach to Charleston. Just three weeks earlier, President Jefferson Davis had telegraphed Beauregard requesting that he send any reinforcements he could spare to General Joseph Johnston in the West. Beauregard had quickly replied that he could not send any troops from his department "without losing the railroad and country between here and Savannah."[11]

As Union gunboats bombarded both banks of the Stono River on the morning of the landing, Federal troops under Terry achieved their objective virtually unopposed. Nothing significant occurred that day except for some minor skirmishes with pickets in the area. The lead regiments of Terry's command penetrated a mile or two inland in a token show of force, but did not attempt much more. Among the troops arriving on James Island on July 9 and 10 were the Fifty-fourth Massachusetts, who mainly concentrated on digging in and awaiting further orders. The main Federal assault occurred the next day on Morris Island without them.

In response to the impending attack, General Beauregard directed his chief quartermaster, Major Hutson Lee, to have trains ready on the Charleston & Savannah Railroad to bring 1,100 troops from Pocotaligo and Adams Run to Charleston as soon as possible. He also directed Colonel Ambrosio José Gonzales to have the siege train ready to move on a moment's notice and petitioned Charleston mayor Charles Macbeth to provide labor from available free blacks and slaves in the city. On the night of July 9, Colonel R. F. Graham of the Twenty-first South Carolina alerted his troops of an imminent attack on the island. When morning broke, his warning was proven accurate. Trees and bushes that previously hid Union batteries on Folly Island had been cleared away, and Federal guns shelled the southern end of Morris Island from that position. Union monitors under Admiral John Dahlgren enfiladed the Confederate position, as two thousand Federal infantry commanded by Brigadier General Truman Seymour and led by Brigadier General George Strong were ferried across the inlet to Morris Island.

Seymour's force was opposed by only about seven hundred Confederates. Beauregard could not spare any reinforcements from James Island until he discerned the full extent of the Union movements there. Eventually three hundred men from the Seventh South Carolina Battalion were sent, but they arrived too late to offer any effective aid. Facing overwhelming numbers, the Rebels gave way and hastily retreated up the beach to Battery Wagner, leaving cooking fires burning, food on tables, and even uniforms in tents. The Northern troops chased the Confederates toward Wagner, aided by the steady fire of Dahlgren's monitors, but the Federals fell back when they got to within range of the gunners in the Rebel battery. The

Federal assault achieved such rapid success that General Gillmore had not provided enough support and preparation for his troops to take Battery Wagner. Having seen what they had accomplished that day, Gillmore decided to delay further attack on the battery until the following day, and Seymour's troops spent the remainder of July 10 digging in and fortifying their position on the island. The Confederates on the island averted total disaster and kept the Yankees out of Wagner, but the enemy was now firmly established on Morris Island.[12]

General Beauregard and his troops defending South Carolina's First Military District used the twenty-four-hour delay to their fullest advantage. Not long after Colonel Higginson's unsuccessful raid, the first train carrying Georgia troops from Savannah had crossed the Jacksonboro Bridge on its way to Charleston. This group of reinforcements included more than five hundred men from the First and Sixty-third Georgia Infantry and the Twelfth and Eighteenth Georgia battalions. They, along with the remains of Graham's Twenty-first South Carolina and Nelson's battalion, became the garrison of Battery Wagner. General Mercer, commanding the District of Georgia, also sent seventy-five artillerists and the first of five ten-inch mortars to Charleston via the railroad. Beauregard sent dispatches to agents at every station along the Charleston & Savannah Railroad to "hurry up the troops from Savannah as fast as practicable." A total of ten companies arrived from Georgia over the next day and a half to reinforce the troops on James and Morris islands.[13]

At daylight on July 11, four companies of the Seventh Connecticut, commanded by Colonel Daniel Rodman, formed ranks to lead the assault on Wagner, supported by the Seventy-sixth Pennsylvania and Ninth Maine regiments. Advancing quietly toward the sand fort with bayonets fixed, their vanguard breached the outer works but began to take terrible fire from both cannon and rifle. Grapeshot and canister tore holes into their ranks. They reached the parapet, but the Pennsylvania and Maine regiments had already broken ranks under fire and were rendered ineffective. At this point the entire garrison was able to focus its fire on the Connecticut troops, who began a hasty and perilous withdrawal to the safety of their trenches, leaving more than half their unit lying dead or wounded on the beach. Several skirmishes occurred on James Island the following week, but the Federals advanced no closer to Charleston. On July 16 a skirmish near Grimball's Causeway forced General Terry to retreat to within the protection of his gunboats, and on July 17 his force withdrew to Folly Island to prepare for another attempt on Battery Wagner.

Colonel Robert Gould Shaw, a Massachusetts abolitionist, commanded the Fifty-fourth Massachusetts, a regiment of African American soldiers later immortalized in print and film. They had initially been part of General Terry's diversionary force on James Island, and Seymour placed them at the head of the planned attack. The Federals bombarded Wagner from both land and sea during the majority of the day on July 18 in preparation for the attack. With darkness approaching, Shaw's Fifty-fourth advanced, supported by the Sixth Connecticut, Forty-eighth New York, Third New Hampshire, Ninth Maine, and Seventy-sixth Pennsylvania. More than seven thousand Union troops tried to capture the fort, manned by only twelve hundred Confederate troops.[14] The Yankees marched slowly until they came within nine hundred yards of the fort, at which time they charged with bayonets fixed. Explosions rang out from all around them as they tried to ascend the steep embankment of the battery. Shaw's regiment was torn apart, and he was killed while charging the parapet. The Union troops were signally repulsed and returned to their position at the south end of the island.

Convinced that Wagner could not be taken by a frontal assault, General Gillmore began planning siege operations for Battery Wagner and the city of Charleston. The Confederacy had lost a significant battle at Gettysburg on July 1–3, 1863, and had surrendered Vicksburg to the Yankees on July 4, but the city of Charleston was still out of their reach. The seemingly insignificant skirmish near Jacksonboro had preserved the railroad and allowed adequate reinforcement of the Confederate troops on James and Morris islands. Recognizing the railroad's importance to the city, and its vulnerability, General William S. Walker ordered Colonel Aiken to speed up completion of earthworks to protect the Edisto River bridge against future raids.[15]

That summer the eyes of the Confederacy had been focused on the farmlands of Pennsylvania and the Mississippi River. Meade had thwarted Lee's Northern invasion at Gettysburg, and Pemberton had surrendered the city of Vicksburg and with it control of the Mississippi River. As the end of August approached, the attention of many shifted. There was an impending crisis in Tennessee, where Union general William Rosecrans and his Army of the Cumberland had marched across Tennessee from just southeast of Nashville, outflanking Confederate general Braxton Bragg, who had fallen back toward Chattanooga and the important railroad connections there. This was more than enough to arouse the attention of President Davis and General Robert E. Lee, who met in Richmond during the last week of August to decide on a course of action. If Rosecrans took Chattanooga, it would open the way to Atlanta. If Atlanta fell, the Confederacy would be split yet again and lose all but the faintest hope of overall victory.

General Lee had initially wanted to send his Army of Northern Virginia back into Pennsylvania for another invasion, but Davis overruled him. While he heard Lee's case, Davis was also listening to other leaders who lobbied for reinforcing Bragg in the West. The Western Coalition—which included General P. G. T. Beauregard, General Leonidas Polk, and Virginia senator G. A. Henry—believed that victory in the West offered the Confederacy its only successful strategy for ending the war. Even Lee's "Old Warhorse," General James Longstreet, saw the benefits of concentrating troops in the West. He was convinced that the western theater offered an opportunity for success and believed that Lee should go on the defensive, retire to a strong position closer to Richmond, and send a corps to Tennessee.[16] Longstreet also offered to command those reinforcements.

On September 5, Davis decided to send divisions headed by General John Bell Hood and General Lafayette McLaws to northwest Georgia under General

Longstreet, leaving General George Pickett's division—still reeling from its infamous charge at Gettysburg—to defend Richmond. This move freed up General Micah Jenkins's brigade to go with Longstreet's corps as part of Hood's division. General Lee stayed in Virginia, where his troops were outnumbered by Union general George Meade's Army of the Potomac. If Meade attacked, Lee might have to fall all the way back to Richmond. Lee gambled on the belief of several staff officers that the Yankees would send troops to counter Longstreet once they learned of his departure from Virginia and leave Lee's army alone.[17]

This most ambitious troop movement by rail was choreographed by Quartermaster General Alexander R. Lawton, a native of Beaufort District, South Carolina, and Major Frederick W. Sims, the new head of the Confederate Railroad Bureau. The move involved transporting part of Longstreet's corps from Richmond to northwest Georgia along with two brigades bound for Charleston to reinforce the defenses there. The routes required as many as eight different transfers, with additional delays because of unconnected or incompatibly gauged tracks. The movement depended on the timely availability of engineers and cars at each transfer point along the way and forced Lawton and Sims to press all available rolling stock into service, straining the already overwhelmed capacity of the Southern railroads. Troops were jammed into everything from passenger and baggage cars to coal cars and boxcars.[18] No matter the size, type, or condition of the railcar, if it could roll, they used it.

Lawton and Sims initially decided that the troops would travel the most direct route from Richmond to Chattanooga via Lynchburg, Virginia, then south toward the Tennessee Valley. Sims had arranged travel on the Virginia & Tennessee Railroad, but Union general Ambrose Burnside had entered the relatively undefended city of Knoxville, Tennessee, on September 3, effectively blocking this route. Sims and Lawton quickly planned a more circuitous route further to the south.

The first of Longstreet's corps left Richmond on the morning of September 8, 1863. The troops traveled light—many taking only their haversacks or blanket rolls—and stuffed themselves into overcrowded boxcars. Many decided to make the trip riding on top of the cars, and some cut holes in the sides of the cars for light, fresh air, and viewing the passing sites. Rail traffic diverged onto two separate routes, a central route and a coastal route, at Gaston and Weldon, North Carolina. The central route passed through Raleigh and Charlotte, North Carolina; Columbia and Kingville, South Carolina; and Augusta, Georgia, then on to Atlanta. The coastal route used railroads that passed through Wilmington, North Carolina, and Florence, South Carolina; from Florence the troops traveled to Atlanta, either through Kingville and Augusta or by way of Charleston, Savannah, and Macon.[19]

To reinforce the Charleston defenses General Longstreet had chosen two brigades of Georgia troops under General George T. Anderson and General Goode Bryan. Because of the Georgians' low morale, General Longstreet did not want to take them into their native state for fear they would desert. On September 10,

General Lee and President Davis suggested that it would be better to send the brigades of General Micah Jenkins and General Henry A. Wise to Charleston. Anderson's brigade was already en route to Charleston, and General Longstreet feared that Beauregard would keep both brigades if he sent Jenkins's to replace Anderson's. Longstreet therefore decided to reassign Jenkins's brigade to Hood's division in Anderson's place; Wise's brigade was eventually sent to Charleston in place of Bryan's Georgians.[20]

Noting the increased activity in the Confederate camps of Northern Virginia, the Yankees stepped up their intelligence efforts. The Confederates routinely sent agents as "deserters" across enemy lines to spread misinformation, but despite such attempts, Union general in chief Henry Halleck felt something was afoot. Longstreet was moving somewhere, and Halleck was going to do something about it. Union forces increased their activities, both overt and covert, along the Southern coastal railroads. They raided the Wilmington & Weldon Railroad, burning or destroying all the water tanks along that route. In the South Carolina lowcountry they sent out a raiding party to intercept telegraph communications between Charleston and Savannah.

Telegraph lines adjacent to railroads often served to make rail travel safer. Operators notified engineers of breaks in the rails, oncoming rail traffic, and other problems along the line. With the outbreak of war, operators also used the telegraph lines to warn engineers of military activity ahead and to coordinate the movement of troops and supplies to critical points along the rail line. The telegraph was extremely important to the Confederate defensive strategy on the Carolina coast because it depended on the rapid concentration of troops at points of enemy movement. As soon as Fort Sumter had been fired on, the construction of a telegraph line was begun between Charleston and Savannah. The system contained three connecting cables that ran the entire length of the Charleston & Savannah Railroad. One cable spanned the Ashley River, one stretched from there to the Ashepoo River, and the third went from the Ashepoo to the railroad bridge across the Savannah River.[21]

A young Canadian telegraph operator named William Forster was sent with a small group of Union soldiers to tap the telegraph wires running along the Charleston & Savannah Railroad. The party was dispatched around September 7, 1863, along with a supporting company of the First South Carolina Volunteer Infantry, who were temporarily bivouacked on Williman's Island. Forster was equipped with a large quantity of wire and all the instruments necessary for intercepting communications. On the afternoon of September 12, he had successfully attached a wire to the main cable when he heard a locomotive approaching.

On the train was the railroad's master of roadway, J. H. Buckhalter, who had gained some measure of renown for his heroics during the Battle of Pocotaligo the year before. When he spotted the wire about a mile south of Green Pond Station at around 3:30 P.M., he sped full steam toward the depot. Lieutenant Colonel William Stokes of the Fourth South Carolina Cavalry, who commanded the subdistrict between the Combahee and Ashepoo, was in his quarters when the railroad agent dashed in to inform him of Buckhalter's revelation.[22] After he and Buckhalter conferred, Stokes ordered fifteen men of Captain W. P. Appleby's company of cavalry to dismount and board the train, which took them to the point at which the conductor spotted the intercepting wire. When the cavalry arrived, the enemy party had fled, and Appleby's men dashed into the woods in pursuit. Colonel Stokes ordered Lieutenant J. W. R. Berry, with the remainder of Appleby's company, to skirmish the forest between Whitehall and Green Pond. Pickets were also placed south of the position along the Combahee and Ashepoo Ferry Road from Colonel William C. Heyward's plantation to the Chehaw Road because Stokes felt certain that the enemy would have to pass that point. His instincts were correct, but a handful of pickets allowed the entire enemy party to pass through their line unchecked. Incensed at this lapse of judgment, Stokes decided to postpone the search because of darkness, but he kept the Combahee River closely guarded overnight.

While walking in Charles Tidyman Lowndes's rice field the next day, Stokes heard the firing of a rifled gun from the battery near Lowndes's mill. He quickly saddled up and rode to the battery, which was manned by Company C of the Eleventh South Carolina. Their commander, Lieutenant J. J. Guerard, reported hearing a noise on the riverbank near Lowndes's mill and decided to lob a few shells in that direction. Stokes then ordered Lieutenant Frank Sineath to take six men and investigate. At the river Sineath found a small raft made from old planks, which the enemy had used to cross the river. Seeing the approaching Confederates, the Federals abandoned the raft and ran. Sineath and his men gave chase, and soon caught Chaplain James H. Fowler, Private Robert DeFoe, and a first lieutenant of the First South Carolina hiding in the marsh near the river.[23]

Captain Appleby had also resumed his search that morning adjacent to Colonel Heyward's plantation, using bloodhounds sent on the train by Colonel Charles J. Colcock. Searching the marshes near the river, they found Forster "almost naked and in a wretched plight from mosquitoes, bugs, etc." The dogs were sent out again on Monday, September 14, resulting in the capture of another black private of the First South Carolina on Colonel Heyward's plantation. They also found approximately one-third to one-half mile of good-quality telegraph wire, which they forwarded to General Beauregard's headquarters. Through questioning their captors, they learned of the encampment on Williman's Island. The Confederates planned a surprise attack for the evening of September 14, but the Yankees had abandoned the position prior to their arrival.[24]

Stokes's men found the telegraph apparatus used by the party two months later near Lowndes's plantation, and discovered their valise soon afterward. It contained two candles, some shirts and pants, and a quantity of dispatch paper and envelopes, most of which were rotten.[25] It is doubtful that Forster's raiding party gleaned any viable information on the movement of Longstreet's corps or the reinforcement of Charleston, but one thing is certain. Whatever information Forster obtained went with him to the grave. After his capture and interrogation, the twenty-three-year-old telegraph operator was sent to Andersonville prison, where he died of dysentery the following summer.[26]

The first troops to leave Richmond for the West were the men in General John Bell Hood's division, the majority traveling the central route via Raleigh, Charlotte, and Columbia. General McLaws's division followed, going through Petersburg to Weldon, and then to Wilmington. After a stop in Wilmington and a steam-ferry ride across the Cape Fear River, they boarded another train bound for Florence, South Carolina, on the Wilmington & Manchester Railroad. The condition of the locomotives on that road had deteriorated to the extent that the Second South Carolina Infantry traveled only forty miles in twelve hours. Eventually their engineer pulled the train onto a siding and telegraphed Florence for another engine, which arrived hours later. At Florence the local quartermaster routed many of the troops to Kingville, the junction of the Wilmington & Manchester and the South Carolina Railroad. To relieve congestion at Kingville, it was decided to send a portion of the corps from Florence to Charleston on the Northeastern Railroad. Those sent to Charleston included Kershaw's brigade of South Carolina troops and a regiment of General Jerome Robertson's brigade.[27]

In spite of the inconveniences and delays some soldiers saw the movement as a dashing and heroic adventure. With news of their brief homecoming in the wind, citizens came from far and near to cheer the troops as they passed. During layovers, "old men slapped their hands in praise; boys threw up their hats in joy, while the ladies fanned the breeze with their flags and handkerchiefs." As the locals cheered their "Boys in Gray," the soldiers inside and atop the cars responded by "cheering and yelling themselves hoarse." Other troops remembered the sumptuous banquets prepared for them and the coquettish women serving them.[28] The romance temporarily blocked out the fear of the impending fight.

While some had their spirits lifted during their return to South Carolina, the combination of homesickness and the proximity to home was difficult for others to overcome. As the first of Kershaw's troops arrived in Charleston, they were brought back to reality by "the booming of the great guns" of the Federal batteries shelling the city. They detrained, marched through the upper portion of Charleston and across the Ashley River Bridge. After resting for an hour, they boarded a train for Savannah. Many of the troops that made it to Charleston on later trains fought their own emotions during their layover. Sergeant Dubose Eggleston wrote that in passing through Charleston "the patriotism of troops never was more severely tested." Lieutenant Colonel Franklin Gaillard agreed that his regiment's time in the Port City was under "the most trying circumstances."[29] Some of the Charlestonians in the regiment spent the night with their friends and families before bidding them tearful good-byes and boarding the trains for Savannah the next morning.

The train from Charleston to Savannah halted only briefly to take on water at Pocotaligo before terminating at the junction with the Central of Georgia outside

Savannah. There trains had been readied prior to their arrival, enabling them to make their transfer immediately and speed off toward Millen. After Millen they passed through Macon and then reached Atlanta, where they boarded trains on the Western & Atlantic Railroad to Ringgold, Georgia.

Longstreet's command arrived at Ringgold's Catoosa Station in detachments, and as quickly as they reached their destination, the troops marched off to join General Bragg. Five small brigades reached North Georgia in time to fight at Chickamauga on September 19–20, 1863. Kershaw's and Robertson's were two of them. The addition of the five brigades to Bragg's Army of Tennessee had helped the Confederates outnumber those in the Federal Army of the Cumberland by 12 percent, and the confident, battle-tested Confederate veterans under Longstreet's command made a decisive contribution to the victory over Rosecrans at Chickamauga.[30] Lawton, Sims, and company had done their best given the circumstances. Layovers, unconnected tracks, and a need for more rolling stock decreased the efficiency of the movement, but it had been accomplished nonetheless. Although the Confederates won the day, had more of Longstreet's infantry and artillery arrived in time, the victory would have in all probability been more decisive.

Of the 12,000 to 15,000 troops who rode the rails to Chickamauga, about 1,200 were diverted over the rails of the Charleston & Savannah Railroad. Exactly who made the decision to send the troops along this route is not known, but it appears to have been a good one, since the trip via the Northeastern, Charleston & Savannah, and Georgia Central was not fraught with the same difficulties experienced on the roads in eastern North Carolina. Though they were busy routing some 900 state militia troops to Charleston and though their depot at Kingville was receiving converging trainloads of soldiers from Florence and Columbia, officials of the South Carolina Railroad took issue with the decision to reroute a portion of the troops to Charleston. Superintendent H. T. Peake, who had been prepared to receive 20,000 troops, immediately telegraphed Major Sims. Peake expressed surprise when asked to furnish two trains to aid in shunting 1,200 men over the Charleston & Savannah route. He took the diversion as "rather an embarrassment than a relief," since "the troops [had] all gone from [Kingville] and Columbia promptly so far."[31]

The choice to use the Charleston route may have simply been to relieve some of the congestion around Kingville, but Peake failed to consider another possible explanation. At that time a large Federal force lay just outside the city of Charleston, and General Beauregard feared an attack there. General Anderson's brigade had arrived in Charleston, but General Wise's troops were hundreds of miles and several days away. Someone in the Confederate chain of command may have decided to divert the troops to give the appearance of a much larger force bound for Charleston. Such a move might mislead General Gillmore into thinking he faced a much larger force defending Charleston than he expected, causing him to delay any planned assault on the city. It might also give the impression that the Confederates planned to send fewer troops to General Bragg in Tennessee than initially thought. Had this been the case, perhaps Peake would have been less insulted; however, since no one has documented the rationale behind the decision, it remains open to speculation. Regardless of the reason, the Charleston & Savannah Railroad was called on and efficiently participated in what George E. Turner called "the outstanding operational feat of the Confederate railroads during the war."[32]

TEN

The Squeaky Wheel

With the change from summer to autumn the atmosphere in and around Charleston was also changing. The Union's fifty-eight-day Morris Island campaign had ended during the first week of September with the Confederate evacuation of Battery Wagner, and the heavy bombardment that characterized the Siege of Charleston had begun on August 22, 1863. On September 8–9 Admiral Dahlgren had attempted a frontal assault on Fort Sumter, but it had been repelled. Charleston's outer defenses had survived another attempted invasion, but could the people of Charleston stand the persistent bombardment from General Gillmore's Marsh Battery and the enemy guns now trained on the city from Morris and Folly islands?

The campfires that dotted the Charleston & Savannah Railroad were now fewer and farther between. As General Beauregard's courier rode the morning train from Charleston, he passed installations along the railroad that were either unmanned or had just a token force. He knew the reply that he carried from Beauregard to the commander of the Third District would not be welcomed news. The cars came to a loud, grinding halt at Pocotaligo Station just after midday, and a weathered-looking cavalry scout escorted the messenger to Brigadier General William S. Walker's headquarters not far from the depot. Walker read Beauregard's dispatch with a mixture of disappointment and frustration. "Yourself and General Robertson," it said, "must make with your present forces, the best show of resistance to the demonstrations of the enemy."[1] Walker neither liked nor agreed with the decision, but he was a soldier, and he obeyed orders.

After the fall of Morris Island on September 9, the majority of the Eleventh South Carolina Infantry, except for a few detachments retained as artillerymen and for bridge defense, were transferred to James Island as part of Brigadier General Johnson Hagood's brigade. In their place Walker was sent a militia regiment of "390 old men and boys"[2] and a battery of light artillery, leaving him a force of approximately 2,200 effective troops and eighteen pieces of artillery to defend the area stretching from the west bank of the Ashepoo River to the Savannah River swamps. A little more than two thousand men to defend batteries and forts designed originally to be defended by a force four to five times that size.

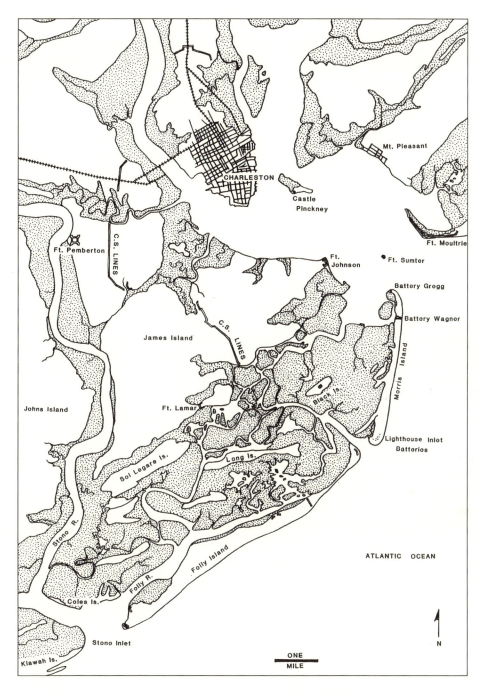

Confederate and Union positions around Charleston, June 1863. From *Gate of Hell: Campaign for Charleston Harbor, 1863*, by Stephen R. Wise, © 1994 University of South Carolina

General Walker possessed a knowledge both of ground and naval warfare and knew that the Federal invasion was taking more time, and costing more money and lives, than the leaders in Washington expected. He felt that the Yankees would grow weary of their attempts to take Charleston from the front and that they would begin to look for alternative approaches to the city. This, he supposed, would lead them into his district, and he was not at all confident that without reinforcements his paltry force could defend against a column of four thousand enemy troops. The earthworks below the railroad were strong as long as there were adequate troops for their defense. With his present force a large Federal force might drive him from the works before reinforcements could arrive from Charleston or Savannah. Walker feared that he would have to retreat toward the Edisto River and abandon the Charleston & Savannah Railroad. This would result not only in further damage to the economic prospects of the region but also in the loss of the shortest line of communication between the cities and a strong line of defense.

The vulnerability of the railroad was magnified when the Union began transferring troops from their bases in North Carolina and on Morris Island to Port Royal. General Beauregard telegraphed Secretary of War James A. Seddon of the concentration. If these Federal troops were intended for offensive operations based out of Port Royal, Beauregard warned Seddon, the railroad communication between Charleston and Savannah would "soon be interrupted, for [his] forces [were] too small to protect it."[3] He quickly had troops in Charleston and Savannah provided with rations, ammunition, and transportation to be sent at a moment's notice to General Walker's district, where an attack was most likely. He also realigned part of his command by subdividing St. Andrew's Parish. General Henry Wise would command the area south of the Ashley River and west of Wappoo Cut, and General William Booth Taliaferro commanded the forces on James Island. Fortunately for the Confederates, no attack was forthcoming.

At around this same time the troops and citizens in and between Savannah and Charleston were excited about a presidential visit. President Davis had departed Richmond on October 6, 1863, for a visit to the western theater, where General Bragg had failed to capitalize on his victory at Chickamauga. There, he met with many of Bragg's subordinates and inspected the troops in eastern Tennessee. On his western tour he made appearances in Augusta, Atlanta, Selma, Meridian, Montgomery, and Macon. He arrived in Savannah on Halloween night, welcomed by a crowd that included the mayor and many of the city's prominent dignitaries. While there Davis toured the forts and batteries along the river with General Hugh Mercer. Later that evening, Davis addressed a crowd from the portico of Pulaski House, where he lodged for the weekend. He attended Sunday services at Christ Church and spent the next day with his staff reviewing reports from Chattanooga, where the Yankees under General Ulysses S. Grant had opened a new supply line into the besieged city. Davis was in a pensive mood as he met General Jeremy F. Gilmer and a group of South Carolina notables to board the Charleston-bound train on Monday morning, November 2.[4]

Along the way, the special train stopped at several hamlets for President Davis to address the soldiers. At Grahamville, Colonel Charles Jones Colcock and his battalion of cavalry received encouragement and thanks from their leader while the cars idled on a siding. General Walker had assembled a regiment of cavalry and several companies of infantry at Pocotaligo to hear a brief discourse from the president. Colonel William Stokes and a column of the Fourth South Carolina Cavalry cheered heartily as the president's train passed Green Pond, and Davis stopped at Adams Run to meet with the officers stationed in the vicinity.

President Davis's visit to Charleston was his first since the formation of the Confederacy. He arrived at St. Andrews Station in the early afternoon of November 2, 1863, and was greeted by a fifteen-gun salute. After a circuitous carriage ride to the Charleston City Hall, he spoke to a crowd, most of whom were getting their first glimpse of their president. As he spoke, "the thunder of the enemy's guns, which sounded in his ears, suggested the present condition and prospects of the city." Davis spoke with a tone of optimism, stating his belief that the city would not fall to the Union army. The difficulties of a water approach, he noted, would be hard to overcome, and Union troops would meet certain defeat if they advanced by land. He also made the prophetic statement that if Charleston ever did fall into enemy hands, he hoped that it would be a mass of rubble. He empathized with the people of Charleston, who had endured the hardships of a Federal siege for over two months. He implored them to have a "spirit of harmonious cooperation" by "casting away all personal consideration and looking forward with an eye devoted singly to the salvation of our country, that our success is to be achieved."[5]

Davis's visit to Charleston was not just to comfort and encourage its populace, however. He also came to meet with General Beauregard and gain firsthand knowledge that would enable him to analyze reports received from the area more accurately. After a night's respite at the residence of former governor William Aiken, Jr., President Davis spent the next day inspecting the defenses of Charleston's harbor and James Island. After a couple of days of speaking, inspecting, and reassuring, he departed the city for Wilmington to do much of the same there.

What Davis had seen on his trip from Savannah to Charleston could not have been encouraging to him. The tired, shabbily dressed old men and boys who had cheered him along his route were all that stood between the Union army and the interior of the state. He had pledged to the people of Charleston that, if Lincoln sent more troops into the region, "reinforcements should not be wanting to the army whose duty it was to defend Charleston."[6] General Walker wondered if President Davis would objectively assess reports sent from his district. The defense of Charleston, her railroads, and South Carolina itself depended on Davis's keeping his promise to reinforce the troops defending Charleston.

As the defense of the railroad became more tenuous, Captain Singletary and Superintendent H. S. Haines did their best to maintain business as usual. With the staggering cost of supplies, a shortage of manpower, and a deficiency of rolling stock, railroads throughout the Confederacy were becoming less reliable channels of commerce. The Charleston & Savannah Railroad had the added difficulty of operating while siege operations were ongoing against Charleston. The company's management feared that hostile enemy activity might partially or completely interrupt their business.

Deficiencies of rail chairs, nails, springs, boiler iron, and wheels existed, and only a select few Southern firms could produce them. The law of supply and demand dictated an unprecedented increase in their cost. The price of lumber suitable for building rolling stock had quintupled since 1861. Other staples, such as nails and illuminating oil, had risen anywhere from ten to twenty times their prewar price. The result was a 61 percent increase in the Charleston & Savannah's operational expenses from 1862 to 1863. In just one year the railroad's operational cost per mile had risen from $12.70 to $21.16.[7] To make up for these increased expenses, the company had no choice but to raise their rates occasionally for both civilian and government passage and freight.

The company could ill afford to reduce expenses by laying off employees. Because of the loss of many workers to the military and the reluctance of local planters to hire out their slaves to the railroad, railway workers during this time were anything but expendable. New legislation introduced in the Confederate Congress threatened an already decimated workforce. The Confederate Conscription Act of April 16, 1862, had provided for the exemption of railroad employees at the option of the secretary of war, and another act passed a week later had specified immunity for persons in actual service on railroad routes. In October 1862 Congress reduced the broad exemptions of the previous act, specifying protection only for certain categories of railroad employees: executive officers, conductors, engineers, station agents, mechanics, and two expert track hands per each eight-mile section of track. On September 5, 1863, however, the Confederate Bureau of Conscription collected a census of all white males engaged in railroad work in Virginia, North Carolina, South Carolina, and Georgia, implying that they might one day be drafted into active military service.[8]

As the military situation became more urgent, it appeared that the exemptions might be tightening up. By the end of 1863 the Confederate Congress was discussing the conscription of all able-bodied railroad workers under the age of forty-five and replacing them with older men and disabled soldiers. Such an action might severely damage the ability of rail carriers to support the war effort. The Charleston & Savannah Railroad already found itself relying on men such as Private Robert S. Walker, who had been detached from the Fifth South Carolina Cavalry for service on the road, and Private E. John White, whom surgeons had deemed unfit for military duty and sent to serve as a route agent on the railroad for the Confederate postal service. The railroad was so strapped for workers that R. L. Singletary asked that the Confederate Post Office Department relieve the company of its responsibility of transporting mail between the depot and the post office. When Singletary and railroad presidents from all over the Confederacy appealed to the Railroad Bureau, Major Frederick W. Sims intervened, and Congress allowed the bill to die in committee.[9]

New legislation, passed in 1864, raised the military age to fifty and tightened exemptions for railroaders. The number of employees of a given road could not exceed the number of miles of track it devoted to military transport. Railroad superintendents specified exempted personnel by name and job description and reported them to the Bureau of Conscription. If these men left their posts, the agency would immediately revoke their status. Strict interpretation of the law would impede rail transport throughout the Confederacy, and railroad managers wanted the Railroad Bureau to have ultimate control over the exemptions.[10]

The need for railroad labor, compounded by rampant inflation, also caused problems with the maintenance of rolling stock. At the close of 1863 Haines reported the company's rolling stock to be in satisfactory condition, but maintaining it was becoming problematic because of the increased cost of labor and materials. The company's machinery had been continually subjected to brutal service. Additional mechanics and a large supply of wheels and axles would be needed for the ensuing year. Because of the lack of manpower, workmen had undertaken nothing beyond "ordinary repairs" along the roadway that year. The decline in maintenance standards translated to a reduction of the engines' speed and thermal efficiency. Before the war the 104-mile trip from Charleston to Savannah took only five and one-half hours. In April of 1863 the same trip took six hours and fifty minutes, and by November it took a full ten hours. At the end of 1863 the company's locomotives could travel only 41 miles per cord of wood, compared to 65 miles per cord before the war began—a 36 percent drop in thermal efficiency. The Charleston & Savannah locomotives could not match the thermal efficiency of the Georgia Central's, which averaged 60 miles per cord of wood; however, they were not as slow as those of the Wilmington & Manchester, which took nearly twenty-four hours to travel the 171 miles between Wilmington and Florence.[11]

Despite maintenance difficulties Superintendent Haines refused to compromise his railroad's safety standards. While breakdowns, wrecks, and injuries were common on neighboring roads, the Charleston & Savannah counted only one derailment in 1863. While the locomotive was damaged, no one was hurt in that accident, but the company did report two deaths during the year. One was a freight conductor, Patrick Hannon, who tried to board a moving train, and the other was a slave of Colonel Charles J. Colcock's, who was killed while coupling a train together. Compared to customers of other roads, Charleston & Savannah customers experienced fewer delays and relatively little accidental loss of life and property. It was a credit to Haines that the loss and damage to freight, baggage, and livestock for the entire year totaled only $4,650.[12]

Matters of maintenance and safety aside, the lack of usable rolling stock continued to hinder the railroad's management. While they had been able to purchase ten boxcars from the South Carolina Railroad and six more through other channels, their numbers of coaches, stock cars, and platform cars remained virtually unchanged from the previous year. The company had to spend $17,000 in 1863 to hire cars from other railroads, an amount that could have purchased much more rolling stock in peacetime.[13] Although they had more boxcars, their relentless use by the military and lack of storage space still created delays in service and clogged depots with both civilian and government freight.

When Brigadier General Johnson Hagood's troops were running out of meat, the railroads were blamed, but even Hagood's staff realized that it was more of a systemic problem than a local one.[14] The Confederate government had the power to do something about the situation, but President Davis and the Congress were hesitant to interfere with private enterprise. Railroad Bureau chief Frederick W. Sims had ideas, but the government gave him no authority to act. Sims ideas fell on deaf ears in Richmond, and the result was a temporizing policy of shifting rolling stock from one place to another. Though both Sims and his predecessor, William Wadley, had tried, no universal system of freight car interchange had been developed. This shortcoming often created problems involving several railroads simultaneously—especially in the transport of government cotton.

Forced to compete with the private sector for railroads and warehouse space, the Confederate Ordnance Bureau had for some time concentrated its blockade-running efforts at Wilmington, North Carolina, giving private blockade-runners nearly free rein of Charleston. After the Union army established a stronghold on Morris Island in July 1863—thus effectively squelching activity in and out of Charleston harbor—the race was on for Wilmington. On October 11, 1863, the Confederate commander at Wilmington, General William Henry Chase Whiting, informed Quartermaster General Alexander R. Lawton that he was unable to store all the government cotton being shipped to Wilmington and that its accumulation in the city created a fire hazard. He requested that the government delay further shipments until he could arrange adequate storage or until a large enough quantity could be shipped seaward.[15]

General Lawton wired his agents to exercise prudence with future shipments, but within a week an outgoing vessel was available, and Lawton contacted Major Hutson Lee, quartermaster at Charleston, to forward government cotton from Charleston to Wilmington as fast as it could be compressed into bales. Lee was unable to obtain the necessary rolling stock from the Charleston & Savannah Railroad; therefore, the Georgia Railroad sent cars from Augusta to Charleston to carry it through. When Lawton requested the use of a train belonging to the Central of Georgia, its superintendent, George W. Adams, balked. When Adams turned down a second request a week later, Lawton appealed personally to the president of the Georgia Central, R. R. Cuyler. "It is of first importance to the Government," he wrote, "that cotton should be promptly forwarded to load ships now at Wilmington, North Carolina, and this can not be done without your cars running through for a few days. I beg that You will send them to Wilmington until the present pressure is removed." Cuyler again refused, but Major Sims of the Confederate Railroad Bureau successfully persuaded Superintendent S. S. Solomons of the Northeastern Railroad to assist in the matter. Although Solomons's cars were occupied by the transfer of portions of Longstreet's Corps from Florence to Charleston, he agreed to ship a thousand bales of cotton as soon as practicable.[16]

Meanwhile, as the supply at Wilmington diminished, bales of cotton—publicly and privately owned—continued to accumulate at depots in Charlotte, Augusta, and Charleston. In mid-October, General Lawton was dismayed to find that the Quartermaster Department had no government cotton available for shipment at Wilmington. When Thomas Sharpe, superintendent of the Charlotte & South Carolina Railroad, asked permission for one of his trains to carry cotton through Columbia to the junction of the Wilmington & Manchester line at Kingville, Superintendent H. T. Peake of the South Carolina refused, stating that he was prepared to ship the cotton from Columbia "on [his] own train." In Augusta more than two thousand bales sat at the depot of the Georgia Railroad awaiting transport over the South Carolina road, waiting for several days without budging an inch toward Wilmington. The flow of cotton slowed to the point that blockade-runners were offering to pay the rail freight if the government would assist them in getting the cotton to the wharves.[17] To the privateers less cotton meant less profit, and to the Confederate government less cotton meant further delays in obtaining essential goods because exported government cotton paid for imported military supplies.

Because the port at Wilmington had taken over the majority of Confederate exportation east of the Mississippi River, the government tried to keep two or three trains operating between points in Georgia and Wilmington to insure the shipment of government cotton. Even though many Southern newspapers gave the impression that cotton was freely flowing out of Southern harbors, the truth was that Wilmington blockade-runners were not expecting to ship more than fifty thousand bales the entire year. Much of the cotton lay damaged and rotting on wharves and at railroad depots throughout the South. The problem was not the inability of ships to make it through the blockade—between October 22 and November 28, 1863, twenty-one ships cleared the port at Wilmington, and all but one made it successfully to Bermuda or Nassau. The problem was the inability to get it to the wharves.[18]

As blockade-runners shifted their operations to Wilmington, private cotton flooded railroad depots at a time when the Confederate government had none to send. By the first week of December 1863, the Charleston & Savannah Railroad had one thousand bales piled up at their Charleston depot, in addition to seventeen carloads of cotton that Major J. M. Seixas refused to receive in Wilmington. In Florence, South Carolina, the Northeastern Railroad had fifty cars laden with privately owned cotton that the Wilmington & Manchester was either unwilling or unable to receive. To discourage further private shipment the Charleston & Savannah line imposed a charge of two dollars per day for each bale stored by the road. Since they were unable to remove the cotton that had already accumulated, the railroad also refused to receive any more government cotton.

The situation worsened when the Charleston & Savannah Railroad refused to unload a train of government cotton sent to Charleston by the Central Railroad of Georgia. Major Seixas asked the Central to forward the cotton directly to Wilmington, but Superintendent George W. Adams of the Georgia Central refused, citing his company's policy prohibiting its cars from operating beyond Charleston. This left local railroad managers and Major Hutson Lee, chief quartermaster at Charleston, with a dilemma. The government needed its cotton in Wilmington, and the Georgia Central needed its cars for the daily transport of commissary stores. A considerable amount of government cotton from Macon was accumulating in Savannah, and unless the situation was remedied, it would paralyze cotton shipments west of Macon and Augusta, Georgia. Further complicating matters, Major Lee and local railroad officials received a dispatch from General Lawton on December 11, 1863, mandating that any train running to Wilmington for private interests must reserve half its capacity for government freight. He authorized military authorities in Wilmington to seize any train found to be in violation of the order.[19]

As soon as several locomotives became available to him, Major Lee decided to take action. He had the government cotton transferred to another train and shipped it, along with a significant amount of private cotton, to Wilmington the following week. General Whiting seized the trains—consisting of rolling stock belonging to the Northeastern and Charleston & Savannah railroads—on their arrival at the Cape Fear River. This outraged the presidents of the two railroads, both of whom appealed to Major Sims at the Railroad Bureau for intercession. Both companies were prepared to send more government cotton as soon as their cars were returned, but they would concentrate on other matters until the military released the cars. "The whole business of the government has been checked," reported Robert L. Singletary, president of the Charleston & Savannah Railroad, while his counterpart, Alfred F. Ravenel of the Northeastern, noted that "the government [was] defeating itself while [he] was trying to serve [it]."[20]

After traveling to Charleston and Savannah to inspect the situation, Major Sims appealed to General Lawton to release the trains belonging to the Northeastern and Charleston & Savannah railroads. At the urging of Majors Lee and Sims, Lawton ordered the trains returned to their respective railroads on December 23, just in time to be worked through the holidays, hauling government freight to relieve the congestion along the line. He also granted both the Charleston & Savannah and the Northeastern an exemption from his earlier mandate that half their trains' capacity be devoted to government freight. Although the government had traditionally afforded the Railroad Bureau little authority over the activities of local quartermasters,

Lawton ordered his agents to defer to any orders issued by Sims regarding shipments over the two lines.[21]

The difficulties encountered shipping government cotton, like the partially successful movement of Longstreet's Corps from Richmond to Chickamauga, brought the inefficiency and disorganization of Southern railroads to light. Such occurrences exasperated Sims and his constituent railroad operators, but the leaders in Richmond remained either oblivious to the problems or unwilling to do anything about them. Difficulties in maintaining usable rolling stock and the increasing demands from commercial and military interests caused cotton and other freight to clog depots at all major railroad junctions. Uncompromising railroad officials often refused to cooperate with neighboring lines and the Confederate government, exacerbating the logjam. The biggest failure of all, however, may have been the inability of the Confederate government to coordinate the effort efficiently. The government attempted to allocate rolling stock to points of necessity, but overall it was unable to achieve adequate communication or cooperation among itself, the various departments of the military, and local railroad officials.[22]

The beginning of the Union army's Morris Island campaign and the Siege of Charleston took its toll on the railroad's business. In reality, though, the problems started with the fall of Port Royal in November 1861 and worsened as the war progressed. In 1861 close connections with Savannah's other major railroad and its steamer companies provided access to markets in Jacksonville, Tallahassee, Atlanta, and Montgomery. The conjoint efforts of the Savannah, Albany & Gulf Railroad, the Pensacola & Georgia Railroad, and a connecting stagecoach company helped to provide the shortest through route to Tallahassee.[23] By the end of 1863 Charleston & Savannah trains required almost twice as much time to travel between Charleston and Savannah as they had in 1861, and Federal control of southeastern waterways made all but the railroad's connection to the Georgia Central worthless.

With the port of Charleston effectively closed to exportation and the constant bombardment of the city from Morris Island, a large volume of business that might have otherwise been bound for Charleston was shunted to interior lines. Freight business was limited primarily by the lack of rolling stock, but passenger business seemed to suffer most from the siege. In April 1863 the Charleston & Savannah road had added a second passenger train, and their business had immediately escalated. Passenger receipts from April to July outpaced receipts from the same period of the previous year by 60 percent. When General Gillmore began operations against Charleston, however, passenger service declined to such an extent that Superintendent Haines discontinued the second passenger train in mid-July. The decline in absolute passenger volume was cushioned by an increase in fares, which the company viewed as "fully warranted by the corresponding advance in the cost of labor and material." Passengers who were paying four cents per mile in January now found themselves parting with six cents per mile—a 50 percent increase. Because of this fare increase, the railroad realized a 25 percent rise in passenger receipts for the year.[24]

Transporting freight along the South Carolina coast was heavily weighted toward Charleston. Of the 30,000 bales of cotton shipped by the railroad, 28,000 were received at the company's Charleston depot; of 160,000 bushels of rough rice, 130,000 were received in Charleston. The company's chief freight agent, Isaac B. Davis, reported that the government sent almost all its freight through Charleston. This necessitated the transfer of many empty boxcars back to way stations and the Savannah terminal, increasing the inefficiency of the railroad's service. Business for the first quarter of 1863 had lagged behind that of the first quarter of 1862, and the company decided it had to raise their freight tariffs. In Treasurer William Swinton's final tally, the corn, rice, cotton, animal hides, and other sundries conveyed along the road netted an extra $97,176 for the year.[25]

Receipts from Confederate government business were also up from the previous year. Previously the railroads had been paying their exorbitant bills at the expense of their civilian customers. It was now time for the Confederacy to ante up. President R. R. Cuyler of the Georgia Central Railroad called for a convention of rail carriers to be held in Macon, Georgia, on November 25, 1863. Major Sims, whom Cuyler also invited, hoped that the railroad managers would address several important issues regarding Southern rail transportation, but the only thing on their minds was an increase in government rates. Following the recommendations of the Macon Convention, the Charleston & Savannah began charging the government three cents per mile for troop transport, up from two cents per mile. Transporting horses was also increased by 50 percent, from eight to twelve dollars for a trip between the two terminal stations.[26] An increase in the volume of military transport and the rate hike provided the road with an additional $124,000 in receipts.

Many people accused the railroads of acting in their own best interests. Even after the war began and the need for a unified system of rails became more apparent, railroad managers were often reluctant to place Confederate concerns over private interests. Southern railroaders had always cast a suspicious eye toward any government foray into railroad affairs. They saw the government's role as one of assistance and viewed centralization and control as meddlesome and inappropriate. Expediency and self-preservation ultimately trumped any patriotic affinity to the Southern cause. In-keeping with the philosophy of antebellum management, a railroad's operation extended beyond the boundaries of its home state only when distant, untapped markets warranted it. Railroad executives were first and foremost professionals whose reputation and livelihood depended on the success or failure of their companies. As noted by Mary DeCredico, these men operated their railroads with a view to increasing the wealth and standing of their respective communities and to ensuring profits and dividends for their directors and stockholders.[27] The Charleston & Savannah Railroad was no different from any other Southern rail company.

There is little doubt that any for-profit corporation would be guilty of self-serving to a certain extent, but no one could accuse the Charleston & Savannah Railroad of being completely unpatriotic. The company was continually active in building up the defenses of Charleston. They provided an engine and fifteen cars to the Northeastern Railroad to transport firewood into the city of Charleston. Charleston & Savannah conductors often collected donations to send to South Carolina soldiers stationed in Virginia, and the railroad shipped contributions from Georgians to their native sons stationed in and around Charleston free of charge.[28] These acts of patriotism notwithstanding, the company enjoyed a substantial profit at the end of 1863.

This profit was quite a surprise for Singletary and the board of directors. The company's net income for 1863 almost equaled its gross receipts for 1862, and—had a large quantity of their business not been shunted to interior lines—the proceeds would have been even larger. The railroad had begun paying the interest on its bonds and was showing signs of improving its financial strength, at least on paper. The company's second-lien bonds, which began the year selling for $71.50, had peaked at $103 before the siege began and were still bringing $90 in December. Captain Singletary felt that the railroad's success—and its escape from enemy harm—was "attributable, in a great measure, to its availability for its own defense." Because of its importance as the base of military movements along the seaboard, he felt that "no exertion should be spared to maintain its present state of efficiency."[29]

Because the railroad meant much to the state and to the defense of the region, General Walker agreed that it was worth preserving. He communicated his plan to General Beauregard in mid-November, requesting that Beauregard send him one thousand veteran infantry and an experienced artillery battery to be stationed with transportation at Pocotaligo. It would be a mobile column, which Walker could dispatch to any point under attack. He also wanted Beauregard to station two thousand to three thousand infantry and two batteries in Charleston near the railroad with ready transportation to serve as a reserve. With his current command, predominantly cavalry, and the mobile unit at Pocotaligo, he could hold back four thousand Union invaders and with the reinforcements from Charleston possibly ten thousand. If the Yankees attacked Charleston, Walker could send his troops there relatively quickly by train. The Charleston & Savannah Railroad, he was confident, would provide a train for the purpose, and if they could not spare enough cars, he thought he could persuade the Georgia Central to provide some.[30] While he watched Federal troops accumulate at Port Royal, he eagerly awaited General Beauregard's response to his request.

Though Walker's plan was a logical one, General Beauregard summarily rejected it. He thought it unfair to provide troops for the Third District if he did not do the same for Brigadier General Beverly H. Robertson's Second District of South Carolina. If he sent equal reinforcements to both, it would severely deplete his already inadequate garrison in Charleston. Beauregard also worried that the Federals would feint an attack on the interior, reembark after reinforcements had been sent from Charleston, and attack Charleston with overwhelming numbers. In his dispatch, which reached Walker at Pocotaligo during the last week of November, Beauregard

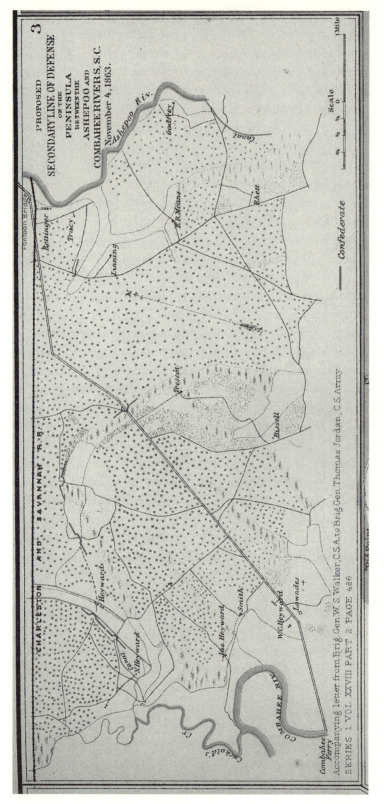

Proposed defenses for an area crossed by the Charleston & Savannah Railroad, November 4, 1863. From *Atlas to Accompany the Official Records of the Union and Confederate Armies*, I, plate 26, no. 3; courtesy of Francis Lord Civil War Collection, Rare Books and Special Collections, University of South Carolina explained, "No work is sufficiently strong to resist a determined attack unless properly garrisoned." He felt Charleston was not. "Is it prudent," he asked, "to risk the safety of Charleston for that of the railroad and the country lying between it and Savannah? No one would hesitate in the selection."[31]

General Beauregard countered with a different plan. He intended for Generals Walker and Robertson to make do with the forces they presently had. If attacked in great number, they would retire while hotly contesting every inch of ground. Their objective would then be to guard the country in their rear and protect the South Carolina Railroad from the bridge across the Edisto near Branchville. They would set up a new defensive line from Bee's Ferry on the Ashley River to Four Holes Swamp, then from Givhan's Ferry to the Savannah River along the southern boundary of Barnwell County. To keep the Federals confused as to their number and intent, Beauregard wanted Walker and Robertson's men to beat drums and fire rockets and salutes near their picket lines.

While General Beauregard's plan confounded General Walker, it also raised eyebrows at the highest level of the Confederate government. Secretary of War Seddon had received the plans and forwarded them to President Davis. The prospect of abandoning the Charleston & Savannah Railroad alarmed Seddon, who telegraphed Beauregard to ask him if he had understood him correctly. The railroad was important to the Confederacy, and Davis did not understand why Beauregard could not grant Walker his request for additional troops—especially if he could arrange their immediate return to Charleston. While Davis was sympathetic to Beauregard's predicament, he noted, "Immediate wants must decide the location of the forces we have." When the Federals again began to move troops from the North to Port Royal, Beauregard acceded to the wishes of his commander in chief. He ordered General Johnson Hagood to transfer the Fifty-fourth Georgia Volunteers, the Twelfth Battalion Georgia Volunteers, and the Thirty-second Georgia Infantry to Pocotaligo for "temporary duty."[32]

With the increased enemy activity at Port Royal, General Beauregard began to devote more of his attention to the Third District. Some intelligence reports indicated that the Yankees had given up hope of taking Charleston for the present. Troops were still traveling from the North to Hilton Head, where General Walker estimated some six thousand troops were already encamped. Beauregard sent his second in command, Major General Jeremy Gilmer, to inspect the defenses. General Gilmer suggested several changes, including the combination of the Third District with the District of Georgia. He was given temporary command of the newly formed superdistrict, which contained three subdistricts: Pocotaligo, commanded by General Walker; Grahamville, commanded by Colonel C. J. Colcock; and Hardeeville, commanded by Lieutenant Colonel Thomas. H. Johnson. In addition Beauregard tried to concentrate Walker's forces by extending General Wise's right to the Edisto River, and shifting General Robertson's district to include the area between the Ashepoo and Combahee rivers.[33]

General Gilmer went to work constructing new works and batteries and placing additional torpedoes in local waterways. He also combined the defensive plans of the Third District with that of the Savannah garrison so that Georgia troops could serve in South Carolina. Requests for more troops were sent up and down the line. General Beauregard petitioned for the return of a portion of General W. H. T. Walker's division, which was wintering in Dalton, Georgia, and also considered calling on General Lee for the return of General Jenkins's or Kershaw's brigade to their home state at least temporarily. With the reorganization and the dribbling of reinforcements into the region from the North and West, the increased defensive strength was beginning to become noticeable. "A great many troops," noted L. H. Boineau, "are now placed on the Charleston & Savannah Railroad. At Green Pond alone, a soldier told me they had to feed 3,000 horses belonging to General Robertson's brigade. During Christmas, several trains of cars passed with troops to be placed along the line between Pocotaligo and Savannah. An attack at some point is hourly expected."[34]

General Beauregard did expect something to happen in his department, and he supposed General Gillmore's next move would be up the Broad River from Port Royal. Elsewhere in the Confederacy the Union had secured Chattanooga and was quickly returning troops to the eastern theater for fear that General Longstreet would rejoin General Lee and move on the Army of the Potomac before anything could be done. Since the disjointed Southern railroad system had not yet conveyed Longstreet back to Virginia, some of Meade's reinforcements might be sent south for a winter campaign against Charleston or Savannah. Union general Henry Halleck was beginning to realize the potential of rapid troop concentration by rail. The Confederates had enjoyed this advantage locally but had not been able to exploit it on a national scale. "I suppose," lamented Beauregard, "by the time we shall have no more troops to concentrate, we will learn better."[35]

ELEVEN

Unrealized Gains

An attack was expected along the railroad at any time, but all remained quiet as Christmas 1863 approached. The only military activity on the line was a Federal reconnaissance by a detachment of the First South Carolina Volunteers that reached Pocotaligo Station but went no further. No engagement occurred, but the small Yankee band was able to assess the strength and position of the Confederate camps in the area and carry off a number of Daniel Heyward's slaves.[1] This, however, was before the reinforcements began to arrive.

With the restructured defensive plan and strengthened forces, the mood in the various Confederate camps seemed to be lighter as the holidays neared. Troops stationed in and around Hardeeville enjoyed a holiday tournament and barbecue, and soldiers of the Third South Carolina Cavalry paraded on horseback and on foot to entertain the locals, in return for a fine feast of goose and turkey in the company of local beauties.[2] While the fine buffet, the music, and the revelry could not erase the troubled reality these men faced, it was a temporary boost to their flagging morale.

There were no parades or shows of horsemanship in his camp at Pocotaligo, but General William S. Walker was nonetheless in better spirits. By year's end reinforcements had increased his command to nearly 4,100 effective troops.[3] This strength remained constant at least through January as his charges closely monitored enemy activity at Port Royal and in the Broad River. It is not known whether President Davis's trip over the Charleston & Savannah Railroad contributed to his decision, but he made good on his promise to reinforce Charleston and its environs. Even though he came through at a critical time for the Charleston & Savannah Railroad, he still refused to take a more active stance on Confederate railroad policy in general.

The Davis administration's failure to pay responsible fares to the Southern railroads exposed them to financial ruin. Railroads such as the Charleston & Savannah operated at full capacity, which became more and more occupied by military transport as the war progressed. The government paid the railroad at half its normal commercial rates in Confederate bonds, which were rapidly depreciating. Union quartermaster general Montgomery Meigs justified the generous fares the U.S. government paid Northern railroads by pointing to the excellent and economical services they rendered; however, Southern roads received no such enticement from the Confederacy.

Legislation had given the Confederate government the power to take military control of the railroads, but Davis never used this authority to force their cooperation. He did not, however, protect the majority of the Southern roads from meddling by army officers, and he gave the Confederate Railroad Bureau—headed by Colonel Frederick W. Sims—very little authority to enforce its policies. The Confederate government also allowed the military to consume a majority of the raw materials and imported supplies that would have aided efficient railroad operation. Another failure of Davis's policy was his laissez-faire attitude toward blockade-running.

The Southern states had been purchasing most of their industrial products from the North or from Europe since before the war. They also planned to import weapons and other military stores that they could not produce themselves. The Union blockade targeted ports with deep-water channels, dockyards, banks, and railroad connections. Though the blockade was initially porous, the inefficient Confederate response to it allowed Union forces to close six of the South's best eight ports effectively by the spring of 1862.[4]

Following the lead of other railroads in Georgia and the Carolinas, the Charleston & Savannah explored the idea of blockade-running. Their machinery had deteriorated to the point that it threatened both their civilian business and their essential service to the Confederate army. In the company's annual report in February 1864 President Singletary informed the stockholders that "preliminary steps have been taken which, if successful, will enable us to procure abroad such articles as are essential to the continued successful operation of the road." As early as 1862 Southern manufacturers and railroad executives had asked the Confederate government to take part in a joint-trading venture by which they planned to run cotton through the blockade to pay for machinery, equipment, and other imports. President Davis refused to participate because he felt that it was "not a proper function of government . . . to support private enterprise."[5] The Confederate navy had also shunned involvement on the grounds that it would require shared authority between different departments, which would have been a bureaucratic nightmare.

By this time fewer and fewer ships were making it through the blockade, but the prices paid for cotton in foreign markets made such trips worth the risk. One shipment of 1,300 bales of cotton brought a $200,000 profit. Unfortunately for Singletary, the Union grip on the Confederate coastline had grown tighter. He was able to acquire a quantity of cotton, but only twenty-two bales made it from Charleston to Nassau in 1864.[6] Because of the lack of support and coordination, many of the ships that made it back to Southern ports contained luxury items instead of industrial matériel. This benefited the entrepreneurs and those who could still afford to purchase their wares but did little to further the Confederate cause. President Davis did eventually sign a law requiring blockade-runners to set aside half their shipments for

Brigadier General Truman Seymour, U.S.A. Courtesy of South Carolina Historical Society

government cargo, but clever captains easily dodged this directive. This law proved to be too little and too late—typical of President Davis's handling of matters, according to George E. Turner, who argued that the Davis administration's "temperament, disposition, and capacity; its lack of industrial comprehension; and its inability to deal in a practical manner with the mechanics involved, rendered it unable to comprehend logistics in modern warfare."[7]

One man who saw the importance of the railroads to the Southern Confederacy was Union brigadier general Truman Seymour. The thirty-nine-year-old Vermont native was one of Major General Quincy Gillmore's most skilled subordinates. Seymour had witnessed the first shots of the war from inside the walls of Fort Sumter and had led the ill-fated assault on Battery Wagner the previous summer. He was reputed to be an aggressive, but sometimes rash, commander whose successes were often achieved at a high cost. He wanted to see a quicker resolution to the conflict, and was becoming impatient with the Union leaders in Washington. After several months of studying the South Carolina coastal defenses, the positions of the main Confederate forces elsewhere, and Confederate railroad maps, Seymour had come up with a plan that he felt would topple the Confederacy.

After the fall of Vicksburg the strength of the Confederacy lay between Mobile, Alabama, in the west and Richmond, Virginia, in the northeast. Between those cities ran a fragmented and disorganized assortment of railroads with General Joseph E. Johnston's Army of Tennessee on one end and General Robert E. Lee's Army of

Northern Virginia on the other. The two main rail routes—if they could be called such—between Mobile and Richmond ran through South Carolina. One of those routes included the Charleston & Savannah Railroad, which lay only a few miles from the Sea Islands and intracoastal waterways controlled by the Union navy. It was also only a day's march from the firmly established Union military base at Beaufort.

The Union had two large armies, one under General Ulysses S. Grant in the West and the other under General George Meade in the East. Neither commander had any immediate plans for offensive maneuvers. Both could afford to send several units to Beaufort without appreciably weakening their forces, since their Confederate counterparts would have to divert significant numbers of troops to defend the vulnerable Carolina coast and counter the Union buildup. A powerful diversion would, at the very least, favor either Grant or Meade by weakening the already outnumbered forces in their front.

With their armies idle in the eastern and western theaters and with efforts to take Charleston essentially abandoned, plans were in the works for what General Seymour called "a trifling expedition" to Florida. He opposed this plan, and was insistent that the Union should move against the railroads in South Carolina. He based his strategy on the principle that no decisive battle could be won without cutting the major lines of communication. It would require an army of fifty thousand to launch a decisive strike on the Charleston & Savannah Railroad somewhere between Pocotaligo and the Combahee River. From there, it would only be forty miles from the junction of the South Carolina Railroad's Columbia and Augusta branches at Branchville. By occupying both of the major rail routes connecting Johnston's and Lee's armies, the Union would control a central position on interior lines from which they could operate against either of the two Rebel armies.

If executed properly, the Seymour plan might have led to the collapse of the South Atlantic states. If the Union army could take Augusta, Savannah would soon follow. Likewise the city of Charleston could not have survived long once its supply lines from Savannah, Columbia, and Augusta were severed. It was, Seymour believed, the only way that Charleston could be taken, since efforts to take the city from the sea had proven futile. It was also his opinion that, if General Johnston could be defeated in the West, Florida could be taken almost without a contest. Thus, he argued, continuing to press the conflict in the West would make the Florida expedition unnecessary. It was also Seymour's view that, if his plan were carried out successfully, "South Carolina might be converted into the battlefield of the War, and probably, eventually, Bragg's [Johnston's] and Lee's armies would appear on this field."[8]

Seymour had presented his plan to General Gillmore the previous fall. Though he apparently supported the idea, Gillmore took no action on it. Gillmore was tiring of the gridlock in front of Charleston and wanted to change venues. A campaign in Florida would cut off a vital source of military supplies—principally salt, cattle, and other agricultural products—to the rest of the Confederacy; procure an outlet for cotton, lumber, and other products for the North; and enable the Union to obtain recruits for their burgeoning African American regiments. President Lincoln was enamored with the idea of reestablishing a loyal government in Florida with enough time to elect Republican representatives to Congress, thus bolstering his chances against his Democratic challenger in November. By dismissing Seymour's plan Union leaders were ignoring the plan's greatest advantages: Federal possession of the sea, ample harbors by which to supply the mission, and the ability to separate Lee's and Johnston's armies. Seymour was so adamant that he wrote to Senator Ira Harris of New York requesting that he present the plan to President Lincoln or Secretary of War Edwin M. Stanton on his own.[9] Despite his appeals the decision had already been made favoring the politically motivated incursion into northern Florida.

When a Confederate lookout at Foot Point reported a large number of Union vessels headed out to sea from Port Royal and a noticeable decrease in the activity level in and around Beaufort in the days following, the news confirmed Beauregard's suspicion that something was afoot. He and General Jeremy F. Gilmer quickly gave the order to prepare trains on the Charleston & Savannah Railroad to rush troops to Savannah at a moment's notice. No one knew the Yankee vessels' final destination, but Beauregard guessed somewhere to the south. His suspicion was confirmed on February 7, 1864, when he received a wire from General Joseph Finnegan in Lake City, Florida, that a Federal fleet had arrived in the St. Johns River. The next day Brigadier General Seymour and his expeditionary force landed near Jacksonville, setting in motion the hurried transfer of reinforcements from South Carolina and Georgia.[10]

On Folly Island, Brigadier General Alexander Schimmelfennig received orders to take a force of three to four thousand infantry and a few pieces of artillery and make a reconnaissance in force on Johns Island. His objectives were to threaten the Charleston & Savannah Railroad near Rantowles Bridge and draw attention away from General Seymour's push into Florida. General Gillmore hoped to prevent reinforcements from being sent to Florida, to hit the Confederates at a vulnerable time and disrupt the all-important railroad, or both. For the operation Schimmelfennig's force consisted of portions of the First Brigade of Gordon's Division, under Colonel Leopold von Gilsa (including the 142nd New York, Forty-first New York, Fifty-fourth New York, and Seventy-fourth Pennsylvania); the Second Brigade, led by Brigadier General Adelbert Ames (including the 157th New York, 144th New York, Seventy-fifth Ohio, and 107th Ohio); and Foster's brigade of Vogdes's division, commanded by Colonel Jeremiah Drake (including the 112th New York and 169th New York). They received their marching orders on the same day that General Seymour's troops went ashore at Jacksonville.[11]

With eighty rounds of ammunition and three days' rations, the Union troops were ferried from Folly Island to Kiawah Island shortly after midnight on February 8. They marched through the night and rested the next day at Vanderhorst's plantation on Kiawah. Resuming their march that night, they waded across the shallows to Seabrook Island and continued forward until reaching Seabrook Plantation shortly before dawn on the February 9. Here, the 142nd New York and 157th New York threw out a line of skirmishers, which advanced toward Haulover Bridge connecting Seabrook and Johns islands. At the bridge was a small picket post manned by several young Confederates from a company known as the Cadet Rangers.[12]

The Cadet Rangers were formed in the summer of 1862 by a group of cadets from the Citadel, South Carolina's military college. After the Battle of Secessionville these cadets had chosen to withdraw from school and serve in the Confederate army. Their company became part of Colonel Hugh Kerr Aiken's First Regiment of South Carolina Partisan Rangers, which later merged with the Sixth South Carolina Cavalry. On this cold February night, several of the former cadets had been enjoying some music and dancing in a nearby plantation home before they were surprised by lead elements of the 157th New York The Federals captured three cadets, while the remaining Confederates dashed back to their base camp to sound the alarm that the Yankees were on the island.[13]

When Captain M. J. Humphrey heard that the Yankees had captured several of his young troops, he sprang into motion. Within minutes he and thirty-five Cadet Rangers were saddled up and rushing forward to retake the post. On their arrival at the White Gate, Captain Humphrey observed the Blue Coats establishing their line on an embankment extending from Bohicket Creek to Haulover Creek. He dismounted his men, and they advanced across a field under the cover of a dense morning fog. The cadets fired a volley into a small detachment of the 157th New York, which fell back to the main Federal line. Seeing the handful of Yankees falling back, Humphrey charged his enthusiastic men into the murky mist to face a Federal force that he had mistakenly estimated to be only about four hundred. With Rebel Yells splitting the air the Cadet Rangers drove the enemy in their front from the embankment.

At that moment Confederate major John Jenkins had arrived from his headquarters and ordered Captain James Jennet's company of the Fifty-ninth Virginia forward to support the Rangers. After the Virginians assumed a position to the rear of Humphrey's men, nature intervened. A stiff breeze caused the fog to lift, revealing the tiny force of Citadel men. Seeing them, the Yankees poured heavy fire into the out-manned Confederates, causing first Jennet's and then Humphrey's troops to fall back. Joined by the Seventy-fifth Ohio, the 157th New York regained the embankment and rushed into the open field to exploit the swing in momentum. A volley from the Virginians provided just enough of a distraction to allow the cadets time to reach cover in the woods. The Federal force, now growing larger by the moment, continued to push forward with the arrival of the remainder of the Second Brigade. They proceeded in formation across two open fields in pursuit of Major Jenkins's hodgepodge of Confederate detachments, who hotly contested every acre of ground that they conceded.[14]

On reaching a third field the Yankees found an entrenched Confederate force, now consisting of the Cadet Rangers, Jennet's company of Virginians, Company I of the Third South Carolina Cavalry (Rebel Troop), the Stono Scouts, Sullivan's Cavalry, and a section of the Marion Artillery. This force of about 150 men stopped the Federal advance in its tracks, rallied, and began a counteradvance. Having moved too far ahead of his reserves and assuming the Confederates had more support than they actually did, General Ames halted and began pulling his brigade back. After several spirited charges by Major Jenkins, the Confederates harassed the Federals all the way back to the position they occupied earlier in the morning. The small band of rebels had crossed and recrossed nearly two-and-a-half miles of ground, but had kept their enemy in check. The Federals bivouacked for the night near the Haulover Cut and began to put up breastworks.[15]

Later that evening the small Confederate force cheered the arrival of Colonel William B. Tabb with a battalion of the Fifty-ninth Virginia Volunteers and another section of the Marion Artillery. The next morning Colonel Powhatan R. Page rode in with five companies of the Twenty-sixth Virginia Volunteers. Shortly after their arrival, the First and Second Brigades moved cautiously up Bohicket Road, the most direct line between their position and the railroad bridge at Rantowles. As they extended their line to the right, General Wise arrived on the field and assumed command from Colonel Page. He saw the immediate concern of being outflanked on his left and ordered the withdrawal of his troops to the vicinity of the Cocked Hat. a three-way intersection about four miles from Haulover Cut. At the Cocked Hat he established a new line of defense and called for all remaining reserves at Adams Run and Willtown. The Federal troops did not follow the Confederate withdrawal, choosing to retire to the protection their breastworks.[16]

General Beauregard, with action now occurring on two fronts, had to make some adjustments. He wired General Finnegan with instructions to do what he could in Florida until adequate reinforcements arrived. To repel the Federal invasion of Florida, Beauregard would have to deplete the forces guarding Charleston and Savannah. He was certain that Schimmelfennig's demonstration was a diversion, but his force was large enough that there could be serious consequences if no action were taken. Among other units Brigadier General Alfred H. Colquitt's brigade was ordered up from James Island, but as soon as his troops had boarded the train for Savannah, they were diverted to Johns Island to help General Wise stem the Yankee incursion there. After an all-night march from Adams Run and Willtown respectively, the Fourth and Forty-sixth Virginia regiments arrived in the early morning hours of the February 11, weary but ready for action. As Generals Wise and Colquitt rode forward to survey the lay of the land, an aide came to inform General Wise that the Yankees were on the move.

After strengthening their breastworks the previous day, the First and Second Brigades were moving up Bohicket Road in formation. Their vanguard approached just as the first of Colquitt's regiments arrived in the vicinity of the Cocked Hat. Finding a newly reinforced enemy, the Federals fell back briefly, giving General Wise time to position his troops.[17] He placed Colonel Page's Virginians on his right on Bohicket Creek and extended the line through a field and some woods back to Legareville Road. As they arrived, Colquitt's Georgians took a position on the Confederate left, while the Marion Artillery and Charles's battery commanded Bohicket Road. At about 3:20, the Federal columns re-formed and began another advance up the route leading to Rantowles Bridge. As soon as their guidons became visible, the Confederate artillery opened fire. The Yankee artillery, which had been hidden from view, quickly answered, and for the next hour and a half an artillery duel ensued. When Schimmelfennig's troops withdrew at around 5:00 in the afternoon, Captain Edward L. Parker of the Marion battery had expended all his shells.[18]

Because of the lack of sufficient artillery and cavalry, the fatigue of many of the troops from an all-night march, and the fear of being flanked by enemy gunboats, General Wise did not think it wise to pursue the Bluecoats as darkness approached. That night General Beauregard unleashed a heavy bombardment on the Federal position in hopes that they would leave the island. Neither he nor General Wise were aware that the Yankees were silently being ferried back across Haulover Cut after burning the Haulover Bridge and William Seabrook's house at Haulover Plantation.[19] By the time the Confederates were ready to give chase, Schimmelfennig's troops were safely back on Folly Island, but not for long. Almost as soon as they returned, they—along with many of the troops stationed on Morris Island—were headed to Florida to join General Seymour's expeditionary force.

Scarcely was Confederate success on Johns Island assured before General Beauregard ordered the regiments of Colquitt's brigade to Rantowles to board a train for Savannah. They and a number of other units in and around Charleston were on the move to aid General Finnegan in Florida as quickly as the rails could carry them. In the days following the skirmish, Beauregard sent additional troops, including a portion of the South Carolina Siege Train, the Forty-sixth and Fifty-ninth Virginia, the Eighteenth South Carolina, and the Eleventh South Carolina. The men of the Eleventh South Carolina marched to the depot in Charleston and waited in freezing conditions before boarding the cars the next morning. While traveling so close to their homes, many of the soldiers of the Eleventh jumped off the train at Ashepoo and Salkehatchie, risking court-martial to spend a few precious hours with their families before facing the Yankees in Florida.[20]

The movement of troops to Florida was carried out over the Charleston & Savannah; the Savannah, Albany & Gulf; the Atlantic & Gulf; and the Pensacola & Georgia railroads. The lack of suitable rolling stock on the Georgia and Florida railroads and a gap of twenty-six miles between Lawton, Georgia, and Live Oak, Florida, were cited as the main impediments to his plan.[21] The Federals had attempted another diversion at Whitemarsh Island, Georgia, but this one was not directed at the railroad. Even though some troops were briefly detained at Savannah because of it, a portion of Colquitt's brigade reached General Finnegan in time to deal Seymour a blow at the Battle of Olustee on February 20, 1864. The Confederate victory, however, had not been a rout because Finnegan lacked enough men.

Had the gap in the rails not existed, Beauregard's reinforcements might have arrived in time to drive the Federals completely out of Florida.

Although Union troops were now firmly entrenched in Jacksonville, the expedition had not achieved the success its designers had anticipated. President Lincoln's "trifling expedition" had failed just as General Seymour feared it would, with Seymour himself contributing to its undoing by attacking the enemy against General Gillmore's orders. Had the Union high command heeded Seymour's suggestion, it might have changed the complexion of the war and caused even more devastation to South Carolina. Instead General Seymour returned to Hilton Head with a defeated army and a damaged reputation. He was soon transferred to Northern Virginia, where he was captured during the Wilderness Campaign in May. Later that month General Halleck requested from the frustrated General Gillmore a full report of his troop and artillery strength. The U.S. military had grown weary of attempting to take Charleston, and its leaders felt they might use his troops more efficiently for "operations against some other point of the Atlantic or Gulf coast."[22]

As military and railroad officials were scrambling to rush troops from the South Carolina coast to Florida, the directors and stockholders of the Charleston & Savannah Railroad were convening for their annual meeting in Charleston. The heavy Federal bombardment forced them to move the meeting from the company's office on Broad Street to the Orphan House on Calhoun Street. After William F. Hutson called the meeting to order, President Singletary read his annual report into record. In it he congratulated the staff and directors for increasing the income of the railroad under such adverse conditions, but he pointed out three glaring deficiencies of the road that needed addressing in the coming year. Construction of a permanent bridge across the Ashley River was needed. The temporary bridge across the Savannah River would require either repair or replacement by a permanent structure. They needed to continue their own track into the city of Savannah from the junction with the Georgia Central. These were reasonable goals in peacetime but rather unrealistic ones during a time when the railroads of the state could not even make change for customers who did not have the exact amount for their fares.[23]

Singletary reminded the stockholders that the company's steadily increasing income had enabled the directors to authorize payment of all interest that had accrued on the funded debt prior to March 1863. As an inducement to the creditors to accept the third-lien bonds, the board instructed Singletary to offer the bonds for the interest from the date of their issue—March 1, 1861. Those who had completed work on or furnished materials to the railroad had generally accepted these terms, but some of the largest creditors did not. The opportunity was open to them until the end of November 1863, and by that time the increased income gave the company enough cash for them to pay in currency all those who had not accepted the offer.

After meeting with the directors, Singletary felt that this would be the best course of action, but when he tendered that alternative to the creditors in December, it was not accepted. Singletary also pointed out that "one of the largest creditors" claimed payment in third-lien bonds, but Singletary did not think this was a wise choice under the then-improved circumstances of the company.[24] Because this represented a large portion of the company's floating debt—and because payment was preferred in bonds rather than currency—he was putting the issue before the stockholders to decide. These statements got the attention of stockholder and former director James Butler Campbell. He knew the circumstances of the matter, and he knew who this "largest creditor" was. It was his client (and an early benefactor of the railroad) John S. Ryan. After the report had been submitted as information, Campbell requested to address the meeting.

In December of 1857 Ryan had contracted to build the portion of the railroad between the Salkehatchie and Savannah rivers, and a few months later he was assigned the contract from the Savannah River to the junction with the Georgia Central. As more subcontractors gave up their portions of the line, Ryan also assumed responsibility for the unfinished work. For his efforts he was to be paid one-third in company stock and two-thirds in the company's 7 percent (second-lien) bonds, half of which Ryan was to redeem at par value one year after the project's completion date.[25] Ryan's acceptance of the work was felt to be the company's only hope of completing the road. It was such a risky proposition that the company allowed Ryan the option of forfeiting the contracts with virtually no personal liability—or so he thought.

The first contract between Ryan and Drane & Singletary was made in August 1858 for work left unfinished by several other contractors. Drane & Singletary initially refused, but later accepted, the job provided that Ryan guarantee payment himself. He advanced them a large sum before the work was done and eventually paid them $18,000 at its completion. Since the cost of the work was 20 percent lower than they would have charged the company directly, Ryan felt that he was entitled to a commission of at least half the amount he had saved the company. He also wanted consideration for certain expenses he had incurred directly during the construction. Ryan had also called on Drane & Singletary when another of his subcontractors had given up his portion of the job. With Ryan facing a large financial setback, the railroad agreed to pay whatever rate Ryan could negotiate to have the contract fulfilled. They completed the work for a cash payment of $84,752; however, Ryan was credited with $44,000 in stocks and bonds, which had a cash value of only $26,000. He had lost nearly $60,000 in the transaction and hoped to recoup some of this loss as well.

Another point of contention was the section of the road cut through Heyward lands. Thomas Drayton had taken this portion of the contract from Ryan and given it to William Henry Heyward, and as feared, Heyward did not finish the work on time, not only interrupting the continuity of the line but also depriving the company of state aid, which had been dependent on the road's continuity up to a certain point. Ryan again subcontracted with Drane & Singletary, who demanded higher rates for the job. The company authorized reimbursing Ryan for his payment to his contractors but did not allow him any profit in the deal. Since he had endured a significant opportunity loss in shouldering so much of the railroad's construction, Ryan felt he was entitled to at least some profit or commission in this matter as well.

When the railroad had begun full operation between Charleston and Savannah, it had owed Ryan nearly $240,000 in bonds, a portion at par and a portion at seventy-five cents on the dollar, equal to a market value of $284,000. At times, because the company owed him so much, the directors advanced him currency from the company's cash accounts. Much of this was money he had raised for the railroad by endorsement of their notes or from acceptances secured by deposits of bonds that were of inferior grade compared with those to which he was entitled. Because of these endorsements and acceptances, he appeared in the company's general cash account as a debtor in the amount of approximately $238,000. When the company was nearly bankrupt, Ryan had endorsed a larger amount than was showing on that balance; therefore, he was held liable—and in some instances sued—by the holders of these bonds.

In May of 1863 Singletary and the board of directors offered to issue the company's third-lien bonds to creditors who would accept them. Ryan did not accept the offer initially but did begin paying off the notes that he had endorsed for the company. By year's end he had paid back all the notes and was ready to apply this amount toward the cash balance against him. At the beginning of 1864 Ryan was still responsible for about $150,000 of the company's liability. When this amount was applied to the various cash amounts he claimed were due him, he hoped it would neutralize the cash transactions between him and the railroad.[26]

It was then that President Singletary, with the approval of the directors, stopped issuing third-lien bonds to the creditors, opting instead to pay the company's floating debt in cash. An experienced securities broker, Ryan knew how inflation had devalued Confederate currency. He wanted no part in a deal that would leave him with a large amount of worthless money, and demanded the company pay him in its third-lien bonds instead. Because Ryan had refused this same offer only months before and because he thought it unwise under the railroad's present circumstances, Singletary refused to budge. At an impasse Ryan enlisted the services of Campbell to negotiate with the company and to litigate the matter if required.

When Campbell stood to speak at the February 1864 stockholders' meeting, he explained to the stockholders that Ryan had been acting on the advice of counsel, and he wanted to be certain that the group was thoroughly informed on the issue. To explain Ryan's rationale Campbell proceeded into a lengthy discussion of Ryan's dealings with the railroad, and by the time Campbell finished speaking the stockholders could see that there were two sides to the matter.

It was evident that Singletary still favored paying off the company's floating debt and discontinuing the issue of third-lien bonds. The directors had agreed with him, but after hearing Campbell's explanation, William F. Hutson put forth a resolution referring the entire matter to the board with the instruction to "act in such a way that their judgment was in the best interest of the company" while also "being just and fair to all creditors."[27]

The final order of business that night was the election of the board of directors. Singletary was reelected president, but the board had many new faces on it for the coming year. Gaining reelection were William C. Bee, Henry Gourdin, Theodore D. Wagner, John H. Steinmeyer, Mayor Charles Macbeth, and Charles Taylor Mitchell. The six new directors were Charles V. Chamberlain, Frederick Richards, William F. Hutson, Richard Yeadon, J. K. Sass, and the president of the Cannonsborough Wharf and Mill Company, Charles M. Furman.

The following morning a special meeting was called to take action on a couple of mundane items. There was little public mention of the previous night's proceedings because of the exciting news coming from Charleston harbor. On the night of the stockholders' meeting, February 17, 1864, the Confederate torpedo-submarine CSS *H. L. Hunley* had sunk the USS *Housatonic*. Because of the stir created by the Confederate triumph, the meeting was a brief one. They handled some housekeeping matters, approved a new schedule, and passed a resolution authorizing the company's Committee on Finance and Accounts (Bee, Gourdin, Furman, and Sass) to investigate Ryan's claims against the company and to speak with Ryan to see if they could reach a mutual agreement.[28]

The company's treasurer, William Swinton, quickly began poring over the company's books, examining every account and entry pertaining to Ryan. After a careful audit of the accounts and an adjustment made by transferring his bonds to the accounts of his contracts, the balances were calculated. The cash balance owed to Ryan by the company was $68,137; the bond account balance totaled $205,600 in Ryan's favor; and the stock account showed a debt of $372 owed by Ryan.[29] After charging the stock account to the cash account, Campbell advised Ryan to accept the remainder in cash, which he believed Drane & Singletary would accept for their share of the settlement and any other amounts Ryan owed them for their work as subcontractors on the road.

Given the precarious state of the economy, Campbell knew that it would be better for Ryan to settle the dispute rather than allowing the account to remain unadjusted or enduring a costly legal proceeding. He also knew the railroad was unable to pay off the bond account in full, and any settlement that forced the company to substitute the market value of its bonds for their face value would charge contractors—such as Drane & Singletary—with essentially 100 percent on the amount due. Campbell did not advise Ryan to insist on the specific delivery of the bonds' market value in currency, but he did think that Ryan should avail himself of the opportunity to "secure some indemnity for the embarrassment and damage he had suffered and to reimburse himself for the expenses he had incurred in achieving the settlement."[30] Instead of payment in notes, Campbell suggested that the company pay Ryan at least partially in third-lien bonds—which should have been plentiful—at 25 percent and the rest in cash. In return Ryan would give up legal claim to more than $150,000 in promissory notes. Campbell took this proposal to Bee, who presented it to the directors.

Bee knew that Ryan was entitled to the balance of the bond account in third-lien bonds or their market value in cash. If he took the latter option, Bee did not think a settlement would be practicable. Ryan wrote to Bee proposing a settlement under which he would accept $100,000 in the company's third-lien bonds and $185,000 in currency. Since there was a desire on the part of both parties to avoid litigation, Bee and the other members of the committee endorsed the settlement. Singletary concurred, noting that the bonds were readily obtainable and that they could draw the majority of the cash amount from the Bank of the State at a discount. On the motion of Sass, they authorized Swinton to pay all the interest due on Ryan's claims in cash.[31]

In July the dispute was finally settled. The Charleston & Savannah Railroad paid John Ryan $100,000 in third-lien bonds and the remaining $185,000 in treasury notes newly issued by the treasury of the Confederate States. Ryan returned $151,165 in promissory notes and 170 equipment bonds with a face value of $85,000 to the company.[32] The railroad had avoided a contentious lawsuit with the man who was singularly responsible for its early existence. The railroad had withstood both fiscal and military hazard, and its rails were still intact. Soon, however, the complexion of the war changed. That spring General Ulysses S. Grant took command of the entire U.S. Army; General William Tecumseh Sherman began his Atlanta Campaign (May–September 1864); and a new commander arrived in the Department of the South. These changes factored prominently in the most turbulent months of the Charleston & Savannah Railroad.

TWELVE

Foster Tries Charleston

Robert L. Singletary's first year as president of the Charleston & Savannah Railroad had been as much a test of his stamina as it had of his business acumen. The Yankees' steady bombardment of the city was a constant reminder of an enemy that was looking for any opportunity to sever his railroad. He was keeping it in operation while also serving as the superintendent of labor for the Charleston defenses. He and Superintendent Haines were juggling the commercial schedules around military necessities with little help from the Confederate government. He was doing a good job, but how long could he continue to do so? How long would the Confederacy last?

In the spring of 1864 the Southern railroad presidents had called what proved to be the last of the Confederate railroad meetings. Set for April 13 in Columbia, the purpose of the convention was to discuss rates on government passage and freight and to address tax legislation passed by the Confederate Congress in February 1864. On the day before the meeting, Singletary—the Charleston & Savannah's sole representative—boarded the South Carolina Railroad for the short trip from Charleston to Columbia, which had been made a little longer by the disrepair and sluggishness of the locomotives.

The Columbia Railroad Convention was yet another opportunity for the leaders of Confederate railroad companies to vent their disenchantment with the government. In February the Confederate Congress had expanded the income tax, adding a surtax of 25 percent on profits exceeding one-fourth of the capital investment in railroad, banking, communication, and transportation companies. Railroad officials were appalled. Because of the devaluation of Confederate currency, the railroad shareholders were subjected to a five-to-ten-times-greater tax than that levied on the average citizen. The railroad leaders considered the act discriminatory "class legislation," which unfairly targeted investors—including widows, trusts, estates of the deceased, and charitable institutions—who invested for convenience of money management. They could neither produce goods nor provide services, and they could not impose the depreciation on others.[1]

The railroad executives pointed out that their companies had sacrificed greatly for their country. They had agreed to carry mail, troops, munitions, and supplies at greatly reduced rates and often at the sacrifice of more lucrative civilian business. At the same time the government had monopolized rolling mills, iron works, and other sources of supply; seized and redistributed railroad iron; and impressed rolling stock. As a result, the railroads had to deal with a deteriorating and dwindling supply of rolling stock and a dearth of overpriced iron and other necessities, which restricted their ability to conduct regular business. Government policy had contributed to the crippling—and in some cases ruination—of many Southern railroads. Georgia Railroad president John Pendleton King pointed out that with the devaluation of Confederate money "railroads in reality have made no net profits for the last three years, and are making none now! It would require much more than the dividend paid to place the property in the same condition it was at the beginning of the war."[2] The railroaders felt that their industry, second to banking, had been asked to sacrifice disproportionately for the Confederate cause.

In that same spirit the railroad managers took up the issue of government rates, which they had deeply discounted for the duration of the conflict. After some debate they decided to raise freight rates on Confederate property by 50 percent and passenger fares to six cents per mile. They then appointed a committee to memorialize Congress to denounce the part of the tax act pertaining to railroad companies. They proposed, instead, an amendment to the law that would impose an ad valorem tax on the entire wealth of the country or at the very least adopt a different valuation for associated capital.[3]

The outcry must have had some effect. At the Charleston & Savannah Railroad's monthly meeting several weeks later, the board addressed whether to return the company's capital stock for taxation or leave it for the stockholders to pay. They received word that the Congress had decided that the company, rather than its stockholders, should be responsible for the tax. The directors authorized payment of the tax, and in an effort to appease the railroad detractors of the day they also decided to advertise this payment in the daily newspapers.[4]

When Singletary arrived back in Charleston from the convention, there was a great deal of commotion in and around the railroad yards. Horses, ordnance, military stores, and troops were noisily transferring from one railroad to another, bound for points north. Only a month before, the exodus had begun with the transfer of the Fourth, Fifth, and Sixth South Carolina Cavalry, the Seventh Georgia Cavalry, and Millen's battalion to Virginia in exchange for the First and Second South Carolina Cavalry, which returned home. The units traveling this particular day were the Twelfth, Seventeenth, Twenty-third, Twenty-sixth, and Twenty-ninth South Carolina regiments—General "Shanks" Evans's old brigade, now commanded by Brigadier General William S. Walker—which had been ordered to Wilmington, North Carolina, on April 14. Word was also out that General Lee had asked General Beauregard to return to Virginia to assist with the defense of Petersburg and Richmond.

On April 20, 1864, General Beauregard received his official orders to take command of the Department of Southern Virginia and North Carolina. His successor in Charleston was Major General Samuel Jones, a native Virginian and graduate of West Point. He began his Confederate service as a major in a corps of artillery, and after his promotion to lieutenant colonel he served as General Beauregard's chief of artillery and ordnance at the First Battle of Manassas in July 1861. After serving in Tennessee and Virginia, he was sent to Savannah to replace General Jeremy Gilmer, who had been ordered to Richmond. Beauregard had some concerns about how Jones would perform, but competent replacements were becoming harder to find in the Confederacy—especially in the Department of South Carolina, Georgia, and East Florida, which was becoming known as the "Department of Refuge" because of the quality of commanders who received permanent assignment there.[5]

A week after General Jones's arrival General Beauregard called for General Johnson Hagood's brigade—consisting of the Seventh South Carolina Infantry Battalion and the Eleventh, Twenty-first, Twenty-fifth, and Twenty-seventh South Carolina regiments—to be transferred to Petersburg, while several Georgia regiments were sent to General Johnston's command near Dalton, Georgia. The first week of May, Jones dispatched Wise's and Colquitt's brigades to Richmond, and a day after their departure Jones wired Johnston that he had just sent his last infantry brigade to Virginia. Just how his superiors expected him to defend the coast with such a meager force, he was not sure.

The State of South Carolina began activating its reserve troops, for there were none left elsewhere in the Confederacy to send. General Jones was so desperate for men that he asked Charleston mayor Charles Macbeth to organize the city's fire brigade into companies. He even asked the presidents of Charleston's railroad companies to muster their employees—whom the government had traditionally exempted from service—into companies, promising to call them only "under circumstances of actual and grave necessity." This would be difficult for Singletary, who was already operating his railroad with such a skeleton crew that he was able to staff only half of his way stations with freight agents.[6] For him and his employees working under conditions of grave necessity was becoming standard operating procedure.

The Federal installations at Port Royal, Hilton Head, and other points along the coast of South Carolina and Georgia were in the midst of their own transition. Preparing for what became known as the Wilderness Campaign in May, General Ulysses Grant had summoned General Quincy Gillmore to Fortress Monroe, Virginia. He was to have taken eleven thousand men with him, but instead stripped the department of about twenty thousand. Much to the dismay of interim commander Brigadier General John Porter Hatch, these soldiers were replaced by raw recruits and newly organized African American regiments.

The Union batteries had restrained their bombardment of Charleston during April and May. Brigadier General Schimmelfennig, commanding the troops on Morris and Folly islands, reported that his "main objective was to compel the Southerners to keep a larger force within Charleston than he was maintaining outside the

Brigadier General John Porter Hatch, U.S.A. Courtesy of South Carolina Historical Society

city, this being my only means of cooperating with the more important movements of our armies in Virginia."[7] His method for doing this was to shell at odd intervals, hoping to disrupt railroad traffic and the transfer of men elsewhere. He was not achieving this goal quite as well as he thought.

General Hatch had been paying close attention to the efflux of Rebel troops from the region. Despite departures on both sides, he knew that he had far superior numbers and decided the time was right for another attempt to cut the Charleston & Savannah Railroad. Hatch recalled Brigadier General William Birney from Jacksonville to lead the operation, which targeted the bridges across the South Edisto and Ashepoo rivers and the long trestlework across the swamp between the two rivers. Hatch also enlisted the help of the navy, who sent several gunboats and a squad of marines to support the army's movement and to execute a feint at Willtown.

The expedition left Port Royal on the evening of May 25, 1864, their destination Bennett's Point at the mouth of Mosquito Creek. Troops of the Ninth U.S. Colored Troops under Colonel Thomas Bayley led the way aboard the transport steamers *Edwin Lewis* and *Boston*. General Birney commanded from the army gunboat USS *Plato*. As the two transports slowly ascended the Ashepoo, darkness and uncertainty caused the pilot of the *Edwin Lewis* to sail past Bennett's Point. Acting Ensign William Nelson, pilot of the *Boston*, quickly notified Colonel Montgomery that they were passing the intended landing site. Perhaps thinking plans had changed, or that Nelson was simply mistaken, Montgomery reminded the pilot that his orders were to follow the steamer ahead of it. This decision sealed the fate of the mission.

Despite the protests of its pilot, the *Boston* followed the *Lewis* upriver another six to eight miles past their original destination up to a Rebel post known as Chapman's Fort. Stationed there that night was a portion of the First South Carolina Cavalry, which had just returned from a stint with General Wade Hampton's brigade in the Army of Northern Virginia. Its troops were eager to prove their mettle on their native soil, and Lieutenant I. I. Fox galloped off to alert Lieutenant Colonel John D. Twiggs, who passed the information up the chain of command. Twiggs then ordered Lieutenant W. J. Leak to reinforce the pickets between Chapman's Fort and the Chehaw River. He sent one section of Captain William E. Earle's battery of the Palmetto Battalion to Chapman's Fort and another to Means Causeway on the Ashepoo Road. As word made it back to Charleston, General Ripley loaded two companies of the Thirty-second Georgia Infantry onto a train and sent them to Adams Run.[8]

When Lieutenant Colonel Twiggs arrived at Chapman's Fort, he observed that the *Boston*, traveling in unfamiliar waters, had run aground about three hundred yards below the obstructions adjacent to the fort. He dismounted a portion of his cavalry, who fired on the boat from the riverbank but were out of range. Captain Earle's cannons were not, however, and they quickly found their mark. When the Confederate shells hit the transport, the men of the Ninth U.S. Colored Troops panicked. Dropping their arms and equipment, they jumped overboard. Those who could swim made it to shore; however, a number drowned.

Meanwhile Lieutenant Commander Edward E. Stone was creating a diversion on the South Edisto. After landing marines on Jehossee Island, he had proceeded up the South Edisto with the USS *Vixen*, *McDonough*, and *Hale*, to shell a Confederate position at Willtown. After keeping up a steady fire for two hours, they sailed back downriver and collected the marines, who had been unable to get within range of the Rebel battery at Willtown. The small squadron sailed back toward base, frustrated at their inability to accomplish anything more than a day's march across an abandoned plantation. As they headed south, Stone saw a large plume of smoke in the direction of the Ashepoo and sent a boat through Mosquito Creek to communicate with Lieutenant Commander J. C. Chaplin aboard the USS *Dai Ching*.[9]

The *Dai Ching* had not been able to sail with Birney's main party because its pilot did not know the channels of the Ashepoo. After having a pilot transferred from another boat, Chaplin was a considerable distance behind the other boats when he finally set sail. He was heading toward Bennett's Point when he received word of the *Boston*'s running aground. Pouring on the steam, he arrived to find the Confederate battery firing briskly on the stranded craft and striking her repeatedly. Nearly all her troops were in the water or on shore. Chaplin immediately opened fire with all his guns trained on the fort. At around 10:00 A.M., with another Federal gunboat returning fire and Captain Earle's ammunition nearly depleted, the Confederates ceased firing. The *Boston* had been hit about seventy or eighty times, with two shells going through her boiler.[10]

The *Dai Ching* and other Union ships in the area kept up a steady fire on the Rebels while the *Plato* and *Edwin Lewis* pulled alongside the *Boston* to take on as many of the crew and remaining troops as possible. After making two trips to the *Boston*, General Birney ordered it set on fire to prevent her capture and use by the Confederates. Having lost the element of surprise, not to mention an entire boatload of troops and horses, General Birney recalled the remainder of his troops and wearily sailed back toward Port Royal. The mission had failed, not because of superior Confederate defenses or because of timely placement of troops, but because of Federal ineptitude. When the tired and frustrated Birney made port, he found that he was to report to a new commander.

Major General John Gray Foster was already familiar with the city of Charleston and had very little love left for it. An 1847 graduate of the U.S. Military Academy, he had served in Mexico and was placed in charge of strengthening the fortifications in Charleston harbor in July of 1860. He was in command of the garrison at Fort Moultrie when the Federal garrison evacuated to Fort Sumter and was second-in-command when Major Anderson surrendered on April 13, 1861. After serving the majority of the conflict in North Carolina, Foster was excited to be back in South Carolina. General Ulysses S. Grant wanted the Federal troops around Charleston to tie up any Confederate reserves that might potentially be sent to aid Lee or Johnston. Grant ordered the forces along the coast to act primarily on the defensive with occasional raids to occupy the Rebel defenders. One of Foster's first orders after assuming command was to increase the shelling of Charleston. He knew of the reduction of force in his front and was intensely interested in capturing—preferably destroying—the city that he had grown to hate.

Major General John Gray Foster, U.S.A.
Courtesy of South Carolina Historical Society

With so many Confederate troops being siphoned away from Charleston to reinforce other areas, the city sat ripe for the taking. General Jones ordered Major Motte A. Pringle of the Quartermaster Department to cooperate with the railroad officials and to have trains ready on the Charleston & Savannah Railroad at a moment's notice. The Confederates also held a battery of light artillery on constant alert to travel on the cars to any threatened point.[11] Well aware of the city's vulnerability Foster decided on a decisive assault on Charleston. He expected at the very least to destroy the railroad connection between the Broad River and Charleston, and he hoped to find a weak point in the line of defense through which he could penetrate and gain the city itself.

General Foster had devised a five-pronged attack that involved both the army and navy. His plan was based on the same routes used by his predecessors, but it was larger in scale and required considerably more coordination. The eastern arm of the assault called for the capture of Fort Johnson, an installation critical to Charleston's harbor defense. Before the attack Union troops would land in force at the southern end of James Island to compel General Jones to send troops from Fort Johnson to meet the threat. In support of this movement the navy would send a fleet of gunboats into the Stono River to shell Battery Pringle.

The western arm of the assault targeted the railroad. A force under Generals John Hatch and Rufus Saxton was to land on Johns Island, traverse the length of the island, cross to the mainland, and cut the railroad near Rantowles. If they were unable to do that, they would attempt to destroy the railroad bridge that crossed Rantowles Creek with their artillery. A second force under General William Birney would ascend the North Edisto and push toward the railroad at Adams Run. If possible, they would destroy the railroad bridge at Jacksonboro and as much of the trestlework as possible between there and Ashepoo Ferry. It was a bold plan, requiring the coordination of five simultaneous attacks, but Foster was confident of success.

In the early morning hours of July 2, 1864, Federal troops under General Alexander Schimmelfennig landed on the southern end of James Island. The Fifty-fifth Massachusetts, Thirty-third U.S. Colored Troops (formerly First South Carolina Volunteers), and 103rd New York Volunteers drove in the Rebel vedettes and advanced toward a picket force commanded by Major Edward M. Manigault, supported by troops under Lieutenant Thomas M. DeLorme. The picket force at Battery Wright twice fended off advances by the Yankees, but on the third assault, the Federals flanked them on both sides. Lieutenant DeLorme's withdrawal was so chaotic that he abandoned two artillery pieces belonging to Captain Francis D. Blake's artillery. (These guns were the same ones salvaged by the Confederates after the Union transport *Governor Milton* burned during Colonel Higginson's failed attempt to cut the railroad the previous spring.) The Yankees continued forward with a heavy skirmish line supported by a gunboat that appeared in the Stono. As they approached the main Confederate defensive line, Rebel canister and grapeshot drove them back a short distance to the works they had captured earlier. Major

Joseph Morrison's New Yorkers maintained a position closer to Secessionville until moving to a more secure position opposite Coles Island. Here, they went into camp and secured the captured guns.[12]

Major Manigault's small force, which included portions of the Second South Carolina Artillery, Captain (later lieutenant colonel) Niles Nesbitt's cavalry company, Major A. Bonaud's battalion, the First South Carolina Cavalry, and a company of the South Carolina Siege Train, were spent. Many of their number suffered from heat exhaustion, and their effective force was not large enough to mount any kind of counterattack. Manigault was content to maintain his defensive position and await reinforcements. General William Booth Taliaferro, commander of the James Island defenses, hurriedly began to shift troops from various points around Charleston to bolster his line. To the exhausted force he sent one hundred men from the garrison of Fort Johnson, along with several companies of the Thirty-second Georgia, troops from Sullivan's Island, and the "Bureau" Battalion. General Joseph Johnston, who was facing problems of his own in northern Georgia, was sending a couple of Georgia regiments by rail, but they did not arrive until the following day.[13]

Later that morning Colonel George Harrison's Thirty-second Georgia and a detachment of the First South Carolina Cavalry pushed forward and discovered that the enemy had withdrawn during the night. Finding no one in their immediate front, they established a new line from Rivers's Causeway to Grimball's Causeway. That morning the Confederates welcomed the Fifth and Forty-Seventh Georgia Infantry sent by General Johnston, and also the Tenth North Carolina Heavy Artillery Battalion, sent from Wilmington by General William Henry Chase Whiting.[14] This presented a slightly more formidable front to the Yankees, who were busy securing their position to the south. They heard the firing in the direction of Fort Johnson, but they did not know until later what had transpired.

General Foster's objective was to demonstrate forcefully enough on the south end of James Island that it would compel the Confederates to weaken their eastern line and allow him to capture Fort Johnson. The Rebel commanders had done just as he had hoped; however, things did not go as planned. Another portion of General Schimmelfennig's force, under Colonel William Gurney, set sail from Morris Island at 2:00 A.M. on July 3. They disembarked about an hour too late to benefit from the high tide, and many of the boats, piloted by inexperienced army men rather than sailors, ran aground. The Fifty-second Pennsylvania Volunteers, the 127th New York, and a detachment of the Third Rhode Island Artillery approached the fort in two columns of barges early that morning; their objective was a beach between Shell Point and Fort Johnson.

At daybreak Colonel Henry Hoyt's Pennsylvanians found a good channel, reached shore, and overran a Confederate battery; however, the 127th New York could not reach shore quickly enough to provide support, and Hoyt's unit found itself trapped. Under fire from Fort Johnson and from a smaller picket force near Battery Simkins, Colonel Hoyt was forced to surrender his force of 150 men to Lieutenant Colonel

Joseph A. Yates of the First South Carolina Artillery. Because the prisoners outnumbered their captors, Colonel Yates had to await reinforcements from the Citadel before transferring them to Charleston.[15]

To the west the Union effort to disrupt the Charleston & Savannah Railroad was underway. The troops boarded transports and left Hilton Head Island on the night of July 1, reaching the North Edisto River the next day. After the boats carrying General Hatch's men split off to land on Johns Island, Birney sailed farther up the river. General Foster accompanied the small flotilla, which carried a small company of marines; the Seventh, Thirty-fourth, and Thirty-fifth U.S. Colored Troops; a portion of the Seventy-fifth Ohio; and two howitzers. General Birney's combined force of 1,200 men landed at White Point and camped at Legare's plantation.

They began their march before daybreak on July 3 with orders to penetrate to the railroad at Adams Run, destroy the railroad bridge at Jacksonboro, and destroy as much as possible of the trestlework across the swamps and rice fields between there and the Ashepoo River. After marching a half mile, they came under a smattering of fire from Confederate cavalry sent out as skirmishers. Birney's men continued steadily forward, driving the Rebel scouts back over three or four miles until they came to King's Creek. There the Rebel cavalry crossed a bridge, dismantled it behind them, and joined a battery of light artillery placed there by General Beverly H. Robertson.

The battery was in a good position, commanding the approach to the earthworks and the creek. When their rifled guns began firing at the Federal troops on the road, Birney quickly moved his men into the cover of the woods and sent couriers back to General Foster. While Birney reconnoitered the ground, he moved his howitzers into position and began returning the Confederate fire. General Foster sent back orders for Birney to cross the creek and told him that he should take the gunboats into Dawho Creek to provide a covering fire from the flank.

General Birney knew that this assignment would be no easy undertaking. The swamp in front of him was boggy and deep. It was swept by the guns of a bigger Rebel fort near the Dawho and by the guns on the battery and earthworks in their front. The narrowest point to cross was thirty-five to forty yards but had marshy borders, which he estimated to be about fifty yards on each side. As he continued to assess the situation, shells began falling into his lines, but they did not appear to be from the Rebel battery; in fact they were coming from the Federal gunboats. As soon as Birney realized he was under friendly fire, he sent an urgent message to Foster. Once Foster redirected his fire, the gunboats and Birney's howitzers dropped a few shells on the Rebel battery with little effect.

In his final analysis Birney concluded that a crossing was impossible. He faced six enemy guns commanding a creek surrounded by marsh. His men were on an island and could not flank the Rebel works. He lacked sufficient artillery to cover his crossing, was unable to build a bridge, and was not confident in the ability of his troops to execute the planned assault. These factors rendered useless the boat that General

Foster had provided him for the task. After little more than two hours of skirmishing, Birney signaled Foster for permission to withdraw.

From his command post aboard the gunboat, General Foster tempered his annoyance at General Birney's lack of enthusiasm and hesitance in executing his orders.[16] Birney may have misunderstood the importance of his part of the operation, writing in his report that his troops had performed well and that "the affair was an excellent drill for them preparatory to real fighting." Foster was forced to recall Birney but instructed him to hold his position until nightfall. After they embarked under the cover of darkness, Foster ordered them to James Island, where they joined Schimmelfennig's forces later the following day.[17]

In the meantime the party near Johns Island was encountering more difficulty than expected in getting everyone landed. General Hatch was in command of a force that included his own brigade, General Rufus Saxton's brigade, Colonel W. H. H. Davis's brigade, a battery of the 103rd New York Artillery, and one hundred men of the Fourth Massachusetts Cavalry. The 144th New York Infantry landed first and occupied Haulover Cut on the morning of the July 2, with Davis's brigade and the cavalry supporting them. Because of the shallowness of the water at the dock, Hatch did not have everyone safely on Seabrook Island until Saxton's brigade finally landed on the morning of July 3. As the infantry concentrated near the Jenkins's plantation on Bohicket Creek, the engineers were busy constructing a bridge across the cut to allow passage of the artillery. A small force of Rebel cavalry kept a careful watch of the activities, firing on the invaders occasionally, but falling back swiftly when pursued. Captain Edward L. Parker wisely chose to withdraw his small section of artillery back toward the Wadmalaw River rather than risk capture.

Reacting to the landing on Johns Island, Confederate general Samuel Jones sent telegrams to Savannah requesting more troops. The First Georgia Regulars—encamped on Whitemarsh Island outside of Savannah—were loaded into boxcars and sent on July 3. Colonel Charles Jones Colcock sent a company of Bonaud's Twenty-eighth Georgia Battalion, two companies of the Thirty-second Georgia, and a portion of his own Third South Carolina Cavalry from the southwestern part of the state. After their arrival at Rantowles Station, they crossed the Stono River onto Johns Island and went into bivouac just before dawn on July 4. Later that day the Yankees formed at the Cocked Hat intersection and marched down Kiawah Road to Abbapoola Creek. They were prevented from going further by intense heat and exhaustion of the troops, who welcomed the torrential downpours that came later in the day. The delays took away any element of surprise the Northern troops might have had and allowed the concentration of additional Confederate troops between them and the railroad.

Unable to gain the upper hand, General Foster decided to change the focus of his mission and ordered General Hatch to march across the island to a point opposite Battery Pringle, from which his artillery could enfilade the first line of Confederate defenses and destroy the bridge across the Stono River. Hatch continued operations here for a few days before withdrawing to regroup and prepare for a second assault. His force set out on the morning of July 5. As his command spread out, he left four companies of the Twenty-sixth U.S. Colored Troops at the previous night's camp and two companies of that regiment at Huntscom's Crossroads to guard his rear and prevent the Rebels from cutting his lines of communication. In response Confederate major John Jenkins ordered all of his cavalry to monitor the main body of Hatch's troops and hamper their march. He sent the rest down Edendale and Bohicket Road to get in Hatch's rear, threaten communications, and possibly induce a withdrawal.[18]

The First Georgia had been joined that morning by two companies of the Thirty-second Georgia, two companies of the Third South Carolina Cavalry, a Napoleon gun of the Marion Artillery, and a howitzer from Charles's battery. They marched toward the crossroads through underbrush so dense that it reduced their visibility as much as it hindered their progress. As they emerged from the woods, they came upon the surprised companies of the Twenty-sixth U.S. Colored Troops, who responded with a nervous, scattered volley. Simultaneously the Confederate guns erupted at close range, decapitating a couple of the black troops and injuring several others. Momentarily stunned, the Yankees broke into a full retreat with the Georgia troops in hot pursuit. After learning of the rout, Hatch sent Saxton's brigade back to the crossroads to restore his line of communication. With the large force of Yankees on its way, the Confederates broke off their chase to prevent being cut off from the mainland. Having retreated to safety, Jenkins's command hurriedly marched to Waterloo Plantation to take a position in Hatch's front.

The Federal strength on James Island increased with the addition of General Birney's troops, but the Union forces had not made much progress since July 3. General Schimmelfennig had kept up a steady, slow bombardment on the Confederate lines as his men dug in, and Admiral John Dahlgren sent several gunboats into the Stono River to put additional pressure on the Rebel batteries there. Their fire was answered—and quite accurately—by the guns of Battery Pringle and Battery Tynes, which kept the Yankee ships at bay. The Union infantry on the island made a few small demonstrations on the Confederate works, but they were easily neutralized.

At this point General Foster reevaluated his strategy. Both Birney and Gurney had failed in their missions, and Hatch had been delayed to the extent that the Rebels had time to concentrate enough resistance to protect the railroad. He was pleased in one respect, however, and that was in his harassment of the Confederate defenses from Sullivan's Island to Savannah. He could hear the shrill whistles of locomotives day and night as trains rushed troops, horses, guns, and other supplies to the area. Those train whistles also vexed him; if Birney had met his objective, the railroad would be severed and the train whistles silenced. He might have been walking the streets of a defeated Charleston by now.

Confederate general Beverly H. Robertson continued to funnel whatever troops he could muster to reinforce Major Jenkins's small force on Johns Island. He sent for

Colonel George P. Harrison's Thirty-second Georgia, Bonaud's battalion, and the Forty-seventh Georgia, which were dispatched from James Island. General Lafayette McLaws entrained two-thirds of the Fourth Georgia Cavalry and sent them on their way to Charleston on the evening of July 6. General Robertson stopped the train before it reached Charleston and held the Georgia cavalry regiment, along with Webb's Battery, in reserve at Adams Run.[19] Meanwhile Jenkins's force occupied an earthen fence that extended the entire length of an open field at Waterloo Plantation. It was described as "a splendid earthwork, covered in undergrowth and trees,"[20] and the Yankees held a similar position approximately seven hundred yards away.

After changing his strategy, General Foster sent Montgomery's brigade to reinforce Hatch's troops on Johns Island, and when Saxton's brigade caught up with the main force, Hatch began to entrench. Union ships in the Stono now protected the Federal lines of communication, and General Hatch's force of five thousand troops occupied a strong position opposite Battery Pringle. The causeway he occupied had a marsh in front, the Stono River on the right, and a virtually impassible thicket on the left. From the causeway his artillery could enfilade Battery Pringle. The Rebels' only access to them would be to cross a bridge on the main road or to execute a difficult flanking maneuver. Hatch's position was known as Burden's Causeway, but it later became better known as "Bloody Bridge."

On July 6, Hatch sent a regiment of African American skirmishers to within four hundred yards of the Rebel lines, but the canister and shell of the Washington Artillery easily drove them off. On the morning of July 7, Federal artillery entered the field and began a sharp duel with the Confederate batteries. The firing ceased just before noon, and Major Jenkins left to confer with General Jones in Charleston, leaving Major (later colonel) Robert A. Wayne of the First Georgia Regulars in command. So small was the Confederate force that Wayne had to deploy his men five paces apart to hold the line. He had two pieces of artillery on each flank, and his extreme right was manned by a handful of men from the Stono Scouts. While Jenkins was in the city, the Yankees decided to make a quick strike.

Union sharpshooters opened on the Confederate left but soon withdrew under heavy fire. Later, around midafternoon, General Saxton attacked the Confederate right flank. The Twenty-sixth U.S. Colored Troops advanced under the cover of the woods and emerged about two hundred yards in front of the Rebel lines. Two companies of the First Georgia had been sent to shore up the Confederate line, but—before they arrived—the Yankees overwhelmed the small band of scouts and overtook their works. The men of the First Georgia advanced to meet the enemy in the open field, but the troops of the Twenty-sixth pressed them hard from the front and flank, driving the Georgians back to their original position.

After the Confederates reached the safety of an earthen fence, the black regiment paused briefly to re-form their ranks, as the Yankees prepared to assault the Rebel works. The Georgia troops fixed bayonets and kept up a scattered fire on the enemy. As the Federals resumed their attack, things began to look desperate on the

Confederate right. A column of the black regiment nearly flanked the Confederate position, but at the last moment troops from the First and Thirty-second Georgia regiments arrived to save the day. About one hundred of the Georgia troops and a handful of men from the Second South Carolina Cavalry charged into the enemy with such force that the Federals began to fall back. The black regiment was so stunned that it failed to take advantage of the works they had just captured. A disorganized retreat across the field ensued, and the Confederates soon recaptured their rifle pits.[21]

On July 8 General Robertson sensed another Union attack might be pending and sent the Fourth Georgia Cavalry from Adams Run. They took their place on the Confederate line along with the Forty-seventh Georgia, Bonaud's Battalion, and the full compliment of the Thirty-second Georgia Infantry, which had arrived earlier that afternoon. The troops sent by rail from Savannah and its environs were proving indispensable in opposing the Federal troops entrenched in front of and behind Burden's Causeway. The guns of Battery Pringle lobbed shells across the Stono into Hatch's ranks, but caused more alarm than damage. During the shelling Robertson attempted a flanking maneuver by sending Colonel Harrison's command down a road leading to the enemy's rear. Unfortunately, Harrison's ambulances, medical supplies, and ammunition train had not yet arrived on the island. Any attack on the Federal position had to wait until the next day.[22]

Very early the next morning, Colonel Harrison gave the order for the Confederate attack. His command advanced toward the Union center with the majority of the Thirty-second Georgia leading the attack on the right and the Forty-seventh Georgia and Bonaud's battalion on the left. They were supported by the remaining companies of the Thirty-second Georgia, the First and Fifth Georgia Infantry, and the Fourth Georgia Cavalry, all under Major Jenkins. Colonel Harrison had been instructed to carry the Federal line unless he encountered severe artillery fire, in which case he should not needlessly sacrifice his men.

Harrison advanced the skirmish line across an open field, parading steadily in the face of enemy fire. When they were within yards of the works, he gave the order to charge. A few of the Georgia troops mounted the parapet as their comrades poured over the earthen walls to overtake the Federal position. Cheers and Rebel Yells resounded as they watched a small force of enemy pickets retreating. The majority of the Union troops had already been withdrawn to the second line of defense and had dug in to wait the Rebel attack. Saxton's pickets had retreated to a well-entrenched position beyond Burden's Causeway and had taken up the bridge behind them. The new Union line was flanked by marsh on either side; a strong line of breastworks commanded the approaches, and a dense forest skirted its front.

Harrison's Georgians pressed forward in hot pursuit. When they were within 250 yards of the new Federal line, the Yankees greeted them with a deadly volley. With his troops outnumbered and unable to get closer, Harrison recalled his men, who retreated to the protection of the captured works.[23] The Confederates took a position on elevated ground between the Gervais house and the causeway, but Hatch's force was concentrated and waiting in a strong position of its own. As Colonel Harrison and his command caught their breath, Major Jenkins came up with the remainder of the reserve with orders to press the Yankee position again.

As Jenkins's reserve joined Harrison's, the combined force headed back toward the Federals behind the causeway. Morning fog rising from the swamp mixed with the smoke from rifle and cannon to create an opaque curtain that obscured the Confederates from view. Federal infantry and sharpshooters put up a stiff resistance until Hatch brought up his artillery and swept the onrushing Rebels with canister and shrapnel. The deadly combination of infantry and artillery threw the Confederate line into confusion, forcing Colonel Harrison to break off the second assault and withdraw to his works.[24] That evening a train sat idly on the tracks adjacent to Stono Ferry, waiting to carry the Confederate wounded to Charleston.

General Robertson's duty had been to keep the Federals from establishing batteries that could enfilade the Confederate works on James Island.[25] This he had done, and it was all that General Jones could expect of him. For the rest of the day and night General Hatch gradually moved the Federal force to the rear, within the protection of Union gunboats. General Foster met in a conference with General Hatch and Admiral Dahlgren on the afternoon of the July 9 and decided that they "had done all that [they] intended."[26] By the next day the Federal forces on Johns and James islands had re-embarked, and their nine-day operation was over. Another halfhearted attack on Fort Johnson was ordered, but it was easily quashed by the Confederates.

Foster had begun his tenure with high aspirations but was frustrated in his grand stab at Charleston. Coastal topography, oppressive midsummer heat, and inefficient subordinates had doomed the operation; however, the ability of the Confederate troops to concentrate troops from remote areas by rail could not be discounted. Toward the end of the campaign Foster unleashed what became a protracted bombardment of Fort Sumter, but it did not change the fact that his superior force failed to meet its goal. The city still flew the flag of the Confederacy, and locomotives continued to labor over the Charleston & Savannah Railroad. The Confederate army and the railroad continued to benefit one another, but the wear and tear of the war was taking its toll. Nevertheless the railroad, like the Confederacy, continued the fight.

THIRTEEN

Under Siege

N. J. Bell, a Civil War–era Southerner who spent nearly his entire adult life in the railroad business, once said that there was "no class of men where better men can be found than among railroad men."[1] They were hard-living and hard-working individuals upon whose collective backs rested the burden of Southern supply. They were the men who fired the engines, coupled the cars, and handled the freight. Many were self-made men who started in the most menial positions and worked their way up to higher-paying jobs. Most engineers started their careers as firemen or mechanics, and a majority of conductors began as brakemen or flagmen. They were not rich men; nor did they aspire to wealth—that was for the aristocratic investors. Those who felt called into railroading did their jobs dutifully even in desperate economic times.

As the war progressed, such men became harder and harder to find. The difficulties initially caused by the shortage of raw materials and replacement parts were now being compounded by an equally crippling shortage of human resources. Some blamed the army; many blamed the war itself; others pointed to the low wages. One important element was the rigid pay scale. Until the fall of 1862 there had been practically no pay raise to offset the rising cost of living. By the end of 1863, with inflation skyrocketing, the Confederate dollar had a value of five cents relative to gold, magnifying the need for something to be done.

A Southern railroad conductor in 1861 and 1862 made between $70 and $80 a month on average, with mechanics and baggage handlers making less. The salaries of railroad officers varied widely from one company to the next. Treasurers, auditors, and general freight agents rarely made more than $2,500 per year. Presidents' salaries ranged from $2,000 to $6,000, and superintendents took home just slightly less; however, the superintendent of the Georgia Central Railroad actually made more than its president. In the spring of 1863, the directors of the Charleston & Savannah Railroad had voted to give their employees and officers a raise in pay. The salaries of Superintendent Haines and Secretary-Treasurer William H. Swinton were to increase to $3,000 a year. An increase in employee salaries was left to the discretion of President Singletary. Singletary's salary was approved at $4,500 per year, but he initially declined the raise to help save the company money.[2]

At the end of the summer of 1864 clothing and food in the Confederacy were bought at a premium. Men's hats cost $75; shoes cost at least $125; cloth ranged from $15 to $45 per yard. Tea was $22 a pound; coffee ranged from $12 to $60 a pound; flour—just $6 a barrel at the start of the war—now cost $125 to $500 a barrel. The directors of the Charleston & Savannah Railroad knew something must be done to retain their employees and improve their morale. In June they authorized a 25 percent raise for the company's officers and again voted to give their workers a raise, which would be meted out however Singletary deemed appropriate. Now that he had advocated meeting the needs of his workers, he would have to find the funds with which to pay them.[3]

With the constant threat of military activity in and around Charleston, Singletary watched civilian business on his railroad slowly evaporate. To make matters worse Confederate secretary of war James A. Seddon—with little warning or explanation to the public—placed a restriction on travel to and from Columbia, Augusta, Charleston, Wilmington, and Petersburg in mid-May 1864. The government enacted the policy to prevent congestion on the rails between the two main Confederate forces in Dalton, Georgia, and Richmond, Virginia, and to facilitate the movement of corn and troops—both of which were in shorter supply—to Lee's army in Virginia. The restrictions created some fuss in South Carolina, and in June the Northeastern and Charleston & Savannah railroads coordinated their schedules in an effort to alleviate at least some of the congestion between Florence and Macon.[4] Fortunately for the Charleston & Savannah line, its income was being propped up by the steadily increasing demands of the Confederate military. The pace, however, was becoming consumptive, since the government and military benefited from rates far below what the railroad needed to stay afloat. With revenues declining in the face of astronomical supply and labor costs, Singletary had to make changes on both the credit and debit sides of the ledger.

During the summer, in an effort to increase the supply of ready cash, the directors authorized the sale of a portion of the company's 4 percent Confederate certificates, of which they held more than $200,000 worth. They also voted to purchase 150 bales of cotton from a Columbus, Georgia, planter for $.80 per pound with the hopes that it would bring a higher price in another market. At the same meeting board member Henry Gourdin was instructed to continue his negotiations on the sale of White Hall Plantation, which the company owned but was not using at present.[5]

To expand the company's commercial opportunities Singletary worked with Superintendent H. S. Haines to take advantage of interline cooperation. They tried to attract more business from travelers between Richmond and Montgomery by coordinating their schedule with those of the Northeastern and Georgia Central railroads. With immediate, direct connections they could alleviate the annoying layovers that caused congestion in Florence, Charleston, and Savannah—and the mail would reach Savannah a full twelve hours sooner. Another opportunity presented itself when President Thomas Webb of the North Carolina Railroad contacted Singletary about the establishment of an express service over connecting railroads in the Confederacy. The directors embraced the proposal and were ready to act whenever practicable, but no formal agreement was ever finalized.[6]

Singletary and Haines also looked for ways to cut expenditures without jeopardizing the company's daily operation. Toward that end they decided to remove their freight agents from the depots at Rantowles, Ravenel, Salkehatchie, White Hall, Coosawhatchie, Savannah River, and Monteith.[7] Agents would remain at the stations near important military posts, but freight delivered to the unmanned depots could not be guaranteed once it was taken off the cars. This cost-saving method was also a necessity owing to the dwindling workforce.

When necessary, the railroad augmented its workforce inexpensively by increasing the hire of slave labor. Because Singletary was superintendent of labor of the Charleston defenses and because the railroad was vital to the coastal defenses, the railroad had ready access to African American workers. The majority of black hands performed hard and menial tasks associated with maintenance of way. A number of railroads used them as brakemen or firemen, and many blacks served as skilled mechanics.[8] Measures such as these helped Singletary deal with the attrition caused by the war and the dismal economy and made it easier for him to increase the wages of his paid employees.

In Charleston shortage and sacrifice were becoming the orders of the day. The city and its inhabitants had been subject to a steady bombardment for the better part of a year, and no end was in sight. When the shelling had begun the previous autumn, thousands left the city voluntarily, but thousands also stayed. Those who remained in Charleston were forced to move their businesses well north of Calhoun Street and to seek shelter in safe houses in the northern part of the city.

The once stately houses on the Battery now lay open to the elements, with many of their windows, roofs, and doors battered by Union ordnance. There had been a slackening of the barrage during the spring, but that changed once General Foster assumed command of the Yankee forces. The area south of Broad—now known as the Shell District—was mostly deserted and overgrown with weeds. Soldiers scavenged lead and other useful metal items from the houses; looters carried off much of the rest. The storefronts of Charleston's business district were in shambles. Shattered masonry and mangled iron lay everywhere. The Charleston & Savannah Railroad office at 34 Broad Street was not spared the wrath of the Federal guns. Major Edward M. Manigault noted, as he surveyed the damage that summer, that "just where I used to sit making out estimates, returns, monthly d<u>o</u>, requisitions, &c. . . . a shell had knocked away the wall. If I had been sitting there at the time, I would have had no further trouble in this life."[9]

Day and night the Union batteries on and around Morris Island kept up their fire—at times a shell every fifteen minutes—using the church steeples as their landmarks. Because of the bombardment, very few farmers were interested in shipping their goods to Charleston. The resulting food shortage necessitated the establishment of the Charleston Free Market. City officials traveled the region and purchased produce directly from local farmers and planters. With the cooperation of the city's three railroads and the military, the goods were shipped back to Charleston and sold at greatly reduced prices. Their cooperation with this effort added to the burdens Charleston's railroads suffered for the war effort.

The bombardment of Charleston and restriction of travel to and from the city largely choked off civilian travel and commerce over the Charleston & Savannah Railroad. During the summer a plan formulated by General Samuel Jones increased the transportation of a different sort of passenger on the railroad—the Yankee prisoner of war. As damage to property and endangerment of women and children continued, General Jones felt an urgency to take some of the pressure off the city. In June he requested that approximately fifty Federal prisoners be brought to Charleston "to be confined in parts of the city still occupied by citizens, but under the enemy's fire."[10]

A week after this request, President Davis approved the transfer of five Union generals, among whom were General Truman Seymour, and forty-five field officers to Charleston. They arrived in the city on June 14, 1864, and were billeted in a home at the western end of Broad Street. This move angered General Foster, who did not slacken his fire; instead he sent a request to Washington for the transfer of a similar number of Confederate prisoners to his command. The Union prisoners remained in Charleston for a few weeks until they were exchanged in August for the Rebel prisoners being sent to Foster. The Confederate leadership followed the situation in Charleston earnestly. In its end result they saw a potential avenue for future prisoner exchanges through Charleston, as well as a way to relieve some of the overcrowding of Confederate prison camps.

In late July the Confederates transferred six hundred additional Union prisoners from Andersonville to Charleston with the hopes of exchanging them outside Charleston harbor. Among this group was Captain John Adams of the Nineteenth Massachusetts, who was part of a group of officers who had plotted an escape if the opportunity presented itself. The intent was to seize control of the guards in each car, disarm them, leave them on the train, and set it in motion again. Captain Adams and the men in his car had emptied their guard's cartridge box and awaited the signal— but it never came. Captain Adams was sure that one of his own men betrayed the conspirators, but the plan was likely foiled by the Confederates' placement of a large number of guards on the engine and the first car.[11] When the prisoners realized their plan had been thwarted, a number of them jumped from the train. Many, however, were soon captured and reunited with their comrades in Charleston's Shell District.

When the Confederates proposed an exchange for these prisoners, General Foster refused. Instead he again played tit for tat and had six hundred Confederate prisoners sent south and placed on Morris Island under the fire of Rebel artillery. They remained there until, after two months of futile negotiation, the Union prisoners were

Train transporting prisoners from Savannah to Charleston, sketch by Alfred R. Waud. Courtesy of Library of Congress, DRWG/US Waud No. 793

sent to Confederate prison camps further inland. By this time General William T. Sherman had broken through Confederate defenses in northern Georgia and taken Atlanta on September 1. His presence there—and their uncertainty about his next move—alarmed the commanders at Andersonville prison. When they contemplated the damage that more than thirty thousand liberated Yankee captives might inflict, they began transferring more prisoners to other quarters. This resulted in a steady flow of Union prisoners riding the rails between Savannah and Charleston for the remainder of 1864.

After they left Andersonville, most of the prisoners were sent first to Macon and then to Savannah. If they were not housed in Savannah jail, they were put aboard trains bound for Charleston. Sixteen miles outside Savannah, their train crossed the rickety bridge over the Savannah River. Picking up speed after departing Monteith Station, the cars swayed and the bridge creaked and shook so much that the prisoners feared they might crash down into the muddy river. Once across the river, the scenery that greeted them was strikingly different from that which they were accustomed. Compared to "the famine-stricken pine-barrens of Georgia," the country was more pleasing to the eye. Despite the destruction in other parts of the state, the foliage surrounding the railroad remained lush. The landscape was decorated with clusters of laurel, palmetto trees, and great live oaks adorned with Spanish moss. The swamps and rice fields, with their abundant waterfowl, added still more interest to the journey. From their cramped quarters the prisoners often caught glimpses of grand plantation houses surrounded by majestic oak trees in the distance.[12]

As pleasant as the environs outside the train were, the condition of its cars certainly were not. The wear and tear of persistent use was evident, as was the shortage of usable rolling stock. By September the Charleston & Savannah Railroad was transporting prisoners in rickety freight cars that the railroad typically used for hauling coal. Forty or more men rode in each car, the bottom of which often contained an ample glaze of coal dust that the Rebel guards refused to sweep out.[13] The wheels and axles of the cars lurched and squealed from lack of lubrication, and the maintenance-starved locomotives wheezed and sputtered at speeds of ten to twelve miles per hour.

These slow speeds afforded many of the prisoners the opportunity to tempt fate by jumping from the trains—especially those traveling at night. They usually fell victim to harsh treatment from the elements. Often, escapees ended up starved, sleep deprived, and tormented by hordes of mosquitoes until they appealed to local slaves for help. Some escapees did receive aid, but others were handed over to Confederate pickets stationed along the way. After a meal and sometimes an exchange of pleasantries, Rebel guards took their prisoners to the nearest depot to board yet another train for Charleston.[14]

General Jones, the overseer of these prisoner transfers, was concerned for the future of Charleston. As conditions deteriorated, he ordered the construction of the Florence Stockade and began shunting prisoners there, as well as to similar facilities in Columbia and Salisbury. The forces he commanded were undermanned and ill equipped for the defense of the city. His patchwork command could not concentrate on guarding large numbers of prisoners and resisting the advance of thousands of Union invaders. He was relieved of his stresses and command of the department when President Davis replaced him with General William Joseph Hardee at the end of September. In mid-October Hardee officially took command of the newly formed District of South Carolina.[15]

The official order from Richmond came on September 28, but General Hardee did not assume command of the department until October 5, 1864. A native Georgian and a West Point graduate, Hardee had commanded a corps in the Army of Tennessee through the Atlanta Campaign, taking part in fighting at Resaca, Kennesaw Mountain, and Peachtree Creek. He led the Confederate attack on Union forces at Jonesborough and was defeated on the day before General John Bell Hood evacuated Atlanta. Before the war he had been best known for devising a system of infantry tactics for the U.S. Army. Nicknamed "Old Reliable," he was reputed to be one of the most efficient field officers in the Confederate army. He inherited a department fraught with problems, not the least of which was lack of manpower. He could field only 12,446 effective troops, many of whom lacked adequate weaponry and uniforms. He spent that autumn recruiting and training artillerists, building wharves, obstructing Charleston harbor, shoring up forts in the area, and inspecting his troops. He was even able to effect the removal and exchange of Federal prisoners from Charleston, having complained after his arrival that "there [were] already in the Department more prisoners than [he could] properly guard."[16]

Not only was Hardee a good field commander, he also recognized the importance of factors ancillary to the army. He aided port collectors, secured necessities for the navy, and strengthened lines of communication and transportation. He recognized the importance of the railroads to his cause, helping to regulate their use by the military and sending out work parties to collect unused and endangered railroad iron to replace damaged or worn rails on more vital lines.[17] Even at this late juncture, Confederate railroads did not easily accept regulation and sacrifice. If they did either, it was usually with great reluctance—particularly where it concerned the Confederate government. General Hardee, like his predecessors, realized the importance of the Charleston & Savannah Railroad to achieving his objectives. He hoped that the railroad would reciprocate.

The railroad's directors remained externally calm about the fall of Atlanta and the Siege of Petersburg, which had begun in June 1864. The minutes of the company

Lieutenant General William Joseph Hardee, C.S.A. Courtesy of South Carolina Historical Society

that fall contain no direct mention of the war other than the company's generous contribution to the Soldiers' Wayside Home. The directors deliberated over the execution of bonds, the raising of freight and passenger tariffs, and monthly business statements, barely acknowledging the dire circumstances in Charleston and elsewhere in the Confederacy. At the October board meeting Singletary once again petitioned the directors on behalf of his employees. He pointed out that the pay of certain company employees was much too low, lower than other companies were paying for the same services. He made a motion that they increase the salaries of those employees by a full 50 percent. After a brief discussion, the board gave its assent.[18] At this same meeting they also addressed the ongoing saga of the Ashley River Bridge.

The problem of crossing the Ashley River had plagued the company since Thomas Drayton's ill-informed decision to terminate the tracks on its western bank. The board objected to paying the Charleston Bridge Company for the nonmilitary use of the present bridge and regretted the continued need of steamers for their commercial business. When they realized they would make a profit—albeit it a falsely inflated one—for 1863, the directors decided to explore their prospects for a permanent bridge of their own. Such an opportunity presented itself in November of 1863, when Captain Singletary received a letter from Charles M. Furman, the president of the Cannonsborough Wharf and Mill Company.

The Charleston & Savannah Railroad had purchased or leased property on Doughty Street from Furman's company in 1859 with the agreement that they would make certain improvements on the land. The term of the contract was about to expire, and those improvements had not yet been made. Assuming that the railroad company intended to forfeit the contract, Furman wanted to know if the company wished to remain on the property. If so, he wanted to renegotiate their agreement. Singletary brought the letter to the monthly board meeting; however, they lacked a quorum and could take no official action. Those present asked him to delay his reply as long as possible.[19]

In March at the monthly board meeting, the directors wanted several questions answered as they weighed their options on the Doughty Street land. Was it appropriate for the company to carry out the contract with the Cannonsborough Wharf and Mill Company? If so, was execution of the contract still practical for the railroad? If it were in the railroad's best interest to abandon the project, what was the best alternative for crossing the river and establishing a depot elsewhere? The board also wanted advice on the subject from John McCrady, the company's attorney. After reviewing the necessary information, McCrady told Richard Yeadon, chairman of the railroad's Committee on Road and Machinery, that there could be no forfeiture under the previous contract but that an action might be brought against the railroad to enforce the terms of the contract. In April the directors nominated a special committee consisting of Singletary, Charleston mayor Charles Macbeth, and Henry Gourdin to explore the company's alternatives and handle any future negotiations with Cannonsborough Wharf and Mill Company.[20]

In June one of the mill company's principal owners, James Butler Campbell, contacted Singletary with a suggestion of his own. Campbell offered to sell the railroad his company's land bordering Spring Street on the north and the Ashley River on the west. Railroads were not the only businesses hit hard by the poor economic conditions, and he felt that a railroad terminus and depot at this location would be mutually beneficial. If the railroad would determine how much land it required, Campbell promised that his company would deal favorably with the railroad and that the Doughty Street property would no longer be an issue.[21]

Focusing on his main goal, which was to determine the most cost-effective route that would best accommodate the company's future business, Singletary went back to the board and recommended a line beginning on their existing track a half mile from the St. Andrews wharf, crossing the river just below the present bridge, and terminating on the south side of Spring Street next to Chinquapin Street. This route provided the shortest way to the crossing with the least expense in acquiring right-of-way. It also gave them ample space for their depot, storage, and maintenance shops. The board agreed with him and authorized him to continue negotiations with Campbell.[22]

Meanwhile the railroad had temporarily suspended travel across the existing Ashley River bridge, and the steamers were again busy. Concern had arisen when engineers found a cracked wooden pile during a routine inspection of the bridge. This prompted a more thorough investigation by Superintendent H. S. Haines and Captain Robert H. Lucas, a Confederate engineer assigned to the city's defenses. Piles from several other locations were examined, tested, and eventually found to be sound. They concluded that the defective pile was likely cut long before it was driven into the riverbed, thus causing the bark to fall off prematurely. Without protection from the elements the pile wore down faster than expected, and it was easily injured. Haines and Singletary promptly restored freight and passenger travel, but not before all the bridge's piles that were not covered with bark had been replaced.[23]

As the summer passed, Campbell decided to take the lead in negotiating for the Cannonsborough Wharf and Mill Company. He did not want the sale to fall through, but he did want to be certain he got the right price for the land. Before the war started, the property had been appraised at around three thousand dollars per acre, and Campbell felt it was worth at lest five to six times that amount in the present currency. He had a certain amount in mind, but he assured Singletary that his company had no designs on selling to anyone other than the railroad. On hearing this the Charleston & Savannah board authorized Singletary to close the purchase of the property whenever they arrived at an agreeable price. Since the planned bridge would be used in large part for government transport, the directors hoped to receive financial assistance from the Confederate government for the project. They instructed Singletary to petition Secretary of War Seddon in person for the aid, but Singletary was never able to make the trip to Richmond.[24]

The events of the summer had called into question the structural stability of the temporary bridge the railroad was presently using. Even though it was now passable, Singletary did not know how long it would remain so. He also knew that the steamer used to carry the railroad's freight across the river was becoming unreliable. The contract with the boat's captain would expire the following June, and Singletary felt it would likely be impossible to obtain another one by that time.[25] Further complicating matters, Sherman was now in control of Atlanta and its many railroad connections. Not even Confederate military leaders were sure of Sherman's future plans.

To outside observers the decision to pursue plans for a permanent bridge was made from either misguided confidence or utter denial. Singletary knew how much business had dropped off, and he knew how little cash the railroad had on hand. He also knew that both major Confederate armies were facing overwhelming odds in north Georgia and in Virginia. There was plenty of timber for a bridge, but the industrial iron and the labor required for the task were in very short supply. To Singletary, though, the choice was easier. The company needed a permanent bridge, whether or not it was built immediately. If by some remote chance the South was able to prevail, the influx in business and rebounding economy would sustain the venture. If Sherman or Foster took Charleston, at the very least the company would have the land at its disposal once it was able to resume operation. Singletary knew that what little cash the railroad possessed was nearly worthless, but he had 150 bales of Georgia cotton sitting in storage. He offered Campbell $200,000 for the property, the entire amount payable in cotton. The railroad had purchased it earlier in the year at $.80 per pound and would evaluate it at $1.30 per pound for paying Campbell. The difference in price would serve as the railroad's fee for shipping it from Columbus, Georgia, to Charleston. Campbell accepted the offer on September 26, and the deal was closed on October 19, 1864.[26]

A month after the sale, the wisdom of Singletary's use of cotton rather than currency for the purchase became evident. The value of Confederate currency continued to plummet, and after examining the company's annual financial report, Singletary informed South Carolina governor Milledge L. Bonham that "the financial condition of the company does not warrant the recommendation that a dividend be declared on the capital stock of the company."[27] Military developments in the coming weeks soon made any further efforts toward building a permanent bridge pointless.

FOURTEEN

Honey Hill

Even in wartime Savannah maintained its reputation as a city of charm and beauty. Its wide streets were laid out at right angles and were lined with live oaks, bay trees, willows, and magnolias. The parks at the centers of its squares provided a peaceful haven. Brick and wood-framed houses had large yards decorated with multicolored flowers, ornamental shrubbery, and flowering trees. Locals and visitors could also venture out to the oak-lined avenues of Bonaventure for an afternoon of sanctuary.

Savannah was a commercial center whose shipyards, sawmills, and iron works played a significant role in the Confederate war effort. It was also an important railroad hub. The Savannah, Albany & Gulf and the Atlantic & Gulf connected the city with south Georgia and Florida. It was connected with Millen, Georgia, and points west by the Georgia Central and with the Northeast via either the Augusta & Savannah or the Charleston & Savannah. Since the fall of Fort Pulaski in April 1862, military activity in Savannah had consisted mostly of shunting troops in various directions via the city's connecting railroads, building ironclads for the Confederate navy, and reinforcing its coastal fortifications.

The Confederate army constructed a strong battery on Wassaw Island to guard the entrance to Wassaw Sound and built a fort several miles upriver on a bluff facing the Wilmington River. To protect Ossabaw Sound, they erected a battery on Green Island, guarding the Vernon River, in addition to Fort McAllister, a substantial earthwork on the southern bank of the Great Ogeechee River. Fort McAllister defended the entrance to the river and protected the bridge of the Atlantic & Gulf Railroad a short distance upstream. It had, according to Derek Smith, "come to symbolize Rebel defiance throughout the war."[1] Infantry attacks on Savannah had been expected from time to time but had never come. In 1863 Union monitors had entered Savannah's harbor but had inflicted very little damage. On February 22, 1864, the Rebels repulsed a diversionary Union assault on Whitemarsh Island coincident with Seymour's Jacksonville Campaign, but otherwise Savannah had been relatively quiet.

In Atlanta, however, General William T. Sherman was planning to strike out for the coast in the direction of Milledgeville, Millen, and Savannah. On October 9, 1864, he telegraphed General Grant from Atlanta to ask how many Rebels were likely to oppose him on this route and whether his army could live off the land in that region. He spent a few weeks chasing General Hood's Army of Tennessee into northern Georgia, but as the Confederates moved farther away from Atlanta, Sherman decided not to take the bait. He returned to Atlanta with two-thirds of his army, leaving General George Thomas to shadow Hood. After initially expressing reservations, General Grant finally approved Sherman's plan but asked him to wait until President Lincoln was safely reelected before launching such a bold mission. Sherman hoped to reach the coast by December 1, 1864, but he had not yet set Savannah as his target. He intended to march toward either Savannah or Port Royal, but he had not ruled out Pensacola, Florida, as a possibility either. Sherman's army—sixty-two thousand strong—left Atlanta on November 15 with a force consisting of two columns, the left commanded by General Henry W. Slocum and the right under General Oliver O. Howard. They were on their own, cut off from communication and supply lines. When Sherman's army set forth, Confederate generals Beauregard and Hood thought he was still following Hood into Tennessee. By the time they learned of his whereabouts, he was well on his way to Macon.[2]

When Confederate general William J. Hardee assumed command of the Department of South Carolina, Georgia, and East Florida, he had 12,446 effective troops, mostly clustered in and around Charleston and Savannah.[3] He depended a great deal on the assistance of General Samuel Jones, whom he had just replaced, and General

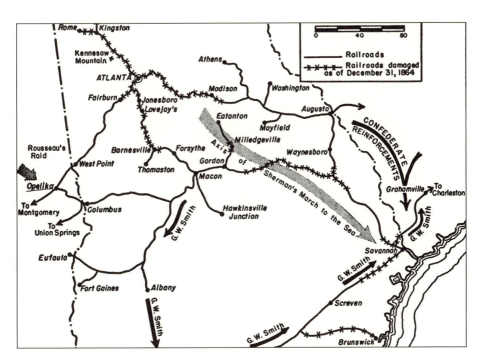

Sherman's march through Georgia, September–December 1864. From *Railroads of the Confederacy*, by Robert C. Black, © 1952 by the University of North Carolina Press; renewed 1980 by Robert C. Black III. Reproduced by permission of the publisher

Lafayette McLaws, a veteran of Longstreet's Corps with a checkered past. Many of his other subordinates had little combat experience. As Sherman stormed through Georgia, the Confederate government ordered Hardee to consolidate his command under General Braxton Bragg in Augusta, thinking Sherman would attack there.

General Bragg, who was convinced that Sherman's progress could not be slowed, concentrated his efforts on the defense of Augusta, allowing Hardee to operate independently in Savannah. He made sure the Savannah River was made impassable to Union vessels by shoring up the obstructions and the powerful network of guns there. He also began moving some of his heavy artillery to the western side of the city. When General McLaws pointed out the incomplete defenses covering the Charleston & Savannah Railroad's bridge across the Savannah River, he sent crews to build more works there. Anticipating an eventual attack, Hardee ordered all able-bodied men in Savannah to report for duty.[4]

The citizens of Savannah braced for the worst. The city's squares were crowded mostly with women and children, because every male capable of bearing arms was away at the front. With rumors flying the people hoped against all hope to avoid a siege or invasion, praying that Sherman would choose some other unfortunate city on which to unleash his punishment. For months Savannah had seen its troop strength dwindling. When Atlanta and Petersburg needed reinforcements, the trickle of troops out of the area became a stream. With Sherman on the move and McLaws counting barely two thousand effective troops, Hardee was forced to do anything to increase his numbers. Out of desperation, he activated everyone from soldiers at home on furlough to those convalescing from injuries or illness. He released soldiers who had been confined to quarters for minor offenses and sent them back to their units. Hardee expected every available man to fight, including Chatham County's reserve militia—almost all of whom were between the ages of fifty and sixty-five. Despite these measures, the total force mustered at Savannah was still fewer than ten thousand men.[5]

The largest single addition to the garrison was approximately two thousand Georgia Militia and State Line troops under General Gustavus Woodson Smith. Smith was a graduate of West Point, an able engineer who had resigned his commission in the regular Confederate army in 1863 and relocated to Georgia. Since that time he had aided in the construction of some of the state's defensive works and was later appointed by Governor Joseph E. Brown to command the state's militia.[6] With Sherman on the move Smith again readied himself for the business of combat. On November 17, 1864, the troops under General Smith were assigned a defensive position on the eastern side of Macon, Georgia. Four days later, Bragg ordered them to move as soon as possible to Augusta. They made it as far as Griswoldeville on November 22, where they were involved in an intense engagement with Sherman's rearguard and suffered six hundred casualties. After losing one-fourth of his command and several of his best field officers, Smith and his Georgia reservists limped back into Macon. After assurances that Sherman's army was safely away from Macon,

Major General Gustavus Woodson Smith, C.S.A., Georgia State Militia. Courtesy of South Carolina Historical Society

the Confederate commanders ordered Smith's force to join General Hardee in Savannah. Unfortunately for Smith, the main body of Sherman's army sat squarely on the most direct route to Savannah, and huge segments of the Georgia Central Railroad had been decimated. He had to take the scenic route to the coast.

Their journey began at the depot of the Southwestern Railroad in Macon, where Smith's four brigades were loaded aboard the cars on the morning of November 25. They made the 104-mile trip to Albany, where they spent the night, and started their march to Thomasville the next morning. Covering the 60 miles in fifty-four hours, they arrived in Thomasville at noon on November 28. There, General Smith was exasperated to find only two trains waiting for him on the Atlantic & Gulf Railroad instead of the five that he had requested. He was forced to leave three of his brigades behind and was unable to get underway with the First Brigade until after dark. The dilapidated engines took twice the usual time to reach Savannah, crawling into the depot at 2:00 A.M. on the November 30.[7] General Smith and his men were tired, hungry, and eager to find a place to rest. As he sat on the train, a courier entered his car with an order from General Hardee telling him to "proceed at once with the first two trains of your troops . . . to Grahamville and Coosawhatchie on the Charleston & Savannah Railroad, which places are being threatened by raiding parties of the enemy." In the past week Smith's men had fought a hotly contested battle, marched sixty miles in just over two days, and been forced to endure the inadequacies of the state's railroads. Now General Hardee was ordering Smith's troops across the state line, which was a violation of the statute calling them into service.[8]

Smith stormed over to Hardee's headquarters and demanded a conference. Smith was not about to take his men into South Carolina unless he felt there was a good reason for doing so. He awakened Hardee, who used maps and several recent dispatches to demonstrate his point. The Yankees had come up the Broad River and landed a few miles below the railroad near Grahamville. If the Confederates could not muster adequate opposition, the Yankees would likely cut the railroad shortly after daybreak. He needed Smith's men to help hold the line until 2:00 P.M., when reinforcements from Augusta and Charleston were expected to arrive.[9]

If the Federals took control of the Charleston & Savannah Railroad, Hardee said, they would cut Hardee off from supplies and reinforcements from Charleston, and he would be left without a line of retreat if pressed by Sherman. The Union army would also acquire a base from which to move on Charleston or Savannah from the interior.[10] Smith was satisfied. Despite protests of some of his officers and men, they would go. They arrived just in time to take part in the Battle of Honey Hill.

The Union invasion up the Broad River grew from a seed planted by General Sherman as early as November 11, 1864, when he telegraphed General Henry Halleck from outside Atlanta. He had just burned the foundries in Rome, Georgia, and was about to set out for his "grand raid" within a few days. He requested that General Foster break the Charleston & Savannah Railroad somewhere near Pocotaligo by December 1. The idea had crossed General Foster's mind as well. He had been observing the exodus of troops from the line of the railroad to reinforce the troops opposing Sherman in Georgia. Assuming he was still under orders to "stand strictly on the defensive," he still wanted to take advantage of his numerical superiority and "scrape together a small force of three thousand men and attack and capture, if possible, some point on the railroad." He had conferred with General John P. Hatch, who agreed with the plan. Hatch favored a raid up the Broad River, which would allow the use of all available resources from Beaufort and Hilton Head. Foster then ordered Hatch to select his three strongest regiments, four pieces of artillery, and all the cavalry he could spare for the mission. Garnering extra winter clothing, ample ammunition, and five days rations, Hatch's force left Morris and Folly islands under the cover of darkness on the November 27 to rendezvous with troops at Hilton Head to form what would become known as the Union's Coast Division.[11]

The military wharf at Hilton Head was crowded on the night of November 28, 1864, full of supplies, guns, horses, and soldiers who had not yet boarded their transports. The atmosphere was intense, but not chaotic, as the soldiers prepared for their short trip up Broad River to strike the railroad. The First Brigade, led by General Edward E. Potter, was made up of the Fifty-sixth, Fifty-seventh, 127th, and 144th New York Infantry; the Twenty-fifth Ohio; and the Thirty-second, Thirty-fourth (formerly the Second South Carolina), and Thirty-fifth U.S. Colored Troops. The Second Brigade consisted of the Fifty-fourth and Fifty-fifth Massachusetts Infantry, and the Twenty-sixth and 102nd U.S. Colored Troops, commanded by Colonel Alfred S. Hartwell. The force also included batteries from the Third Rhode Island and Third New York Artillery, two squadrons of the First Massachusetts Cavalry, and a naval brigade of sailors, marines, and a battery of boat howitzers.[12]

The signal to launch was given at 2:30 A.M., but just as they were set to push off, a heavy fog enshrouded the harbor. With visibility reduced to only a few yards, the expedition was delayed for a couple of hours before the expedition finally set sail. The murky conditions slowed travel up the Broad River that morning, and many of the transports ran aground owing to the low tide. Some simply decided to drop anchor and wait for improved conditions after daylight. Several ships, including the *Canonicus*, which carried engineers and supplies for constructing a landing, mistakenly wandered up the Chechesee River.[13]

The skillful pilots of the shallow-draft navy vessels were able to proceed cautiously upriver, and shortly after sunrise they entered a creek that led to Boyd's Landing. Through the curtain of fog, they saw a grand plantation home, livestock grazing in fallow fields, and moss-tinseled oaks and pines that bordering the creek. Not far from the manse, a handful of Rebel pickets from the Third South Carolina Cavalry huddled together, warming themselves in front of a small fire. As the fog began to lift, revealing the Union transports as they approached the landing, the startled vedettes quickly saddled up, a handful hustling off to alert headquarters of the enemy's approach. The remainder withdrew to a safer position from which to monitor their movement.[14]

The double-enders anchored at 8:00 A.M. near Boyd's Neck, where a dilapidated old dock was all that could potentially serve as a landing. From there, a twisting, rutted wagon road led the way to the Grahamville depot some seven or eight miles distant. When General Hatch's transport arrived, they began a slow debarkation of the naval brigade. It was already 11:00 A.M. General Potter's brigade arrived at around noon, and Hartwell's not long afterward; however, the engineers had not arrived to repair the wharf so that cavalry and artillery could be safely landed. The troops were taken ashore in smaller boats and scrambled up the muddy bank. Soldiers had to push many of the horses overboard, forcing them to swim ashore.[15]

The naval brigade, under Commander George H. Preble, was first to land and advanced in a skirmish line to scare off the Confederate pickets. They paused at an intersection near the landing and erroneously took the right fork, chasing the Confederates toward Coosawhatchie. Without reliable maps or guides they moved about two miles to the right—watched closely by the Rebel cavalry—and began to entrench. They were soon joined by the Thirty-second U.S. Colored Troops, which had been sent to support them. Meanwhile, the Fifty-fourth Massachusetts Regiment was sent out a half-mile to dig in and cover the landing.[16]

By 4:00 P.M., as more troops were landing, General Hatch decided to push forward and attempt to seize the railroad. General Potter rode ahead to find the naval brigade and caught up with them as they were beginning to set up camp. After conferring, he and Commander Preble ascertained that they were on the wrong road. They were soon on the move, retracing their steps back to the first intersection. Fatigued from the marching and from dragging their howitzers over the sandy road by hand, they were ordered to camp there for the night.[17] General Hatch and

General Potter then led a portion of the First Brigade further down the road in the opposite direction.

On reaching the crossroad at Bolan's Church they received misinformation from their guide that contributed to the undoing of the raid. Instead of turning right on the Honey Hill Road, they maintained their course southward, heading toward Savannah. After taking this route for four miles without encountering any resistance, they realized their mistake. After careful review, an embarrassed and thoroughly frustrated Hatch ordered a countermarch back to Bolan's Church. Worn out from the landing and marching debacle, they reached the church and bivouacked for the night. It was now 2:00 A.M. on November 30.[18]

As commander of South Carolina's Third Military District, Colonel Charles Jones Colcock had been fifty miles away the previous day, superintending the construction of field works along the Savannah River. After leaving there, he planned to marry Agnes Bostick, who would be his third wife, on November 30, 1864. His second in command, Major John Jenkins, had been in Charleston on official business and was transported by a special train back to Grahamville after receiving word of the Federal landing. After assessing the situation, he wired General Jones in Charleston to request that additional troops be sent to him that night. He was assured that the Forty-seventh and Thirty-second Georgia would be sent right away.[19]

After couriers relayed the news to Colonel Colcock at Mathews Bluff, he detoured by the Bostick's plantation, apologized to his fiancée, and left immediately for Grahamville. He arrived there at 7:00 A.M. and was immediately handed a telegram informing him that General Smith was on his way. Before Colcock's arrival, Major Jenkins had detached a small force from Companies B and I of the Third South Carolina Cavalry to monitor any Federal movement up Honey Hill Road. At about 8:00 A.M. General Potter's brigade broke camp and began its march toward Grahamville with the 127th New York spread out in a skirmish line, closely followed by the Twenty-fifth Ohio and a battery of the Third New York Artillery. The Fifty-fourth and Fifty-fifth Massachusetts and the naval brigade were coming up from Boyd's Landing and additional troops were not far behind.[20]

In Savannah, General Smith and his troops were embarking on an unexpected leg of their journey. After the general had returned from Hardee's headquarters, he had ordered the trains carrying the Georgia Militia to be switched from the Atlantic & Gulf Railroad to the Charleston & Savannah. At the depot he gathered a handful of his junior officers and briefed them on the situation. The officers' responses were nearly all the same. If the general felt it necessary, they were willing to go. A few minutes later, with bravado awakening the groggy soldiers, laughter could be heard emanating from the cars. After being properly watered and fired, the first of the trains inched out of the depot bound for Grahamville at 6:00 A.M.

The trains consisted of old and battered passenger cars and platform cars; the engines were rusty and leaky. Looking out their windows, Smith's men observed scores of slaves working on the earthworks bordering the railroad outside the city.

They were crooked structures and only three feet high. The swampy ground in front of the works would have to serve the purpose of abatis.[21] Passing the outskirts of Savannah, the tracks made a wide sweeping turn before approaching the river. From here Savannah's river front came into view. In the distance Confederate banners waved proudly and defiantly above the Rebel installations along the river. General Smith made his way through each car, relating to his men the gravity of their current mission. The city's supply line—its very survival as part of the Confederacy—rested on them. As they rattled over the Savannah River Bridge into South Carolina, Colonel Colcock awaited their arrival nineteen miles away at Grahamville Station.

General Smith arrived with the first group of Georgia troops at 7:30 A.M. and was greeted by Colonel Ambrosio Gonzales, who introduced Smith to Colcock. As the three men spoke, a scout rode up to inform Colonel Colcock that a column of Yankees was moving up the Honey Hill Road from Bolan's Church. Colcock quickly dispatched a twelve-pounder Napoleon gun of Captain J. T. Kanapaux's Light Battery and Company K of his cavalry regiment under Captain William B. Peeples to meet the enemy. As Colcock prepared to leave for the front, General Smith requested that he select a position for his Georgians. Smith would wait and come up with the men on the second train as soon as they arrived.[22]

As the Federal column marched up Honey Hill Road, they met their first resistance about a half mile up the road. A handful of Confederate vedettes exchanged fire briefly with the vanguard of the 127th New York and then fell back across a causeway. As the Federal troops fanned out into lines of battle, the small detachment of Rebel cavalry retreated to join two companies of Colcock's Third South Carolina and two fieldpieces across the causeway. The position had an impenetrable swamp on the left and a wide field on the right intersected by canals and ditches. Captain Peeples's company dismounted on their arrival and deployed as skirmishers across the field.[23]

As soon as the lead elements of the Federal column came into view, the Confederate guns opened fire. The procession came to a temporary halt as the two forces exchanged scattered fire. The Thirty-second U.S. Colored Troops, under Colonel G. W. Baird were marching near the rear but were called on to advance up the causeway to capture or drive off the Confederates. On the double-quick they attempted a flanking maneuver along the causeway and into the field. The Rebel guns, firing from close range, had a "deadly effect" on the center and rear of the black regiment.[24] To counteract the flanking movement, Colonel Colcock ordered Captain Peeples's men to set fire to the field. As the dry, crackling broom sedge ignited, a prevailing wind carried the flames rapidly toward the Federal troops, temporarily arresting their advance. Men from the Union pioneer corps moved forward to beat out the flames, but another thirty minutes were lost in the process. Meanwhile General Smith was marching forward with his second brigade of militia. With the small band of cavalry holding the Yankees in check, Colonel Colcock rode back to advise Smith on the ground and to help position the troops.

The place Colonel Colcock had chosen for his defensive stand was two-and-one-half miles below Grahamville at a group of old breastworks at Honey Hill. Colonel Thomas L. Clingman's North Carolina troops had constructed the works in 1861, and Generals Roswell Ripley and Robert E. Lee had ordered them fortified early the following year to protect the railroad.[25] The breastworks sat atop a crescent-shaped hill extending two hundred feet on either side of the road to Grahamville. It had parapets for light artillery and trenches for infantry and commanded the main approach to the railroad from Boyd's Neck. The ground in its front had been cleared of trees but was covered with a thick flooring of vines and underbrush. The Confederate left extended into a pine forest with little protection and the right sat on the edge of a swamp. A shallow, but relatively wide, stream lay 150 yards from the foot of the hill. The only practical approach was by the narrow road from Bolan's Church, which made such a sharp turn after passing the swamp that the earthworks were effectively invisible until a person was in very close proximity to it. Until this day General Hatch and his brigade commanders had been completely unaware of the earthworks' existence.[26]

As the Rebels retreated to the battlements at Honey Hill, Colonel Colcock ordered them to dismantle the bridge crossing the stream in front of the works. He sent another detachment to a position behind an old dam. The colonel and General Smith had at their disposal the Georgia Militia and State Line troops, three companies of Colcock's Third South Carolina Cavalry, five guns of the Beaufort Artillery, and two guns from Earle's battery of the Furman Artillery and Kanapaux's battery of the Lafayette Artillery. Two battalions of reservists from Augusta and Athens, Georgia, formed the reserve, and a large portion of the Forty-seventh Georgia Regiment were being rushed to the scene by train from Charleston. Against Hatch's 5,500 men the Confederates counted on an effective force of approximately 2,000.

The Union column, headed by the Fifty-sixth New York, proceeded up the road to Grahamville in pursuit of the Rebels. Just ahead they saw a small dirt road—fittingly referred to by many historians as the Woods Road—branching off to the right into a dense forest that paralleled the stream. It was at this point that the main road made a sharp turn to the left toward town. When the lead elements of the column rounded the curve, they came into full view of the Confederates, who unleashed a heavy fire of canister and musketry upon them.

With their troops thrown into a temporary state of chaos, the Union field commanders tried to restore order and began a hasty deployment. General Hatch ordered Potter's brigade to form a line of battle parallel to the Confederate position. The Fifty-sixth and 157th New York regiments manned the extreme left, and the 127th New York took a spot just left of the road about two hundred yards from the enemy works. To the right, Potter sent the Thirty-second U.S. Colored Troops, the Twenty-fifth Ohio, the 144th New York, and the marine battalion from the naval brigade. A section of artillery occupied the intersection of the Woods Road and Honey Hill Road. As they moved to the right, the Thirty-second Colored Regiment was stunned by a volley from the Rebels positioned behind the dam. As the men of the Thirty-second began to waver, the Twenty-fifth Ohio charged forward and easily dispersed the enemy. That accomplished, they continued down the Wood's Road to extend the Federal line in that direction.[27]

The first Federal unit to charge the Rebel line was the Thirty-fifth U.S. Colored Troops, led by Colonel James C. Beecher, brother of abolitionist author Harriet Beecher Stowe. Ordered to charge the enemy center, they advanced obliquely and were firing wildly by the time they reached the turn in the road. The Confederate guns were in close range, and the absence of trees made the African American troops easy targets for the now-entrenched Georgians. Beecher's men went in with a cheer, but because of the swamp and the murderous enemy fire, they never reached an effective position for their assault. With losses mounting, the twice-wounded Beecher ordered his men to fall back.[28]

Hearing the first volleys fired, Colonel A. S. Hartwell directed the lead companies of the Fifty-fifth Massachusetts to hurry forward with the intent of supporting the Thirty-fifth. As the eight companies of the Fifty-fifth prepared for battle, two companies of the Fifty-fourth Massachusetts under Lieutenant Colonel Henry N. Hooper moved to shore up the Union left. The Fifty-fourth had just arrived in the field when Hooper was met by Colonel William T. Bennett, who repeatedly motioned toward enemy lines and ordered him to charge. Seeking clarification, Hooper asked him "Where?" The increasingly agitated Bennett could manage no reply other than "Charge!!"[29] Hooper quickly realized the futility of attempting such a maneuver with only two companies and very little support. He moved his men to the left of the artillery and attempted to enter the woods, struggling through snarls of vines and briars. They formed a line on wooded ground sloping to the creek, and lay down to avoid enemy fire.

Just as the Fifty-fifth Massachusetts was forming on the right, they were met by the men of the Thirty-fifth, who were making a panicked retreat to the rear of the artillery. Left with no one to support, Colonel Hartwell's men were now pinned down and under a heavy fire. The colonel and his company commanders fought to keep their troops' attention, barking orders above the roar of the artillery and the screams of the wounded. As his men began to fall, Hartwell knew he could not stay in this position. Faced with wondering what to do—and having no order from Hatch to fall back—he gave the order to advance. Artillery wagons and a confused bolus of wounded soldiers and stragglers clogged the narrow road ahead of them. As the regiment passed this snarl, the columns became disorganized and three companies were mistakenly shunted down the Woods Road to the right. Once past the intersection, the five remaining companies re-formed and continued their advance; however, owing to the terrain and the hail of enemy musketry, they gained only about forty feet before having to fall back.

On the left Lieutenant Colonel Stewart L. Woodford of the 127th New York reported to General Potter that he would charge the Confederate position with part

Corporal Andrew Jackson Smith of the Fifty-fifth Massachusetts Infantry, who received the Congressional Medal of Honor for distinguished service at the Battle of Honey Hill. Courtesy of C. Douglas Sterner at www.houseofheroes.com

of his regiment, provided that a simultaneous charge could be made on the road to his right. As he waited, Colonel Hartwell, who had been injured minutes earlier, was forming the Fifty-fifth for another charge. The colonel waved his hat confidently as he gave the order, "Forward!" They charged down the narrow road four men abreast, with Hartwell's shouts—"Follow your colors!"—barely audible above the din. It was, as one soldier put it, "like rushing into the very mouth of death going up this road."[30] As they made the final turn toward the fort, the grape and canister began to tear huge holes in their column.

Simultaneously, the 127th New York charged forward almost perpendicular to the Fifty-fifth Massachusetts. They crossed the creek, and advanced into the marsh on the Confederate right. The ground was relatively open with isolated trees and low brush; however, they were slowed by the bogginess of the soil and ten to eighteen inches of standing water. Just as they were in a position to flank the Confederates, the Forty-seventh Georgia Regiment—just marching up from the Grahamville depot—was sent immediately to stem the charge.

Even though news of the Federal landing arrived in Charleston at 10:00 A.M. on November 29, the Forty-seventh Georgia had not received marching orders until 5:00 P.M. that afternoon. They did not reach the St. Andrews depot until 9:00 P.M., and were instructed to await their rations at the station. At midnight they left for Grahamville without ever receiving the promised victuals, arriving just after daybreak on November 30. They waited at the depot for another two hours without orders, but finally started a slow, sullen march through town to the earthworks at

Honey Hill. As the firing in their front increased, their steps became livelier. Arriving at the earthworks, they reported to Colonel Colcock and General Smith and were immediately rushed into battle.[31]

The 127th New York held a position seventy yards from the enemy line for ten minutes until the Fifty-fifth was forced back with heavy losses. With the fire in their front now directed solely on them, the New Yorkers pulled back to within the protection of their artillery, where they remained for the rest of the afternoon. As the Federals withdrew, the Confederates pushed slowly forward from their rifle pits to a point on the Union right.

The density of the woods masked the strength of Hatch's force, but it was clear to General Smith that his men were outnumbered. As it was reported later, "the noise of battle at this time was terrific, the artillery crashing away in the center, while volley after volley of musketry ran down both lines and reverberated from the surrounding forests."[32] The Federals had extended their attack, and the deployment of the Forty-seventh Georgia left very few troops in reserve. At times Colcock and Smith had to improvise, changing the position of companies and regiments as well as extending intervals to prevent being flanked. As Colcock positioned the troops, Smith anxiously gazed in the direction of Grahamville looking for any sign of reinforcements.

The previous day Major Jenkins had asked that General Samuel Jones send the Thirty-second Georgia to Pocotaligo as quickly as possible. General Hardee had also been assured by General Bragg in Augusta that a thousand South Carolina reservists under General James Chesnut were on their way, and two thousand North Carolina troops under General Lawrence S. Baker were about to be sent. Hardee was understandably upset when he received a telegraph from General Smith that the fight was still progressing and the troops still had not arrived. Hardee urgently wired General Jones to ask if the Thirty-second Georgia and the troops from Augusta had left Charleston. If not, he wanted them hurried to Grahamville as soon as possible. An accident blocked the tracks of the South Carolina Railroad 112 miles above Charleston, however, preventing the arrival of Bragg's troops from Augusta. The Thirty-second Georgia was delayed and did not arrive in Grahamville until 4:30 P.M.[33]

While the fighting intensified on the Union left, the right wing spread out adjacent to the Woods Road. The Thirty-second U.S. Colored Troops held the extreme right, the 144th New York sat on the left and rear, near the Honey Hill Road, and in between them was the Twenty-fifth Ohio. Lieutenant Colonel Nathaniel Haughton led the Ohio regiment into what appeared to be an impervious thicket a few hundred yards in front of the Confederate earthworks. Although visibility was limited, he could hear heavy fighting on his left but hardly anything in his front.

Perceiving a weakness on the Confederate left flank, Haughton sent a small party to reconnoiter the position; just as he thought, their report was favorable. Haughton advanced his unit eighty yards to the edge of the swamp, where they were met by a strong force of Confederates. The Thirty-second Colored troops were ordered to

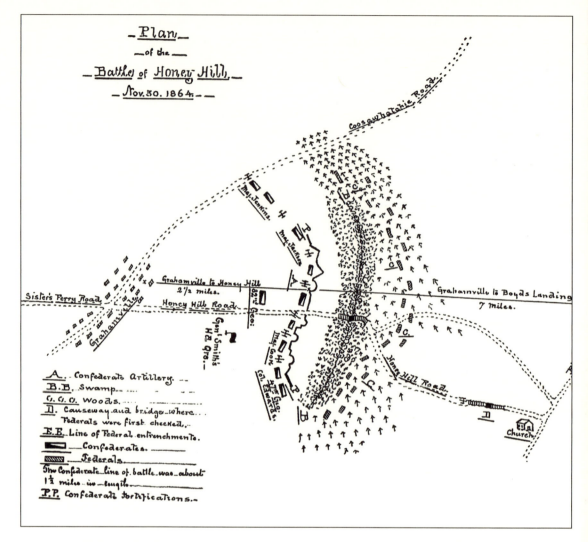

The Battle of Honey Hill. Courtesy of Duke University Special Collections Library, Durham, North Carolina

carry the works from the right, but the marsh, underbrush, and heavy enemy fire halted their progress. They came up too late and did not swing around enough to the right to support the Twenty-fifth Ohio, which was now heavily engaged.[34]

Whereas an organized assault might have carried the enemy flank, the Union troops became disorganized. The 144th New York had not advanced far enough to support the Ohio regiment, and the Thirty-second was suffering heavily on the right. The three "lost" companies of the Fifty-fifth Massachusetts wandered by mistake through the woods and swamp until they happened upon the fighting. Reaching the stream, they found the Twenty-fifth Ohio firing wildly at an invisible enemy. Not knowing if the gunfire on their left was from friend or foe, the Fifty fifth lay down without firing. When the marines formed ranks to the right of the Thirty-second

Colored Regiment, the Federals now had portions of five different units acting independently in the woods to the right of the road.

Noticing the buildup on his left, Colonel Colcock ordered Company B of the Third South Carolina Cavalry to extend their intervals to thirty feet in order to protect the flank. Although the Confederate line was dangerously thin, the Union was unable to penetrate it. The ferocious Rebel fire stymied the Twenty-fifth Ohio in front of the stream under a heavy fire for an hour, until their ammunition was almost exhausted. With no support and his men now scavenging cartridges from wounded and dead comrades, Colonel Haughton was compelled to withdraw. The regiments fell slowly back to the intersection of the Woods Road with the main road, where they replenished their ammunition. They then took a position behind the dam and remained there until the close of the battle.[35]

The last of the frontal assaults was attempted by the Fifty-fourth Massachusetts, veterans of a similarly perilous maneuver at Battery Wagner the year before. When the regiment had left Boyd's Landing that morning, Lieutenant Colonel Hooper detached the four companies as pickets at the Coosawhatchie crossroads under the charge of Captain George Pope. Here, with the help of two howitzers of the naval brigade, they engaged and drove off a small squadron of Confederate cavalry later that morning. Another regiment relieved them later, and they proceeded to march to the front. Halting briefly at the church for rations, they heard heavy firing in the distance and were soon moving forward at the double-quick.

As they approached Honey Hill, their column was strung out by the ambulances, caissons, and wagons that blocked the road. They arrived near the intersection, where General Hatch conferred briefly with Colonel Bennett. After his exchange with Hatch, Bennett led Captain Pope to the intersection and there ordered Pope to take his men to the left of the battery and charge. As they stood in formation, the sights and sounds of battle engulfed the men of the Fifty-fourth. The concussion of artillery was deafening; the Rebel Yell was louder than many had ever heard. In the field before them they could see the dead and wounded of the Fifty-fifth Massachusetts. Shot and canister tore down the hill and up the road as they moved to the left around the artillery. Their charge was over before it began. Because of the congestion in the road, many in the column halted, uncertain of where to go. Captain William H. Homans took command of the bewildered companies and led them into the woods on the right. Once past the battery, Captain Pope was surprised to look behind him and find a lieutenant and only eight men. His only option was to lie down and seek cover.[36]

The Union defeat was total. The Rebels had parried every thrust made upon their left, right, and center. The infantry regiments were at a stalemate—many seeking cover behind the dam—and were running low on ammunition. By 4:00 P.M., the two batteries of the Third New York Artillery had been depleted of shells and had to be replaced by two howitzers from the naval brigade. One battery had lost all its horses and most of its men. Two companies of the 102nd U.S. Colored Troops were sent a hundred yards forward to retrieve the guns—which they did successfully—but they lost a company commander and several officers in doing so.[37]

By that time the firing from both Union and Confederate lines was beginning to slacken, with neither appearing to make any more aggressive overtures. General Beverly H. Robertson arrived at 4:30 P.M. with the Thirty-second Georgia, a battery of artillery, and a company of cavalry. They were too late to participate in the battle but served as an effective reserve. General Hatch had received reports that Confederate reinforcements were arriving by railroad, and he became increasingly convinced that his force could not take the Rebel fort. Even though he had also received reports that the Confederate right might be easily flanked, Hatch ordered General Potter to prepare for withdrawal at dusk.[38]

The Twenty-fifth Ohio and 157th New York moved a half mile to the rear to cover the retreat, while the 127th New York, the 102nd Colored Regiment, and the naval brigade occupied the intersection at the Woods Road. The naval battery kept up a slow fire until 7:30 that night, while the regiments slowly retired back toward Bolan's Church, which the Federals converted into a hospital. For the North the operation had been a disaster from the outset. From the difficulties in navigating the Broad River to the erroneous maps to the series of uncoordinated frontal assaults, the responsibility lay with the commander. The Union had suffered more than 750 in killed, wounded, or missing. By comparison the South counted only 50 total casualties among the units who reported. The number of Federal wounded so overwhelmed their ambulances that the Fifty-fourth and Fifty-fifth Massachusetts were divided into squads to transport them to the church on makeshift stretchers.[39]

For the South it was a triumph of timing, location, and fighting spirit. It was a battle of many "what ifs." What if the weather had allowed Hatch's troops a smoother trip up the Broad River? What if the lead elements of the force had not completely botched the landing and the march on the twenty-ninth? What if they had chosen a different route for the attack on the railroad? What if Captain Peeples had not courageously delayed the Federal First Brigade that morning? And, most important, what if General Smith's militia had been unavailable or unwilling to cross the border into South Carolina?

Only a portion of the Confederate reinforcements bound for Grahamville actually made their destination before the Battle of Honey Hill, but as fortune would have it, they were not needed. The earthworks and the terrain the Confederates commanded allowed a much smaller force to turn back Hatch's Coast Division in convincing fashion. There were many who had scoffed at General Lee's decision to construct the earthworks atop Honey Hill, but after today, not many questioned his wisdom.

The Battle of Honey Hill, like the construction of the railroad, was an exercise in cooperation between the states of South Carolina and Georgia. The troops rushed to Grahamville from Savannah and Charleston had been mostly Georgia troops. The small squads of cavalry that delayed Hatch's march were South Carolinians.

Even though General Smith was the senior officer in the field, he deferred the command to Colonel Colcock, who knew the ground much better than he. Complimenting the effort, General Smith wrote, "I have never seen or known of a battlefield upon which there was so little confusion, . . . and where a small number of men for so long a time successfully resisted the determined and oft-repeated efforts of largely superior attacking forces." The *Charleston Daily Courier* was equally gracious to the Georgians, stating that they had "shown a noble example in defending herself on the soil of South Carolina." The author also submitted that "the pledges given by Bartow, and Jackson, and others, to and for Charleston, on the celebration of the completion of the Charleston & Savannah Railroad, have been fully redeemed so far, but are still recognized by the survivors."[40]

Once again the railroad had played a material role in its own defense. Even in its present state of decline, an inefficient railroad was much better than no railroad at all. Its trains still possessed the capacity to transport much-needed reinforcements at critical times. No one knew how much longer this would last, however, since this victory was different from the others in one important way: the Federal troops did not reboard their gunboats and sail back downriver. As Confederate reinforcements flooded into the area, Hatch's division was busy entrenching at Boyd's Neck to await further orders from General Foster and to anticipate the arrival of General Sherman. The coming weeks were the most tumultuous yet for the railroad and for the defense of Savannah and southwestern South Carolina.

FIFTEEN

Shermanʼs Neckties

It was just after midnight on the morning of December 4, 1864, when the sound of the bugle split the night air. As the last notes of "Assembly" echoed through the barracks, the student-soldiers sprang into action, moving briskly into the quadrangle. They formed by squads, and once all were counted present their commander barked out the order. They must prepare to march in thirty minutes.

After retrieving arms and accoutrements, the corps reassembled. Because of the unseasonably mild weather, many left their overcoats and blankets behind. They stood at attention, organized now by companies. Cadets from the Citadel made up Company A, and cadets from the Arsenal Academy in Columbia, where students at the South Carolina Military Academy spent their first year, comprised Company B. They looked crisp in their state-issued gray uniforms; each cadet sported a muzzle-loading Enfield rifle. When the company commanders reported all present, the battalion paraded out of the sally port and through the torch-lit streets to the tune of fife and drum. They proudly unfurled the colors of the South Carolina Military Academy as they marched from Marion Square to the Charleston & Savannah Railroad depot at St. Andrews.[1]

The Citadel cadets had been stationed at Orangeburg performing home-guard duty when the Yankees struck inland at Honey Hill. On the day of the battle Major James B. White had received orders to "proceed with the Cadets . . . to the nearest Confederate general for service in the field."[2] As their train stopped briefly at Branchville, Major White was handed a telegram from Governor Milledge L. Bonham with orders to aid General Samuel Jones in the defense of the Charleston & Savannah Railroad. From Branchville they were redirected to Charleston, where they spent a few days drilling at the Citadel. They were joined by cadets from the Arsenal Academy in Columbia, and the entire battalion anxiously awaited their marching orders.

Units such as the Cadet Battalion were becoming increasingly important to the defense of the Charleston & Savannah Railroad. While Sherman moved ever closer to Savannah from the interior and Hatch threatened the railroad from the coast, Confederate troops were pouring into the area. The already taxed railroad was pushed to its capacity over the coming weeks. Superintendent H. S. Haines, who had just returned to work after a protracted illness, and Major Hutson Lee coordinated

Major General William Tecumseh Sherman, U.S.A.
Courtesy of South Carolina Historical Society

the transport of thousands of soldiers and their supplies between Charleston and Savannah.

After the Battle at Honey Hill the Confederates—not knowing if the Yankees would launch another assault—remained safely within the confines of their earthworks. One set of works was situated at Bee's Creek and guarded the route to the railroad at Coosawhatchie. The other was at Honey Hill and monitored the way to the railroad via Grahamville and Gopher Hill. As the Rebels celebrated their decisive victory that night, they also cheered the arrival of additional troops. Brigadier General James Chesnut arrived at Grahamville near midnight with his brigade of

South Carolina reservists. Just before daybreak on December 1, Brigadier General Lawrence S. Baker reached the village with 860 of his North Carolina brigade; the remaining 1,100 made it to Coosawhatchie by 9:00 that evening. Also that day General Hardee traveled from Savannah to Grahamville to assess the situation. While there he granted General Smith permission to return his weary troops to Georgia within forty-eight hours.[3]

After the Georgia Militia entrained and was safely on its way back to Savannah, General Hardee telegraphed Superintendent Haines to request locomotives and enough rolling stock to convey an additional 1,200 men from the Grahamville area to Savannah as soon as possible. Further adding to the traffic, 1,300 militiamen under Colonel Wilmot G. DeSaussure had been loaded aboard cattle cars and boxcars and sent from Augusta on the evening of the December 2. Hoping the other brigades had cleared the rails, General Hardee ordered General Samuel Jones to arrange for their transportation from Charleston to Savannah.[4]

Trains of thirty or more cars rocked along the worn-out track with one engine in front and one in back, pulling and pushing the soldiers to their destinations in Coosawhatchie, Pocotaligo, and Savannah. The trains were loaded "thick as peas" with troops, many of whom rode on top of the cars. As they passed stations along the way, some shouted and whooped to the locals as others sang and played banjos to pass the time and take their minds off the battles that lay ahead. They boasted of "cutting the Yankees' throats" and making their enemy tremble. William Rose, then a slave on a plantation near Ashepoo Junction, recalled watching the soldiers as they crossed the trestle over the Ashepoo River: "They going to face bullet, but yet they play card, and sing and laugh like they in their own house. . . . All going down to die."[5] He stood and watched as the train pulled across the trestle and slowly disappeared from sight. The last thing he remembered hearing was the distant sound of the soldiers' laughing and singing.

While the troop strength in the region was increasing, these were by no means seasoned units. Many of the soldiers were inexperienced men more than fifty-five years old and boys younger than seventeen. Some of the militiamen were so poorly equipped that Hardee had to order "600 arms and accoutrements" sent to Grahamville. It appeared, as one observer put it, that "the Confederacy was 'robbing the cradle and the grave' to put men in the field."[6]

Exhausted and outnumbered, the Southern troops had not pursued the retreating Federals after the fight at Honey Hill. The Yankees quietly withdrew to a defensive position close to the spot of their initial landing at Boyd's Neck and began to construct fortifications in anticipation of further orders. The day after the battle was quiet as the Federals entrenched at Boyd's Landing. Early on the morning of the December 2, Confederate artillery opened fire on the Union troops, forcing them back to the safety of their newly built earthworks. The Fifty-fourth Massachusetts Regiment worked feverishly to extend the works for the next two days, as little more than brief skirmishing took place between the two forces. On December 4 the

Twenty-fifth Ohio was taken aboard transports up Whale Branch Creek, where they marched inland six miles and captured two pieces of Rebel artillery at Church Bridge.

The next day—after five days of recuperation, digging, and minor demonstrations—the Union high command was ready to attempt another assault on the railroad. As a diversion, the Yankees sent out to reconnoitering parties on December 5. The Fifty-fifth Massachusetts was sent out with two guns on the extreme left of the Union line, and a larger force under Brigadier General Edward E. Potter went out from the right. It was Potter's force that drew the most attention as it moved on the Confederate works on the way to Coosawhatchie. General Chesnut telegraphed Brigadier General Lucius Gartrell that the Federals were moving in considerable force on the breastworks at Bee's Creek. General Potter advanced his pickets to within two hundred yards of the Rebel works but retired when fired upon by the Confederate battery. At 7:00 P.M. Gartrell reported that all was quiet and that the enemy had retired three miles beyond his outer defenses.[7]

On the day of this minor action General Hardee had placed General Jones in command of the entire line of the Charleston & Savannah Railroad and requested that he set up headquarters at Pocotaligo. Jones left Charleston on the morning of the fifth, but because of slow train speeds and increased rail traffic, he did not arrive at Pocotaligo Station until around sunset. At the time he had no idea of the force under his command—their number, description, or location in the area. He was not even sure how many troops would be required to hold the line of the railroad. As he began to set up his headquarters and process the reports coming to him from Gartrell and Chesnut, he also took stock of the troops at his disposal.

His command included the Fifth and Forty-seventh Georgia regiments, a battalion of the Thirty-second Georgia, several batteries of artillery, a portion of the Third South Carolina Cavalry, and Kirk's squadron. The remainder were for the most part loosely organized reservists and militia from South Carolina, North Carolina, and Georgia—many of whom were without arms. They occupied positions along the railroad extending from Pocotaligo to the Savannah River. The soldiers based at Grahamville were under the immediate command of General Chesnut, and those at Coosawhatchie were under General Gartrell.[8]

The dispatches from Gartrell seemed to indicate that the Federals intended to move on Coosawhatchie via Bee's Creek and the Coosawhatchie River. At the suggestion of Hardee, Jones ordered Chesnut to place the Forty-seventh Georgia and a section of artillery on a train and have them ready to move at a moment's notice. Major Hutson Lee had authorized a train to be kept at Coosawhatchie expressly for the purpose, but the order was not promptly executed. This omission proved costly the following day.[9]

Though Gartrell reported "all quiet" in his front, the Union camp at Boyd's Neck was abuzz with activity. After withdrawing from Bee's Creek that afternoon, the majority of Potter's brigade proceeded to Boyd's Landing, where a dozen transports awaited them. Through the night the soldiers, artillery, and their supplies were loaded aboard the transports, which set sail early the next morning. Moving upriver through the fog, the first of the steamers entered the Tullifinny River at around 8:00 A.M. on the sixth.[10]

Navigation was tricky because of the low tide, and the steamers carrying the naval brigade ran aground. Those transporting Potter's infantry reached the upper landing of Gregorie's plantation first and disembarked without any Confederate opposition. After landing, the 144th New York and a detachment of the Twenty-fifth Ohio set to work securing the landing, while the 127th, 157th, and Fifty-sixth New York; the remainder of the Twenty-fifth Ohio; and a battalion of naval infantry were sent on a reconnaissance in force. At 10:00 A.M. General Gartrell telegraphed General Jones that the Yankees were landing three miles from Coosawhatchie. He requested that the Forty-seventh Georgia and a section of artillery be sent via the railroad as quickly as possible. Unfortunately for him, they were not prepared to move.[11]

The Federals moved north up the road to Coosawhatchie, their right wing marching in an open field and their left in a refused position to protect the flank. The only unit Gartrell had in the area to meet them was the Fifth Georgia. Led by Colonel Charles P. Daniel, the Georgia regiment pressed forward and hit the right of the Federal line about one mile from Tullifinny Crossroads. The naval infantry, which occupied the extreme Union right, were swept by the onrushing wave of gray and were driven back nearly 150 yards before the Confederates became overextended and disorganized. Heavy fighting ensued as the navy howitzers and marines rushed to the scene of the fighting. Colonel Daniel requested reinforcements, but the only troops available were the two hundred men of the First Georgia Reserves.[12]

The Union right was unprotected for a moment, but then the navy guns were put into position and began to open a deadly fire. With the aid of the artillery, the 127th and Fifty-sixth New York regiments were able to turn on the thin line of Confederates. The New Yorkers, under Lieutenant Colonel Stewart L. Woodford, nearly encircled the Confederates, who retreated in a panic, leaving their colors in the field. Some of the retreating Rebels fell back into the woods above the turnpike while the others retreated westward under the cover of their artillery. As the Georgia regiment crossed the Coosawhatchie River, the men partially destroyed the bridge before retiring to their earthworks near the railroad.[13]

One of the units sent belatedly to support the Fifth Georgia, was the Cadet Battalion, now under the command of Major John Jenkins of the Third South Carolina Cavalry. The cadets had spent the previous day distributing provisions to the other troops, and today they were ready for action. Major Jenkins was sent ahead to ascertain the Federal position and led the cadets forward at the double-quick. They were en route when the Confederate line was broken, and by the time they reached the field the Federals had disengaged and were setting up camp on the edge of the battlefield. After the skirmish Major James B. White marched the Cadet Battalion back to Tullifinny trestle in the rain. They remained there on alert, and tried to get some rest in a nearby cornfield. There would be very little sleeping on this night. As they lay in the corn rows, many used canteens to shield their faces from the rain. A few miles away in the Federal camp, tensions also remained high. Scattered firing from pickets caused troops on both sides to form battle lines during the night.[14]

During the night the noise from a twenty-car train loaded with reinforcements from Augusta awoke the exhausted troops. Regiment upon regiment was heading to Savannah as quickly as the railroad and the quartermasters could arrange transportation. Single tracking and the lack of sidings were particularly troublesome obstacles to this endeavor. On December 6 four trains traveling from Savannah to Charleston blocked progress on the railroad, as two trains loaded with troops from Wilmington and Augusta sat waiting on the eastern bank of the Ashley River.[15] With Sherman occupying the route from Augusta to Savannah, maintaining the flow of reinforcements between Charleston and Savannah was never more important than now.

After two-and-a-half hours of fighting on December 6, the enemy now had four regiments positioned along both sides of the road from Old Pocotaligo to Coosawhatchie. The Fifth Georgia had returned to their works at Coosawhatchie protecting the railroad. General Gartrell was frustrated with the results of the day. He had had difficulties with the telegraph operator throughout the day, and Chesnut's failure to have a mobile reserve had been critical. Gartrell was also concerned about his troops' lack of experience, stating "I must have some old troops; the new ones won't stand."[16] He felt that a larger force sent to threaten the Federal rear on the road to Old Pocotaligo could possibly cut the enemy off. General Jones began to gather all the troops he could muster, including the Seventh North Carolina Reserve Battalion, which had just arrived from Charleston. Colonel A. C. Edwards of the Forty-seventh Georgia Regiment would lead the attack the next morning from the left, and Gartrell would demonstrate from Coosawhatchie at the first sound of Edwards's guns.

As the sun rose on December 7, Colonel Edwards called his men to formation. He sent out a skirmish line consisting of the Citadel cadets and three companies of the Fifth Georgia Regiment. Behind them marched the Forty-seventh Georgia, a battalion of the Thirty-second Georgia, the Arsenal Cadets, and a unit of South Carolina militia under Lieutenant Colonel Edwin H. Bacon of the Thirty-second Georgia. They swept quickly over a frost-covered field of broom grass and into the edge of a stand of trees on the Union right. Covered by the dense vegetation and a hazy mist, the cadets advanced to within sixty yards of the main Federal line before they encountered their first resistance.

The Confederate skirmishers hit the Thirty-second U.S. Colored Troops, who were positioned near a bend in the road from Old Pocotaligo. Taken by surprise, the black regiment gave way, exposing a battalion of marines, who had to be withdrawn quickly to avoid capture. The hungry, shoeless Georgians joined the youthful cadets in driving the Yankees back across the road. As the Georgians and the cadets pressed forward, they were joined on the left by Lieutenant Colonel Niles Nesbitt's First

South Carolina Cavalry. The rebel line shifted until it crossed the road at right angles and engaged the entire Union line.[17]

At the same time General Gartrell ordered Colonel Daniel to move with the main portion of the Fifth Georgia from Coosawhatchie. They hit the Union left, but were unable to inflict much damage and were repulsed easily. The Confederates under Colonel Edwards were able to maintain their position for a time, but when Colonel Daniel's advance was neutralized, the Federals were able to shift more of their troops to the right. As the Confederates formed for another assault, a battery of navy howitzers sent a deadly hail of grapeshot into their ranks. At the first discharge an entire battalion of Georgia reserves turned and fled. The tide began to turn, and the artillery—along with a healthy push from the reinforced infantry—forced the Rebels back across the road.[18]

Colonel Edwards had a spot decision to make. The Union line now extended beyond his left, and their artillery commanded the road. The only way he could unseat the enemy was an attack in force. Fielding mostly raw, inexperienced troops and in danger of being flanked, he determined that such a move would be foolish and decided to pull back. As the Confederates retreated to the safety of their earthworks behind the railroad embankment, the enemy chose not to pursue them aggressively, although they did follow them for some distance.[19]

After fighting all morning, the Citadel cadets were nearly out of ammunition and had to be relieved by the Arsenal cadets. When called upon they marched—to the brief amusement of the Georgia veterans—"as if on dress parade" and took their place next to the Forty-seventh Georgia in the rifle pits. The Federal line drew closer to the works with bayonets fixed and appeared to be readying for a charge. The cadets kept popping their heads above the breastworks to get their first glimpse of the enemy. Above the gunfire, shouts of "Down Mr. Hagood! Down Mr. Hayne!" could be heard, prompting one Georgia soldier to remark that the Cadet Battalion's officers were the "politest officers to their men I ever struck up with in the army."[20]

All politeness aside, as the Federals emerged from the swamp into the open field, Major White of the Cadet Battalion rose up and gave the order: "Attention battalion! Ready! Aim! Fire!" On his command their rifles sent a "staggering wave of lead" toward the enemy, who soon broke off contact and retreated back to their works on Gregorie's Neck.[21] The Confederates cautiously pursued the retreat and regained the road to Old Pocotaligo, but the enemy still held a strong line of earthworks within a mile of the railroad.

December 8 was a quiet day on both sides of the line, as both the Southern and Northern troops strengthened their defensive works. Brigadier General Beverly H. Robertson arrived from his subdivision that day to take command of the troops between Bee's Creek and Pocotaligo. He took immediate charge of building up the works near the railroad. Later that day, he sent the Cadet Battalion to the left, where they spent the day constructing breastworks parallel to and east of the tracks. The

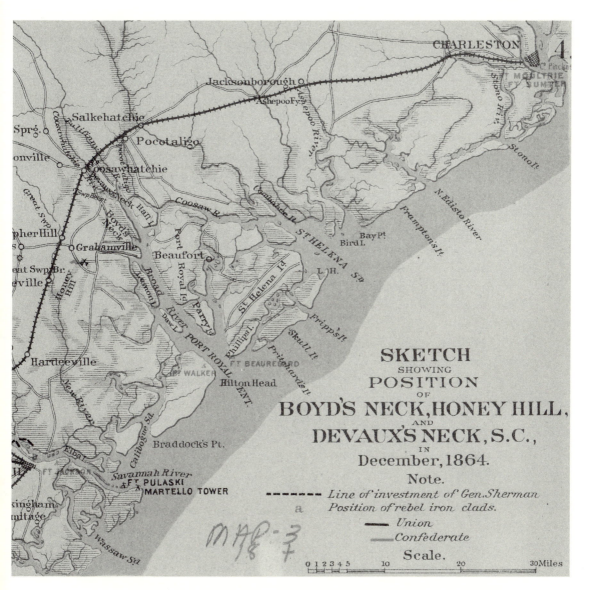

Union navy movements in the vicinity of DeVeaux's (Gregorie's) Neck, December 6–9, 1864. From *Sketch Showing Position of Boyd's Neck, Honey Hill, and Devaux's Neck, S.C. in December, 1864*, detail; courtesy of South Caroliniana Library, University of South Carolina

work continued all along the line throughout the rest of the day and into the next morning.

As the cadets dug in, a train crept slowly down the track, depositing soldiers at regular intervals along the embankment. One by one, they were dropped off, their eyes scanning the woods in their front for any sign of a blue uniform. Many of these were boys of sixteen or seventeen, whose predominant thoughts likely were not on the railroad they were defending or of sustaining the Confederacy. Most were probably like Private George M. Coffin of the Arsenal Cadets, who remembered "standing for hours . . . watching for enemy sharpshooters in the tall pines and thinking that I was only seventeen and did not want to die yet." These youthful pickets stood their posts the entire night but never encountered a single Yankee soldier.[22]

Less than a mile away Confederate pickets could hear the sounds made by spades and axes coming from the Union camp as the Federals spent the day fortifying along the Old Pocotaligo Road. Rain fell intermittently on the Federal engineers as they surveyed the area, and the infantry built up the breastworks. The day before General Hatch had ordered the Federal position at Boyd's Neck nearly abandoned, leaving only one regiment to guard the works there. A small detachment of his force was now landing across the Tullifinny at Mackey's Point. Northern troops were now close enough to the railroad that they could easily hear the rumbling and whistling of the locomotives through the forest. Having been frustrated in their efforts to break the railroad, the Federals now focused their strategy on either destroying or commanding the railroad with their artillery. They also intended to keep as many Confederate soldiers in their front and as far away from Savannah and Sherman as possible.[23]

Word of the enemy's movement was also creating unrest in other parts of the lowcountry. Plantation overseer L. H. Boineau had retreated to the property that his employer, E. B. Heyward, referred to as the Pine Lands. As he wrote a letter to Heyward, Boineau could hear the echoes of heavy artillery fire from the direction of Boyd's Landing and Page's Point. Boineau was making plans for the evacuation of Combahee Plantation in the event that the portion of the state to his south and west was overrun. "We can not," he wrote, "be at any rest until Sherman is disposed of in some way."[24]

Because military communication almost completely monopolized the telegraph lines, the people of Charleston relied on whatever news could be gleaned from soldiers, travelers, and railroad employees. The passenger train from Savannah had been unable to pass on December 7, but a late-arriving freight train brought word of the fighting near Coosawhatchie and Pocotaligo. That night General Beauregard arrived in the city from Augusta, but he did not tarry long. As soon as Haines readied a special train, he was sped to the scene of the day's hostilities.[25]

The springlike temperatures of the previous week had given way to a cold front, and an icy wind greeted the general as he made his way from Pocotaligo Station to General Jones's headquarters. The two poured over maps of the area as they deciphered reports coming in from their subordinates. General Gartrell's dispatch stated that the Federals were entrenched behind a strong line of works on the state road where it intersected the Gregorie's Road a mile from Coosawhatchie. Though Gartrell did not think small artillery fire could dislodge the enemy, he thought he might accomplish it by attacking from the direction of the Tullifinny Bridge.[26] Beauregard agreed that the enemy's position was much too close to the railroad and felt that an attempt to drive them from Gregorie's Neck was imperative.

General Jones knew that such an endeavor would be difficult. The Union First Brigade held a strong position on a peninsula that was less than a mile and a half wide. The Tullifinny and Coosawhatchie rivers and their surrounding swamps protected the enemy's flanks. Jones agreed with Gartrell that the only practical way to unseat the enemy would be an attack on their rear from the Tullifinny side. He had the means to bridge the stream; however, concentrating enough troops for the task would be problematic. More than half of his command consisted of militia or reserve troops, and he had just sent the entire Augusta Battalion and all the men he could spare to Hardee in Savannah. Further complicating matters, the Yankees had landed two regiments at Mackey's Point, and he had to leave enough troops at Pocotaligo to monitor their activity. All Jones's planning and hypothesizing was put on hold the next morning, when Potter's brigade hit the left center of the Confederate line.

The morning of Friday, December 9, was cold and raw. The Confederate troops were up and trying to soak up every degree of warmth possible from their campfires when they heard the first blast of enemy artillery. At 9:00 a.m., the Federal batteries began a furious barrage to clear the woods in their front. After fifteen minutes the ten guns ceased firing, and Colonel William Silliman advanced with a line of skirmishers toward the portion of the Rebel line commanded by Colonel A. C. Edwards.

On the right of the Union line marched a battalion of U.S. Marines; several companies of the 157th New York took the center; and the 127th New York manned the left. Advancing through the trees, they were followed by the main line of battle, made up of the 144th and Fifty-sixth New York regiments, the 102nd U.S. Colored Troops, and the naval infantry. In all they numbered about 1,100. Behind them 500 woodsmen from the Twenty-fifth Ohio stood with axes ready to open a wider lane through the forest, exposing the railroad to the Federal batteries on the right.[27]

At about 10:00 A.M. the skirmishers encountered Confederate pickets in the woods about 350 yards from the railroad. After firing a few shots, the pickets fell back on their reserves. The troops under Colonel Edwards included the Thirty-second and Forty-seventh Georgia regiments, the Seventh North Carolina, and the South Carolina Cadet Battalion. As the pickets fell back, some of the older reservists began firing excitedly before the pickets had gotten back to the works. They had to lie down until the firing slackened, and then retreated to the safety of the railroad embankment.[28]

When the Yankees advanced into the clearing, "they were a pretty sight in brand new uniforms, with a colonel riding at their right wing."[29] The Confederates opened a heavy fire on them as the Blue Coats continued forward. Early in the fight, Colonel Silliman was severely wounded and command fell to Lieutenant Colonel Stewart Woodford of the 127th New York. Pressing his troops, he moved to within two hundred yards of the railroad, where his men became dangerously exposed to enfilading artillery fire from the right.

The guns of Captain William K. Bachman's battery of the German Light Artillery were positioned on the Confederate left and hit the Federals with grape and double-shotted canister. The marine battalion under Lieutenant George S. Stoddard approached the battery and tried to flank it but became mired in a thicket and came under a deadly fire. They were able to advance to within fifty yards of the prized rails, but were stalled by the muck, brambles, and canister. They held their forward position until the 157th New York and their reserves were forced back. The Rebels emerged from their works to pursue the retiring regiment and nearly cut off and captured the unsupported marine battalion. To avoid capture, they hastily retreated via the bank of the Tulifinny back to the crossroads.[30]

To their rear the noise of battle mixed with the sounds of axes and falling trees. As the fighting continued in front of them, the pioneer corps of the Twenty-fifth Ohio busily cleared a swath through the woods. Through midday and into the early afternoon the woodsmen kept up a grueling pace, even as the gunfire began to slacken. At the Federal camp the men of the Fifty-fourth Massachusetts were shoring up the rifle pits and constructing a battery for the guns that would eventually use the opening cut by the Ohio regiment. Potter's and Hatch's plans were slowly coming to fruition. Even if the brigade failed to cut the railroad, they would have the means to fire upon it easily from a fortified position.

By 3:00 P.M. the Ohioans had completed a fire lane five hundred yards long and between twenty and thirty yards wide. The entire Union line was about to be withdrawn in order when the troops heard a muffled roar coming from their left. Trying to capitalize on their defensive success of the morning, the Confederates had decided on a counterattack from their right. The men of the Fifth Georgia and the First and Third Georgia Reserves moved from Coosawhatchie, advancing with volleys of musketry and Rebel Yells that sounded to one Union officer like "a mobbish scream."[31]

As the Confederates charged, Colonel Nathaniel Haughton of the Twenty-fifth Ohio ordered his men into formation. The woodsmen sheathed their axes, unslung their rifles, and proceeded to repulse the onrushing Southerners. The Union line quickly reorganized and parried several attempts by Gartrell's men as they slowly moved back to the safety of their works. The action continued until about dark as the entire Federal force moved back into their trenches south of the road to Old Pocotaligo. Without sufficient numbers with which to charge a well-fortified enemy, the Confederates were recalled to a point almost a quarter mile forward of their position when the day began.

Though they had gained some ground, it proved to be of no real consequence. General Jones had been unable to concentrate enough experienced troops to oppose the Federals on Gregorie's Neck. The total losses in killed, wounded, and missing for the North were 629; for the South about 400. The Confederates had been unable to drive the enemy from the region, but they had been able to prevent them from breaking the vital railroad. Conversely, while the Yankees had not achieved their initial goal, they had succeeded in their "Plan B." They now held a fortified position from which their artillery could easily shell the locomotives traveling to and from Charleston and Savannah. They decided, at least for now, to abandon any further attempts on the railroad with infantry.

The special train carrying Confederate General P. G. T. Beauregard crossed the Savannah River and wheezed into Savannah at around 7:00 A.M. on December 9. "Old Bory" had received an admonition from General Samuel Cooper, who hoped the city could be defended but felt that "the effort should not be to protracted, to the sacrifice of the garrison. The same remarks are applicable to Charleston."[32] Generals Jones and Hardee had been unable to travel to Charleston; therefore, Beauregard had decided to come to them. He met with General Jones for several hours at Pocotaligo, but Hardee and the Savannah defenses required his presence more urgently. He had rested only briefly and then left for Savannah early that morning.

In Savannah, Beauregard was escorted to General Hardee's headquarters, where the two spent the rest of the day reviewing the situation at Coosawhatchie, the position of Sherman's army, and the defensive preparations in and around the city. As they spoke, one fact became glaringly obvious to both men. The fate of Savannah depended entirely on maintaining the integrity of the Charleston & Savannah Railroad.

Opposing Sherman's army of sixty thousand, Hardee commanded a "mongrel mass"—a term coined by his adversary—of ten thousand men of all arms. Confederate engineers under General Lafayette McLaws had selected his advanced line of defense with an eye toward retaining the Charleston & Savannah Railroad and protecting the company's bridge across the Savannah River. The line meandered southward from the bridge, keeping Monteith Swamp in its front. The left was protected by the Great Ogeechee Swamp. Detached fieldworks had been placed at salient points and batteries of light artillery had been hurried into place. Hardee also employed the navy, ordering the CSS *Macon* and the CSS *Sampson* upriver to protect the railroad bridge and to discourage the Union left wing from crossing into South Carolina.

On December 8, Hardee's cavalry commander, General Joseph Wheeler, intercepted an enemy dispatch that revealed Sherman's intention of surrounding Savannah. No longer worried about being cut off from the northern bank of the river, Hardee began to move all available men into the trenches on the west side of the city. He also began to strengthen his main defensive line about three miles from town. This line of works began at Williamson's plantation on the Savannah River and extended southward along a rim of high ground to a point crossing the Central Railroad. From there it ran through farmland, along the borders of rice fields and marshes, and ended at the Atlantic & Gulf Railroad bridge over the Little Ogeechee. The line arced over a span of ten to fifteen miles and stretched the troops so thin that they "amounted to little more than a skirmish line strengthened at intervals."[33]

General William T. Sherman's Union army seemed to be gaining momentum as they marched virtually unopposed toward the coast. Just days before the Federal right wing under Major General Oliver O. Howard had crossed the Ogeechee and forced back several Confederate outposts. Now the left wing lay only six miles from Hardee's intermediary line protecting the railroad. This force, commanded by Major General Henry W. Slocum, was marching in three separate columns down Middle Ground Road, River Road, and the Savannah & Augusta Railroad.

Major Alfred F. Hartridge had been ordered by General Hardee on November 29 to take command of the Confederate line at Monteith. His force consisted of 250 men from the Twenty-seventh Battalion, Georgia Volunteers, 350 men of General Lawrence S. Baker's North Carolina battalion, 150 men from Howard's local battalion, and Captain Abel's battery of two Napoleon guns and two howitzers. Across his lines, which extended from the Charleston & Savannah Railroad bridge to Colonel George P. Harrison's plantation, ran the three most direct routes between Sherman's army and Savannah. Confederate engineers had thought the swamp impassable; however, the major had been able to ride through it in several places. The other glaring weakness was that his only earthworks had been erected at the intersection of the three roads. The position was certainly not as strong as advertised, but Hartridge had been ordered to hold the line as long as he could and was determined to make the best of it.[34]

He set to work immediately to find ways to impede the enemy's advance. With black laborers on loan from local planters, he had trees felled across the roads leading through the swamp. All along their front, bridges were destroyed and rice fields flooded. He extended his line by placing the Twenty-seventh Georgia Battalion on the Augusta Road, Baker's North Carolinians on the road through Harrison's plantation, and Howard's battalion in the center. He divided Abel's battery, sending Baker's and Howard's troops each a howitzer, and placing the two Napoleons on the Augusta Road.[35] Hartridge also had several earthworks constructed, but had neither the time nor the manpower to make the line as strong as it needed to be.

Between the Little Ogeechee and Savannah rivers the ground was mostly swampland and rice fields with stretches of flatlands in between. Running through the area were many canals that were used to flood the rice lands. To some Union commanders the terrain would have seemed formidable, but not General Sherman. He knew the strengths of the Confederate defenses and its commanders; however, he also knew the weaknesses. Nothing—neither the swamps nor the strong Rebel positions —would deter him from his objective.

The day before Beauregard's arrival in Savannah, lead elements of the Federal XX Corps were encamped at Ebenezer Church, just a few miles south of Springfield. At 8:30 on the morning of the December 9, Brigadier General N. J. Jackson's division broke camp and began their march down the road to Monteith. Where the road passed through it, Monteith Swamp was two miles wide. Felled pine trees littered the entire width of the road, hampering the advance of the Union infantry. As they made their way through the obstructions and approached the opposite side, they made their first contact with the enemy at around noon.[36]

On the eastern side of Turkey Roost Swamp, Hartridge's troops had constructed two redoubts with flanking rifle pits. The forward work contained a single howitzer that commanded the road and prevented removal of the fallen timber. When the Confederates opened fire, the Federals threw out several regiments on each flank while the First Brigade under Colonel James Selfridge held the center. To the right of the road was an expansive rice field, which extended to the left of the enemy works. The Rebels' gun completely swept the road, blocking the most practical passage of the swamp.[37]

While Selfridge's brigade demonstrated in front, General Jackson deployed Colonel Ezra Carman's Second Brigade to the right of the road and ordered the Third Brigade under Colonel James S. Robinson to the left. While Carman had difficulty finding suitable ground, Colonel Robinson had better luck on the other flank. Feeling his way on the left, he found a point about a mile from the main road where the swamp was only a quarter mile wide, and decided to cross there.[38]

The Thirty-first Wisconsin and Sixty-first Ohio Volunteers led the way, commanded by Colonel Francis H. West of the Wisconsin regiment. As they entered the swamp, the officers were forced to dismount because their horses could not maneuver in the bog. They negotiated the four hundred yards of marsh slowly but peacefully until they came within the range of the Rebel howitzer. As they shells came crashing through the trees, the Federals had one of two choices—to move forward or to lie down in two feet of slimy, cold water. Choosing the former option, the Union commanders ordered their men forward at the double-quick, out of the swamp, and into an open field bordered by woods in which the Confederates took cover.[39]

To the right of the road Carman's brigade had trouble finding a point from which to launch an assault. He had formed the Third Wisconsin and 115th New York in two lines and prepared to send them out as skirmishers when Colonel West's troops emerged from the woods on the far left. They watched with surprise as their comrades, after an exchange of only two or three volleys, drove the Rebels from their position in a frantic retreat. The Confederates in that area were commanded by Major Samuel L. Black, who had been detached from General Hardee's staff to aid Major Hartridge. The Federals struck his line in the right and rear, driving back the North Carolina Battalion and Howard's battalion. In their hasty withdrawal the Southern troops left behind clothing, arms, and other supplies but were able to carry off their artillery pieces. Major Black fell back until he reached the railroad at Monteith Station while the Federals halted and went into camp near the redoubts at Harrison's plantation.[40]

Even before Beauregard had left Savannah, the intermediary line of defense was crumbling before Hardee's eyes. An entire Union corps was closing in on the thinly spread line of Confederate defenses, which were at little more than brigade strength.

The Yankees had been able to remove the obstructions easily, cross a swamp previously deemed to be impassable, and flank Hardee's detached outer works. After Major Black's troops were driven in, Hardee ordered Major Hartridge to withdraw toward the river, where gunboats would take him back to the city. Not wishing to abandon his guns and knowing that the Yankees rarely attacked with darkness approaching, he held his position on the Augusta Road until nightfall. At 8:00 he and the Twenty-seventh Georgia withdrew to Monteith Station, where he received orders to hold the railroad until General Beauregard's train passed through on its way to Charleston.[41] After that, he was to fall back to the main defensive line four miles from Savannah and join the command of General Gustavus W. Smith.

As Beauregard and Hardee assessed the situation that evening, prospects for the city's defense looked grim at best. Hardee had only thirty days of provisions left for his army; his supply line was in danger; and his men were outnumbered five to one. Hardee expected the enemy to reach the front of his lines by the next day and gave orders for all extra trains on the Atlantic & Gulf and Charleston & Savannah railroads sent into South Carolina as soon as possible. Before leaving General Beauregard asked Hardee about his plans for evacuation and was unpleasantly surprised at Hardee's answer. At present Hardee had no concrete plans for evacuating, since he planned to rely on his gunboats in Savannah's harbor to ferry his troops to South Carolina if necessary.

Somewhat annoyed by this oversight, Beauregard instructed him to begin immediate construction of pontoon bridges across the river. He also cautioned Hardee on the importance of maintaining his communication with South Carolina via Screven's Causeway. His final message to Hardee was quite clear: "It is my desire . . . that you shall hold this city as in your judgment it may be advisable to do, bearing in mind that, should you decide between a sacrifice of the garrison and city, you will preserve the garrison for operation elsewhere." As General Beauregard prepared to leave the city, he received intelligence that the Union army had approached to within three miles of the railroad at Monteith. Deciding that his train would be in danger, he decided to ascend the river on a steamboat and board the train at the bridge. Accompanied by several aides, he finally arrived at the bridge at 1:00 A.M., safely boarded the train, and was off to Pocotaligo to confer briefly with General Jones. By 5:00 P.M. on the evening of the December 10, he was safely back at his headquarters in Charleston.[42]

As Sherman's army carved its path of destruction through the heart of Georgia, it laid waste to the state's economic resources. It was the railroads, however, that seemed to attract special attention from the Northern invaders. From Atlanta to Savannah tracks were torn up; bridges and depots were razed; and rolling stock was destroyed. Sherman had destroyed about 140 miles of the Georgia Central Railroad and had taken up approximately 300 miles of track in the state, leaving nothing behind but the embankments.[43] In fact Sherman's chief engineer, Captain Orlando M.

Sherman's troops destroying the rails, photograph by George N. Barnard. Courtesy of Library of Congress, lot number 4164A

Poe, had invented a machine whose sole purpose was twisting the rails. This invention was soon employed on the tracks of the Charleston & Savannah line.

When general call sounded on the morning of December 10, little stood between the Union left wing and the railroad. As they began their march, General Jackson's division again took the lead. Setting out at 7:00 A.M., they covered the distance to the railroad in just three hours. As column after column of the XX Corps arrived at Monteith, they began dismantling the railroad. Their efforts were assisted by a battalion of the First Michigan Engineers and Mechanics, which had been detached from another corps. After the rails were dislodged, they were heated over fires made from burning cross ties. Using Captain Poe's device, troops placed a hooked clamp under the rails and harnessed a horse to an oblong link at the opposite end. When force was applied by the horse, it gave a rotating motion—similar to the log-turning device in a sawmill—and the rail was dislodged. Many of the still-hot rails were twisted around tree trunks to form Sherman's signature "neckties," or warped to the point of being useless except as scrap iron.[44]

Tearing up the tracks was hot work, but the Federal soldiers took to it enthusiastically, each regiment working on a separate segment. With the other units urging them on, the Eighty-second Ohio set fire to the stationhouse at Monteith, cheering with gusto as it was consumed. They soon resumed their rail-twisting efforts, toiling well into the afternoon. As they completed their portion of the work, the XX Corps was ordered down the line toward its junction with the Georgia Central. Here, the troops dug in and prepared for siege operations against Savannah.[45]

As the other units invested the city, the XIV Corps took over the railroad demolition between Monteith and the river. Corps commander General J. C. Davis ordered Brigadier General Absalom Baird of the Third Division to take his men and destroy the railroad bridge across the Savannah River. On the morning of December 11, Baird advanced on the bridge but was stymied by the wetlands bordering the river. As his men halted, he sent scouts to reconnoiter the area, and their reports disappointed him.

The swamp in his front was impassable except by bridges, which would take days to build even if the Federals could obtain the necessary materials. An extensive line of trestlework, towering fifteen feet above the swamp, led to the river. A gun emplacement occupied by a small Confederate battery guarded the far end of the trestle, which spanned such a distance that the Rebels were out of range of Baird's Napoleon guns. Although the Confederate battery could not reach the Federal position, its guns swept the trestle, which was the only approach to the bridge.

Wanting to make the best decision, Baird rode forward to view the ground for himself. Through his field glasses, he surveyed the trestle and the Rebel battery. Near the battery, he saw a locomotive and a platform car equipped with an artillery piece—a primitive form of today's armored vehicle. Colonel Ambrosio J. Gonzales had first introduced the idea of mobile artillery batteries to the Confederate Army in the latter part of 1861. At that time Gonzales had been assisting with the reinforcement of Charleston's defenses and proposed the construction of railroad tracks running behind, and parallel to, the eastern line of defensive works on James Island. Along this track, locomotives or horses could convey a battery of artillery to any point desired on platform cars that could be moved either together or separately. Although it was an intriguing idea, it was deemed impractical at the time and was never carried out.[46]

Colonel Gonzales revived his idea on November 10, 1864, when he proposed to General Hardee that he employ a similar strategy in defense of the country between Charleston and Savannah. Gonzales negotiated with Charleston & Savannah Railroad president R. L. Singletary for the use of two locomotives and suggested that the unit be comprised of two batteries and two thousand men taken from the Washington Artillery and Captain Hal Stuart's Beaufort Volunteer Artillery. Guns and ammunition were to be kept on board the cars, and special ramps would facilitate unloading at any threatened point on the line. He would station the "flying artillery" column at Green Pond, from which point the batteries would be thirty minutes to two hours by rail from any of the points most vulnerable to a large enemy landing force between the port cities. During General Sherman's envelopment of Savannah, his troops "marveled at the Confederate battery that had been mounted on a railroad car, it was moved from one front to another with celerity, firing accurately into exposed groups of men and generally harassing the Union army."[47] Soon, the operations of the flying artillery west of the Savannah River would come to an end.

Through his glasses General Baird could see the South Carolina side of the river, where another small train sat idly, but he could not see the Savannah River Bridge itself. The only approach to the bridge was by a railroad track under the constant fire of the Confederate guns. It would be impossible to get near the river from his current position without risking heavy losses unless he could find a way through the swamps—and he was assured that there was none. Losing daylight with each passing minute, Baird decided to make the best of a bad situation. He ordered his men to burn as much of the trestlework as they could without exposing themselves to the enemy guns. In addition they took up three-quarters of a mile of the track before retiring. If the Federals hoped to use the trestle for foot passage in the future, they would have to build over the burned portion on top of the piles.[48]

The possibility that Sherman would use the bridge for a large-scale crossing was of great concern to General Hardee and his staff. With the Union army closing in around Savannah, the passage of a significant enemy force into South Carolina would easily cut off Hardee's line of retreat. Union troops had already destroyed much of the railroad between the river and the city, and the bridge really could be of no use to anyone but his enemy. From Hardee's point of view the decision was clear, and on the morning of December 11 he sent a dispatch to Flag Officer W. W. Hunter of the Confederate navy, ordering him to "effectually destroy the railroad bridge over the Savannah River by burning or cutting it."[49]

Hunter ascended the river with a small squadron, which included the flagship CSS *Sampson*, the gunboat CSS *Macon*, and a small transport steamer, the CSS *Resolute*, loaded with supplies. They began preparations for the bridge's destruction but had to wait on the withdrawal of the Confederate outpost, now engaged in monitoring the actions of General Baird's troops on the trestle. Later that evening, after the Confederate battery and locomotive were safely on the South Carolina side of the river, Flag Officer Hunter set fire to the bridge that he had so dutifully defended.

After completing this task, Hunter had orders to return to Savannah to provide naval support for Hardee's right flank near Williamson's plantation. He got underway the next morning, and as he approached Argyle Island, he began to see smoking ruins on the Georgia bank. Proceeding further downstream, he was surprised to see that the enemy had established a battery on a high bluff near the mill at Colerain Plantation. As they boldly tried to pass, the rifled guns of Winegar's battery, First New York Light Artillery, opened up on the Rebel steamers and had soon damaged all three. Hunter ordered the vessels to turn about and head upriver to safety, but the *Resolute* collided with one of her sister ships and became disabled. The immobilized craft and her crew drifted ashore at Argyle Island and were captured by troops of the Third Wisconsin Volunteers, while the *Sampson* and *Macon* fled back upstream. Because of the high water level, they were able to pass the obstructions in the river and escape to Augusta, where they spent the remainder of the war.[50]

The interruption of communication with Savannah and the suppression of all military dispatches reaching Charleston left Charlestonians in the dark as to the military happenings in and around Savannah. The *Mercury* assured its readers that all was quiet at Coosawhatchie except for long-range shelling of the railroad. General Hardee was reported to have fallen back to the junction of the Charleston & Savannah and Georgia Central railroads, but no further information was available. "For the present," the paper stated, "the trains will cease to run through between the two cities."[51] The *Courier* was more forthcoming about the trouble in Georgia, briefly recounting the fighting at Monteith and reporting that the railroad had been torn up between there and its junction with the Central.[52]

The railroad had proven too vulnerable on the Georgia side of the river, but that was not General Hardee's fault. For the better part of three years, the Confederates had focused their efforts in and around Savannah on preventing an assault from the sea. They now had neither the time nor the manpower to counter a massive attack from the interior. Even in South Carolina, where the road had survived numerous amphibious expeditions, it is doubtful that the coastal defenses would have survived such numbers. Hardee tried to defend the railroad, but its bridge crossed the Savannah so far from the city that his network of fortifications could not adequately guard it. His troops were already stretched so thin that he was left with little choice but to pull back the scattered units of his intermediate line. By withdrawing Major Hartridge's command guarding Monteith, he had essentially sealed the fate of the Charleston & Savannah Railroad. Left with no other alternative, he had also sealed the fate of the city of Savannah.

SIXTEEN

Running the Gauntlet

As the Charleston-bound locomotive passed through Coosawhatchie, its driver inched the train forward in an attempt to escape detection by the Union artillerists a half mile away. At the first shot from the Federal guns the engineer ran at full throttle on a harrowing six-minute jaunt as shells exploded all around him. His eyes focused on the tracks in front of him, but all eyes in the Rebel earthworks were transfixed on the train in what had become a morbid sort of entertainment for the troops. Confederate soldiers dubbed the tracks between Coosawhatchie and Pocotaligo the "gauntlet," but to the railroaders the two-mile stretch seemed to be the longest segment on the line.

Ever since the Federals under General John P. Hatch had established batteries on Gregorie's Neck, their rifled Parrot guns had the railroad in range from a half mile away and routinely lobbed shells onto the tracks and into the Rebel camps alongside them. Hoping to trap as much rolling stock as possible west of Coosawhatchie, the Federals allowed trains headed in the direction of Savannah to pass without incident. Those traveling east, however, were under a steady fire over the entire range. Often the trains passed with the engineer and fireman hanging out of the off side of the engine to avoid harm. Robert Herriott of Captain William K. Bachman's battery, which was stationed along the railroad, observed, "It was very exciting, especially at night, each train passing along under fire for two miles, while the cross ties often were struck together with the trees near the track."[1]

General Foster was pleased that General Hatch had stepped up his bombardment of the railroad, but Foster was frustrated by what little damage it had inflicted. He wanted Hatch to destroy the railroad by force and suggested that he cross the Tullifinny and strike a point near Pocotaligo. If Hatch could not accomplish this, Foster expected him to secure a position from which he could destroy any train attempting to pass. The Federal batteries were prevented from getting any closer to the tracks, however, by a Confederate battery situated four hundred yards in its front. Even from this distance the Yankee guns were able to cut the rails in several places, but Rebel work crews quickly repaired this damage during the night. At no time was transportation on the railroad interrupted for more than a few hours.

The only train damaged by the Union artillerists was the *Isundiga*, which was hit by a ten-pound shell on the night of December 13, 1864. The shot struck the train on the right side of the boiler just in front of the engineer's seat, putting it out of commission. General Hatch reported that the explosion killed an engineer and fireman on the train, but his intelligence was faulty. Nonetheless the persistent barrage adversely affected the morale of railroad employees, who were becoming increasingly displeased about having to pass the enemy batteries. Despite their protests, running the gauntlet continued as long as the Confederates had troops to supply at the end of the line.[2]

President R. L. Singletary and Superintendent H. S. Haines of the Charleston & Savannah Railroad had foreseen the catastrophe in Georgia, and their quick action prevented the loss of a significant portion of the company's rolling stock. With Sherman's approach, they received permission from Confederate military authorities to remove their rolling stock to the east bank of the Savannah River. All their property was safely in South Carolina when the bridge was destroyed. Now, with not only their cars but the lives of their employees threatened, they wished to withdraw their trains east of the Coosawhatchie River to prevent their loss if that bridge were destroyed. This time their request was refused, and they were issued a specific order to keep the line open to Hardeeville at all hazards.[3]

Commercial activity on the railroad was nonexistent, as evidenced by the quantities of rice and other goods that sat rotting at way stations along the line. Haines, Singletary, and the Confederate army were managing the affairs of the road with little input from the board of directors. In fact, when the monthly meeting was held in December, only one director, Charles M. Furman, was able to attend. Still these men were able to keep the trains running with no loss of life. Even more remarkable was their ability to coordinate the evacuation of other companies' trains as well as their own with little to no loss of rolling stock. Their service was indispensable to the troops headquartered in Hardeeville, whose job it was to guard the Savannah River and keep open the line of communication between Jones's and Hardee's commands.[4]

Patrolling the bank on the South Carolina side of the river was a division of cavalry under Major General Pierce M. B. Young, who had recently returned to the theater from the Army of Northern Virginia. On December 12 he confirmed to General Jones that the railroad bridge had been destroyed and that he had posted a guard of fifty men near what remained of the structure. They had been on full alert since a small party of Yankees landed at Heyward's plantation, but after burning some houses in the area, the Federal troops had withdrawn without a skirmish.

As Sherman's troops continued to envelop Savannah, General Hardee's only line of communication with Charleston was via Screven's Ferry and then the Charleston & Savannah Railroad. Young sensed that Sherman knew this and would do all in his power to disrupt it. Hardee needed all the troops that could be spared from Charleston and Pocotaligo as well as a few more pieces of artillery. Late in the evening of

December 12, he asked General Jones to "urge upon the railroad men the necessity of keeping at least two trains this side of Coosawhatchie."[5]

The next day, 450 Confederate reinforcements were rushed by rail from Charleston either to be sent directly to Savannah or to be detained by General Young if they were needed east of the river. Later that day a Federal force under Brigadier General William B. Hazen took Fort McAllister, removing the last obstacle between Sherman and the Union navy. General Young had also noted movement from the Union left and was planning a demonstration against a Federal unit on Argyle Island. With the situation on Hardee's flanks deteriorating, General Beauregard warned Hardee that under no circumstances should he be cut off from joining forces with General Jones. On the fifteenth, Beauregard sent Hardee a brigade of dismounted cavalry under General Samuel Ferguson—the last reinforcements that would, or could, be sent.[6]

Hardee was increasingly concerned that Sherman would send a sizeable force across the Savannah River on rice flats to threaten his line of communication. He decided on December 16 to call his lieutenants together for a council of war, where he asked General Joseph Wheeler to cross into South Carolina and set up headquarters near Hardeeville. Known throughout the army as "Fighting Joe," the native Augustan was charged with defending the line east of Screven's Ferry Causeway and along the New River to Hardeeville. To further hinder the enemy, rice fields in the vicinity were ordered overflowed and roads near the river obstructed. Hardee also sent Jones an earnest communiqué, which read: "Our occupation of Savannah depends on your ability to hold the railroad. Whenever you are unable to hold it, I must evacuate."[7]

The reality was that evacuation was already necessary. With the fall of Fort McAllister, General Sherman had gained complete control of the Ogeechee River, giving him free communication with his navy and a secure supply line. Time was of the essence for the Confederate garrison and its harried commander. General Hardee, met with his subordinates, almost all of whom agreed that they should evacuate Savannah as soon as pontoon bridges were constructed. Hardee ordered General Young to gather as many rice flats as possible from plantations along the Savannah River and delegated the bridge construction to the able hands of Colonel J. G. Clarke and Lieutenant Colonel B. W. Frobel.[8]

After assembling a combined detail of Georgia militia and sailors, the Confederates began construction on pontoon bridges that extended a thousand feet from the end of West Broad Street to Hutchinson Island. A plank road was laid across the island, and a second pontoon bridge was to be built from there to Pennyworth Island. Another road would cross that island, and a third pontoon bridge would cross Back River to the South Carolina shore near Screven's Ferry. Wood planking for the bridges was taken from wharves on the Savannah waterfront. General Young's cavalry handed over the rice flats they had collected to Frobel, who had them lashed together with ropes and stringers and anchored into place with the wheels of railroad

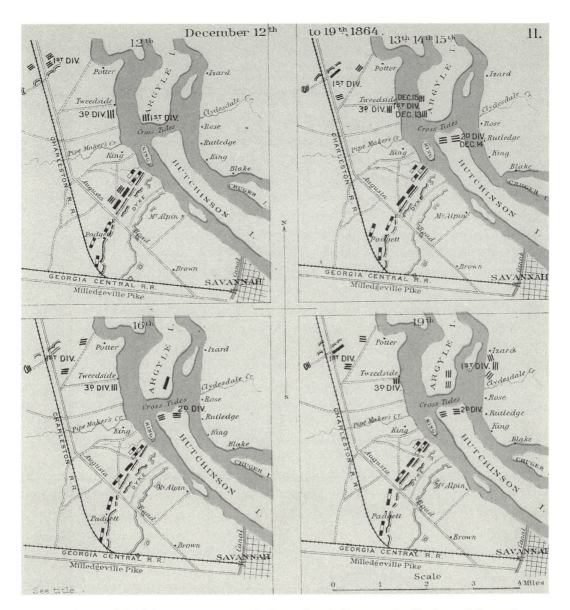

Union and Confederate positions outside Savannah and adjacent to the Charleston & Savannah Railroad, December 12–19, 1864. Courtesy of South Carolina Historical Society

cars, probably from the shops of the Georgia Central Railroad. By the evening of December 17, the first bridge to Hutchinson Island was complete.[9]

On the night of December 16, General Beauregard arrived in Savannah to help supervise the preparations. The next day, he received word that the War Department in Richmond approved of his intention to sacrifice the city in order to keep its garrison intact. His greatest concern was that Hardee might not have enough time to save most of the public property and destroy what would likely fall into enemy hands. Assured that construction of the pontoon bridges was progressing nicely,

Beauregard turned his focus to strengthening the line of retreat. He ordered Hardee to construct earthworks at Turnbridge and Morgan's landings east of Screven's Ferry and ordered creeks leading to the landings obstructed with pilings or torpedoes. He then penned a memorandum advising Hardee on the disposition of his troops once they were safely out of the city.[10]

While the Confederates planned their evacuation, Sherman's troops were busy preparing for a full-scale assault on the city. On December 14 Sherman held a shipboard conference with General John Gray Foster, who had come over from Hilton Head, and Admiral John Dahlgren. The men agreed that Dahlgren would provide boats suitable for navigating the Great Ogeechee and increase his shelling of the Confederate batteries along the river. Foster agreed to send Sherman siege guns from Hilton Head to aid in the bombardment of the city and would have Hatch press his movement on the Charleston & Savannah Railroad. After a couple of days' thought, Sherman cooled on the idea of an attack. In a December 16 dispatch to General Grant, Sherman optimistically informed his commander of his plan to ask for Hardee's surrender. "If General Hardee is alarmed, or fears starvation," he wrote, "he may surrender; otherwise, I will bombard the city, but not risk the lives of our men by assaults across the narrow causeways by which alone I can now reach it."[11]

On December 17 Sherman composed a letter to General Hardee demanding the surrender of Savannah and its defensive works. He gave Hardee a synopsis of his obvious predicament and offered liberal terms to the inhabitants and the garrison. After reviewing the document with General Beauregard, Hardee sent his reply later that same day, politely refusing. Sherman immediately had his men resume preparation for an attack, tentatively set for December 21. The Federal troops were soon occupied with building scaling ladders, foot bridges, and rafts with which to storm the Rebel defenses. Within several hours, however, Sherman reconsidered, reasoning that "the ground was difficult, and as all former assaults had proven so bloody, I concluded to make one more effort to completely surround Savannah . . . in case of success, to capture his whole army."[12] There would be no great attack until he had exhausted all his other options.

For the moment Sherman chose to concentrate on the common link to Hardee's last remaining line of supply and retreat—the railroad at Coosawhatchie. If the railroad and the telegraph wire could be cut in the direction of Charleston, it might bring swift resolution to his efforts at Savannah, but Hatch's advance seemed to be stagnating. Sherman suggested to General Foster that the Confederate force in the area might not be as strong as it appeared and that a bold thrust from the south could break the road. If not, a large demonstration at Coosawhatchie might make it possible for a small detachment to cross the Tullifinny and accomplish the task. Regardless of what happened, General Sherman wanted more activity along that front. In a December 18 dispatch he told Foster that even if nothing else could be done, "let them whale away with their 30-pounder Parrotts and break the road with cannon balls."[13]

On the night of December 19 the Thirty-third U.S. Colored Troops and the Fifty-fourth Massachusetts Regiment marched to landings on Gregorie's Neck, where they boarded transports, sailed up the Tullifinny a short distance, and disembarked at Mackay's Neck. It was a cold, rainy night, and the two black regiments slogged two miles through the muck until they reached the high ground on General Micah Jenkins's plantation. Also in the vicinity were the deserted plantations of the Hutson, Mason, Stuart, and Howard families; in their immediate front were creeks, wetlands, and dense woods. They stacked their arms and bivouacked for the night near the encampment of the Twenty-sixth U.S. Colored Troops.

Because of the steady rain, the Federal march was delayed until the following afternoon. At about 3:30 P.M. Lieutenant Colonel Charles T. Trowbridge and three hundred men of the Thirty-third U.S. Colored Troops made a reconnaissance, supported by two companies of the Fifty-fourth Massachusetts. They marched up the right bank of the Tullifinny two miles in advance of their picket line and then turned east. They planned to hit the Pocotaligo Road a mile beyond Stuart's plantation but came upon a force of mounted pickets stationed between the river and the road.

After halting briefly, Trowbridge sent two companies under Major H. A. Whitney to try to flank the Rebel pickets, but they were discovered and came under a heavy fire. When Whitney's men returned fire, the vedettes—likely a detachment of the Third South Carolina Cavalry—immediately fell back to their reserves, dismounted, and formed their battle lines. Unable to get between their adversaries and the swamp, Whitney pulled his companies back. When they rejoined the main force, Trowbridge formed ranks and charged his command across the open field and into the woods, driving the Rebels back toward the railroad.

Because of the lateness of the hour, the Federals did not press their advantage, and instead decided to return to camp. From his observations Lieutenant Colonel Trowbridge deemed passage through the swamps impracticable except to light troops. He believed that the only way to reach the railroad in force would be by way of the Pocotaligo Road. Colonel Edward Needles Hallowell had enough troops in the vicinity for a strike on the railroad and might have attempted one had not the Confederates evacuated Savannah that night.[14]

The same day that the two regiments of the Coast Division crossed the Tullifinny, Colonel Ezra Carman's brigade crossed the Savannah River from Argyle Island and made a lodgment near Izard's mill. They entrenched on a line between Clydesdale Creek and Izard's plantation but were unable to advance further because of flooded rice fields and the pesky Confederate cavalry. Several days earlier, General Henry W. Slocum had wanted to send an entire corps across the river to entrench and then feel their way toward Union Causeway. Not wanting to risk such a large force, Sherman instructed him to send only a brigade.[15]

Examining his maps, General Sherman thought Hatch's division might be in a better position to move to Bluffton and block Hardee's retreat on the causeway road. Instead of having Slocum move south, Sherman wanted Slocum to head eastward and threaten to flank the movement of any troops attempting to flee from Savannah. Still hoping to trap Hardee in the city, he entertained the thought of a joint army-navy operation near Bluffton, and sailed out of Wassaw Sound to Hilton Head to confer with Admiral Dahlgren and General Foster about his plans. Expecting to be gone for at least two days, he asked Slocum not to take any offensive action until he returned.[16]

General Beauregard wired his superiors in Richmond that a Federal force of 1,500 had crossed into South Carolina, and General Wheeler's cavalry was the only force in the area with which to oppose it. He petitioned President Davis for Hoke's or Johnson's division to be sent to the region until General John Bell Hood's army could arrive. Davis refused, basing his decision on General Lee's fear that such a depletion of his force would jeopardize the security of the Confederate capital.[17] Realistically, however, the already overburdened Confederate railroads could not have executed the movement of two whole divisions in the short amount of time in which they were needed. With Carman entrenched only three miles from his line of communication and the avenue of his retreat becoming smaller every day, General Hardee knew that he had to act now or lose his entire force.

Hardee made sure that the pontoon bridges were nearly finished and prepared to vacate the city. To insure against a threat from Carman's brigade, he sent General Wheeler an additional 200 of Ferguson's dismounted cavalry, 450 infantry, and six artillery pieces. He also strengthened his skirmish lines on the right. While the Confederates evacuated Whitemarsh Island and the heavy guns there were spiked, Hardee had his guns elsewhere ramp up their bombardment of Federal positions. At about 9:00 P.M. on December 19, he received word that the bridges and their connections were complete.

As morning dawned on December 20, 1864, the Confederates resumed their barrage. The artillerists under Colonel Charles Colcock Jones fired at will, knowing that they would likely dump any unused ammunition into the river that night. With the pontoon bridge completed the civilian population of Savannah began to pour across it in wagons and carriages and on foot. To protect the refugees and to protect his flank, Hardee had the gunboat *Savannah* sent upstream to shell Carman's position at Izard's. The ship was to cover the withdrawal, protect the rearguard for two days, and then head out to sea.[18]

During the afternoon Lieutenant Colonel Frobel went to the front to reconnoiter the Union lines. Detecting no enemy movement, he returned to the city, content that everything was going as planned. On his approach he stopped for a moment to watch the long line of wagons and ambulances creeping shakily across the series of pontoon bridges and plank roads to the South Carolina mainland. Behind them more than four dozen pieces of light artillery were rolled by hand to the waterfront to await their appointed time to cross.

As soon as darkness fell, the troops began to leave the outer works and file through the city. The three divisions were withdrawn in order with those having

Evacuation of Savannah (December 1864). From *Harper's Weekly,* January 21, 1865

the greatest distance to march leaving first. From the far left General Ambrose R. Wright pulled his division back at 8:00, followed by McLaws's division at 10:00. As the troops slipped away, they spiked their heavy guns and destroyed their remaining ammunition as quietly as possible. General Gustavus W. Smith's division served as the rearguard, and his skirmishers kept up a scattered fire until they left their works at midnight. His column arrived at the river at 1:00 A.M. and in an hour had cleared the first set of rice flats onto Hutchinson Island.[19]

It was a joyless march, carried out for the most part in silence. Large quantities of rice straw had been spread onto the bridges to muffle the sounds of the retreat, but a low rumble echoed from the bluffs and could be heard by some of the Federal troops upriver. A strong wind fanned the flames in Savannah's shipyard as many Confederate ships were set ablaze. As one Confederate soldier described the scene, "The constant tread of the troops and the rumblings of the artillery as they poured over those long floating bridges was a sad sound, and by the glare of the large fires at the east of the bridge it seemed like an immense funeral procession stealing out of the city in the dead of night." To Sergeant W. H. Andrews of the First Georgia Regulars, the fires were "sad to look at, but at the same time made a beautiful picture on the water."[20]

When the last of Smith's division passed them, Clarke and Frobel mounted up and made their way to the bridge. As they did, a rocket shot into the air behind them, followed by a second—the signal that the evacuation was complete. Clarke and the other engineers were responsible for the destruction of their recent creations, and as a final precaution, they punched holes in the bottoms of the rice flats as they set the makeshift bridges adrift. There was now no one left in Savannah except stragglers, deserters, and those who were too sick or injured to travel.[21]

The march continued throughout the next morning along Huger's Causeway instead of Union Causeway, which Sherman had assumed they would take. The Confederates' route ran northward through the rice fields to Hardeeville—roughly approximating the path of today's U.S. Highway 17—and was barely wide enough to accommodate artillery and wagons. Adding to their misery, a bitter cold rain doused the area, dampening already flagging spirits and saturating the ground, creating a soggy mess. Some of the younger troops began firing their guns to prevent wet powder from clogging their barrels, and tempers flared as the officers tried to stop them. As the day wore on, the edgy, exhausted troops made their way into Hardeeville and bivouacked near the Charleston & Savannah Railroad depot.[22]

The first troops to arrive at the railroad in Hardeeville climbed the water tanks, smashed through a thick layer of ice, and formed bucket relays to pass water to the idling locomotives. Hardee wanted the trains to be ready to move quickly if needed, but he found the amount of rolling stock entirely inadequate for his purposes. General Beauregard bristled at the oversight, placing the blame on the local commanders. He immediately fired off a dispatch to Lieutenant Colonel John M. Otey, informing him that the troops were "almost unprovided with transportation." He wanted Otey to contact the railroad agents and impress sufficient rolling stock for three thousand men at Pocotaligo, three thousand in the Fourth Subdistrict, and six thousand at Charleston. He also demanded that there be three full trains kept on the road from Coosawhatchie to Hardeeville and three or four near Pocotaligo to await troops bound for the Fourth District and Charleston. Superintendent Haines knew that he was not able to supply the needed cars from his stock, and he quickly telegraphed Singletary, asking that he speak to General Beauregard about obtaining them elsewhere. While others tried to iron out the transportation issues, Hardee began to reorganize and assign destinations to the various units as efficiently as possible.[23]

Refugees were an added source of chaos as they crossed the Savannah River ahead of the army. Many flocked to Hardeeville, which possessed the nearest railroad station that was not in enemy hands. The day before the evacuation Captain Edward J. Thomas had found a horse and wagon to convey his wife, son, and three female family members to Hardeeville. On their arrival they were rushed onto what they were told was the last train leaving town that evening. Not wanting to leave without her husband, Mrs. Thomas burst into tears, which got the attention of General Beauregard, who happened to be on the same train. The general generously gave her a note granting her husband permission to accompany her to Charleston and rejoin his unit later. When Captain Thomas arrived, he found his wife near the depot and waited with her until another train bound for Charleston finally rolled in.

To his great relief he was able to get the entire party—including their baggage—on board, and "[accomplish] by luck what seemed at first impossible."[24]

General Hardee was an expert in infantry tactics and had been a corps commander in the Army of Tennessee. He had gotten plenty of experience retreating in front of General Sherman at Atlanta, and it appeared that he would now be doing more of the same in southwestern South Carolina. As he surveyed column after column of troops arriving at Hardeeville, he realized that his work had only just begun. Aside from coordinating the movement of all of his troops, he must find a way to feed them. At this stage in the war, the army—and the region—were ill-equipped for such an enterprise, but fortunately he and his quartermasters were welcomed by local families, who had not yet abandoned their plantations. The cattle, hogs, and sheep donated and barbecued by these loyal benefactors, provided ample—if temporary—fare for the weary troops.[25]

The soldiers defending the railroad at Coosawhatchie were not quite as blessed. For nearly three weeks, General Samuel Jones's men had braved the elements in fields adjacent to the railroad tracks. Some of the Cadet Battalion slept in the open air, while other units had A-tents, which offered little shelter from the rain and cold. Supper was the only cooked meal they received, and no rations were available for breakfast or lunch. The troops obtained water for drinking and cooking from the same ditches they used for bathing, washing clothes, and watering horses. Food ran so low that some troops had to steal corn from the troughs of artillery horses. In tattered uniforms and worn-out shoes, they endured the hunger, disease, and severe winter weather; yet they continued to stand their post.[26]

With the influx of troops from Savannah, Hardee and Beauregard now had more than fourteen thousand troops along the line of the Charleston & Savannah Railroad. The focus now, however, was not to protect the railroad but to impede the progress of the enemy toward Charleston and the rest of the Palmetto State. Their next objective was to fortify and man a succession of defensive lines, the first of which ran from the mouth of the Combahee River to Barnwell Courthouse.

General G. W. Smith's division was the first to move. The Georgia militiamen were sent by rail to Charleston and were ordered to proceed from there to Augusta as soon as additional troops arrived. Colonel E. C. Anderson's brigade and a portion of McLaws's division under Brigadier General Lawrence S. Baker would follow Smith to Charleston for duty on James Island. General Ambrose R. Wright's division and General Chesnut's command of South Carolina reserves and militia were sent to the Fourth Subdistrict and were stationed on a line between Green Pond and Adams Run.

General Wheeler's corps was left at Hardeeville to monitor the landing at Screven's Causeway and the crossings along the Savannah and New rivers. When forced to retire, they would guard the right flank between the Combahee and Savannah rivers. Colonel Charles Jones Colcock's Third South Carolina Cavalry was assigned to protect the coastal flank of the troops retreating from Hardeeville to the

Combahee. General Young's brigade was reinforced by the Seventh Georgia Cavalry and was placed between the Combahee and Ashepoo near Blue House. Colonel Ambrosio Gonzales was charged with the disposition of field-artillery pieces from the Salkehatchie Bridge to the coast.[27]

On December 22 Hardee boarded a train for Charleston to confer with General Beauregard. He left McLaws in command, and instructed him to ship supplies and troops to Charleston as expeditiously as possible. As soon as Baker's command was on its way to Charleston, General McLaws proceeded with the remainder of his division to Pocotaligo and assumed command of the troops from the Combahee to Hardeeville. When relieved by McLaws, General Jones was to assume command of the district in and around Charleston.[28]

On Christmas Eve, General McLaws received the order from Hardee to march "without delay to Pocotaligo, leaving only such guards at Hardeeville as you may deem necessary." He formed his remaining troops into a brigade composed of the First Georgia Regulars under Colonel Robert A. Wayne, Colonel W. R. Simons's regiment of reserves, and the Twenty-seventh Georgia Battalion under Major Alfred L. Hartridge. McLaws assigned command of the brigade to Colonel J. C. Fiser, who marched his new charges eighteen miles to Grahamville through rain and mud, arriving late that evening.[29]

The next day, as General Sherman and his army presented Savannah as a "Christmas gift" to President Lincoln, there was little action across the river in South Carolina. The weather remained cold and a thick ceiling of lifeless gray clouds hung overhead, reflecting the mood in the Confederate camps. General McLaws's men spent Christmas Day at Grahamville and moved out the next day to the Coosawhatchie River, where Colonel George P. Harrison held the Yankees in check. The troops then went on to Pocotaligo, where, they remained on picket duty, establishing their camp in the woods between Old Pocotaligo and the Coosawhatchie River. As the various Confederate units headed to their assigned destinations, their supplies were being hurried from Hardeeville to Charleston by rail. Scouts of the Fifty-fourth Massachusetts closely monitored Confederate troop activities and observed that "such portion of Hardee's army that passed did so on foot, but cars laden with guns and ammunition ran the gauntlet of our fire over the rails." From his camp near Old Pocotaligo, Sergeant Andrews could hear the Union batteries opening up on the locomotives. "I used to think I would like to be an engineer," he recalled, "but not on that train. Think the engineer turns on all the steam and lets her go."[30] Miraculously the trains made it through without injury.

General Beauregard was worried that Sherman would cross immediately into South Carolina to resume his destructive march, but he did not. With the expert assistance of Hardee and McLaws, Beauregard had been able to establish a defensive line up and down the Combahee River and transfer the supplies from Savannah to Charleston, Augusta, and Columbia. Not much was happening in the Federal camp on Gregorie's Neck other than their batteries' shelling the railroad; therefore, it brought great excitement when a family of contrabands revealed to Federal officers on December 29 that few Rebel troops remained in their immediate front. They also reported that a detail of Confederates was beginning to take up the railroad iron.[31]

In a postwar interview General Hardee recalled his evacuation of Savannah as the singular most satisfying event of his military life. There is little doubt that Charleston & Savannah Railroad President R. L. Singletary and Superintendent H. S. Haines felt a similar sense of pride about their evacuation of the railroad's property from the city. The two men had shown great foresight in removing all locomotives and rolling stock from the Georgia side of the Savannah River before the Union army tore up the rails at Monteith and the enemy's presence necessitated the burning of the Savannah River Bridge. The same could not be said of the Central Railroad, Georgia's premier Southern trunk line, which lost a large quantity of rolling stock and equipment during Sherman's conquest. In fact the Yankees captured the president of the Central while he fled the city on an Atlantic & Gulf train.

In contrast Singletary and Haines were successful, not losing a single car while at the same time shepherding the movement of large quantities of other companies' rolling stock over their rails. Meanwhile they successfully kept the line open between Charleston and Hardeeville at great risk to life and property. For three weeks the trains passed within easy range of the Federal batteries on Gregorie's Neck, but the enemy artillery never interrupted service for any longer than it took to fix a damaged rail or two. In fact only one locomotive, one boxcar, and two small sections of track had been severely damaged until after Savannah had been evacuated, the troops and matériel from that city were secured, and command of the region was handed over to General McLaws.[32]

The withdrawal of the troops to defensive positions along the Combahee River necessitated continual contraction of the railroad's operation. The railroad sent out work crews to begin taking up the unused rails to keep them out of the hands of the enemy, but General Hardee had other designs on the precious iron. With the scarcity of Confederate iron, the rails from little needed or little used railroads were being used to replenish those on major trunk lines. The South Carolina legislature had authorized the construction of a railroad connecting Columbia and Augusta, and the company's president, William Johnston, had applied to the War Department for the rails salvaged from several companies including the Charleston & Savannah. Since the operational length of the railway was now only seventy miles, Secretary of War Seddon assented to the request.[33]

General Hardee dispatched Captain William Quick to Hardeeville to cooperate with railroad officials in taking up several miles of track between Hardeeville and Gopher Hill Station near Grahamville. According to Special Order No. 2, the iron would be forwarded to an agent of the Confederate Engineer Bureau in Columbia. General Wheeler was to assist in any way possible by providing work details from his unit and by impressing additional black labor if necessary. When presented with these orders, Singletary immediately called a special meeting of the railroad's board of directors to discuss them. As a former infantry captain and Superintendent of Labor or the Charleston defenses, he certainly understood the needs of the military and the nuances of impressment; however, he must now act in the best interest of the railroad.[34]

As the directors gathered at the office of William C. Bee on January 7, 1865, terms such as *patriotism* and *bankruptcy* were bandied about before the meeting was called to order. After a lengthy discussion the board of directors authorized Singletary to negotiate the sale of any and all, iron removed from the Charleston & Savannah line to any solvent railroad company willing to purchase it. They also asked General Hardee to do all he could to protect the company and its employees during the removal of the rails and to aid them in their mutual effort to increase the efficiency of sister Confederate railroads.

The Confederacy had had a policy of impressing railroad iron from little used or little needed lines since the spring of 1862, but it faced considerable resistance from companies who did not want their roads dismantled. Such was not the case with the Charleston & Savannah, and they wanted to reassure the General Hardee that they in no way meant to hinder his efforts. Citing the State of South Carolina's lien against the company, they did not feel that they had the authority to dispense with that much iron if they could not apply it to the payment of bonds held by the state and the discharge of the lien. On the other hand they also understood Hardee's plight and—better than many of their counterparts across the South—the interdependence of the railroads and the military. To them this compromise made sense and was a win-win situation for the company and the army. The railroad would receive compensation for the loss of the iron, and the interests of the Confederacy would be preserved.

At the January 7 meeting Singletary also addressed his concern for the safety of the company's capital equipment. With Sherman's recent movements and the contraction of services along the line, he did not want to keep any more of the company's machinery and equipment in Charleston than was absolutely necessary to meet their present needs. Agreeing with the president, the board instructed him to remove any excess rolling stock and materials to a safe place whenever he deemed it necessary. The exact place and time would be left to his discretion. As a final gesture the board rewarded Singletary, Haines, and Swinton for their extra efforts in the preceding months with a 50 percent raise in salary—a nice token, but a negligible sum considering the actual purchasing power of Confederate currency.[35]

Singletary got quickly to work and had to look no further than the offices of the South Carolina Railroad for a bid on the recently removed iron. On January 13 he convened another special meeting, this time attracting a majority of the board members. After several days of negotiations, Singletary presented a proposal in which the South Carolina Railroad offered to purchase the rails, spikes, and chairs at a price of fifty-five dollars per ton—their value in 1860—which they thought was a good offer considering the worn condition of the rails. Payment would be in the form of 7 percent bonds of the South Carolina Railroad Company, which would become due twelve months after the end of the war. If the Charleston & Savannah Railroad did not receive payment, the South Carolina Railroad agreed to give back an equivalent number of rails in the same time specified. By unanimous consent the board agreed to the negotiated terms and told Singletary to close the deal.[36] Little did they know that in a few days they would have taken up their last ton of salable iron.

In the days following the fall of Savannah, Sherman's army busied itself disposing of captured property, which included hundreds of guns, a significant quantity of Central Railroad rolling stock, and more than thirty thousand bales of cotton.[37] After being resupplied and enjoying a brief convalescence, however, Sherman was ready to move his juggernaut into South Carolina. His grand plan called for General Oliver O. Howard's right wing to establish a base on the Charleston & Savannah Railroad at Pocotaligo on or about January 15. General Henry Slocum was to cross his left wing over the Savannah and rendezvous near Robertville or Coosawhatchie at about the same time. Sherman did not foresee the torrential rains that plagued his troops throughout the month of January.

On the January 3, 1865, the two corps under General Howard began their transfer from Fort Thunderbolt, Georgia, to Beaufort, South Carolina, by ship.[38] Their movement was going as planned, but floodwaters hobbled Slocum. Federal engineers had constructed pontoon bridges over nearly the same path used for Hardee's evacuation, but they encountered problems from rapid currents and the bogginess of the flooded banks. General W. T. Ward's division of the XX Corps was the first to attempt to cross, but it became bogged down so severely that it took a week for them to complete their task. Seeing Ward's frustrations, Slocum moved the remainder of his corps farther upriver to attempt passage, but when he arrived at Sister's Ferry, he found the river so swollen that nearly three miles separated its eastern and western banks.

At about this same time General John Gray Foster began to recall some of the units from the Broad River to Beaufort and Hilton Head. General Edward E. Potter had informed him that newly established Confederate batteries near Coosawhatchie were beginning to inflict damage on the Federal marsh battery. Not wanting to risk additional needless casualties, General Foster withdrew Commander George H. Preble's marines from Gregorie's Neck. On January 3 the Twenty-sixth U.S. Colored Troops left Mackay's Neck for Beaufort, and the Fifty-fourth Massachusetts Regiment occupied their former camp.[39] While the Federal presence in front of the railroad was—for the moment—thinning, the buildup at Beaufort continued.

By January 11, the Union XVII Corps under General Frank Blair had completed its transfer to Beaufort, and portions of the XV Corps were beginning to arrive. General Ward's division had finally negotiated the Savannah and New rivers and on the seventh had reached Hardeeville, where they were held in check more by the heavy rains than by Wheeler's hapless pickets. Acutely aware of the increasing Federal presence, General Wheeler tried to hurry along the Confederate railroad work crews responsible for taking up the rails and burning cross ties before the enemy advanced.[40]

Several days later, the clouds lifted and the deluge stopped long enough for General Blair to move out from Beaufort with his corps and a brigade from the XV Corps. On the morning of the fourteenth, his force crossed Whale Branch Creek on pontoon bridges and marched toward Garden's Corner. Confederate pickets raced back to Pocotaligo Station to inform Colonel Charles Jones Colcock, who saddled up as many men as he could gather and rushed to meet the enemy. Colcock's cavalry encountered the van of Blair's corps at Huspa Creek between Garden's Corner and Pocotaligo, and a sharp skirmish ensued. Blair pressed forward and successfully flanked the Rebel troopers, forcing them to fall back. The Confederates were able to delay the Yankee pursuit by destroying bridges as they retreated until they at last reached the breastworks at Old Pocotaligo.[41]

The breastworks were situated north of Old Pocotaligo, a short distance from the railroad they were built to protect. In their front was an open field extending to a forest three-fourths of a mile away. Two hundred yards from the emplacements ran a stretch of marsh that recent rains had made virtually impassable. Colonel Fiser's brigade and a battery of artillery arrived at the fort earlier in the day and were just getting into position when they saw Colcock's men falling back across the field. Once they were past the marsh, they rallied, formed company, and strode into the earthworks to join their comrades.[42]

Waiting inside the fortification was Sergeant W. H. Andrews of the First Georgia Regulars. He watched as the blue-clad skirmishers marched across the field in pursuit of the cavalry. Waiting until the Yankees came within close range and halted at the marsh, the Confederate line "belched forth one solid sheet of flame."[43] Taken completely by surprise, half the enemy troops scurried back to the woods and the rest dropped into the grass for cover. Those who dropped eventually made their way back to the woods under the combined fire of infantry and artillery. The two sides exchanged artillery fire until dark, as the Union force bivouacked and planned a flanking maneuver for the next day.

The Yankees had driven Colonel Colcock's cavalry back so easily that General McLaws barely had time to move his men into position, and he knew that he was facing more than just a reconnaissance force. This was an entire Federal corps. McLaws did the only thing he could do, and withdrew his force eastward to the Combahee River. He sent the cavalry several miles north of the railroad up the Salkehatchie River. The infantry and artillery retreated along the railroad, crossed the Salkehatchie Bridge, and burned it before manning portions of the swamp above the bridge.[44]

When reveille sounded the next morning, Blair's planned attack was no longer necessary. His men found the works abandoned, and occupied them without a struggle. Excited Federal pickets on Mackay's Neck heard the fifes and drums of the XVII Corps and immediately informed their superiors. Colonel Edward Needles Hallowell quickly moved in their direction with the Fifty-fourth Massachusetts and Thirty-third U.S. Colored Troops, passing remnants of campfires, debris of Rebel picket posts, and empty earthworks along the way. On their arrival near Pocotaligo, they found Blair's troops in possession of the fortifications there. When they joined the other troops in the works, they became the first troops in Hatch's Coast Division to form a junction with Sherman's grand army. "Thus," remarked Captain Luis Emilio of the Fifty-fourth Massachusetts, "fell a stronghold before which the troops of the Department of the South met repeated repulses."[45]

As General Blair's mud-splattered troops swarmed into Pocotaligo, portions of Hatch's division joined them in a celebratory occupation of the former Confederate stronghold. In their new camps across the Combahee, the ragged Confederates could see the billows of angry gray-black smoke drifting skyward above the distant pines and knew they signified the destruction of the railroad and possibly other houses and buildings in the vicinity. The Confederate army was reeling from Hood's mid-December defeat at Nashville and the loss of Savannah to the enemy. Further darkening their collective mood was the news received on January 17 that combined Federal ground and naval forces had taken Fort Fisher and other Rebel installations at the mouth of the Cape Fear River. A seemingly much smaller victory by comparison, the fall of Pocotaligo opened a wide avenue for Sherman's army. It was now South Carolina's turn to feel "the mailed hand her temper had wrought."[46]

General Howard had carried out his portion of Sherman's plan to the letter and with minimal casualties. In so doing he dealt a severe blow to the South Carolina's defenses and now commanded what had been perhaps the most important point on the Charleston & Savannah Railroad. The Yankees promptly strengthened Pocotaligo as a base and established a depot on the Pocotaligo River several miles from the railroad. Sherman now had a base from which he could support movement in any direction he wished.[47]

On January 18 thunderheads rolled into the area, and southwestern South Carolina was drenched by six straight days of rain. Soldiers on both sides could only wait in anticipation for the tempest that was to come. General Hardee, who had ridden forward to McLaws's new headquarters near Salkehatchie Station, felt that Sherman would move north and cross the Salkehatchie River near Branchville; however, McLaws maintained that Sherman was most likely to strike Charleston. General Sherman did nothing to resolve their disagreement.

While he solidified his plans in Savannah and Beaufort, General Howard sent scouting parties in both directions. A trusted Confederate scout, Sergeant T. M. Paysinger of the Third South Carolina Infantry, reported Federal parties clearing the roads of obstructions in the direction of Robertville and River's Bridge. Howard also demonstrated heavily in the direction of Charleston, sending working parties to reconnoiter and tear up the railroad near Salkehatchie. On January 20, Major

Pocotaligo Depot after the arrival of General Sherman's troops in winter 1865, engraving after a sketch by Theodore R. Davis. From *Harper's Weekly,* April 8, 1865; courtesy of Library of Congress

General Joseph Anthony Mower's division of the XVII Corps advanced to Blountville near the Combahee River but turned back because of rapidly rising floodwaters and the harassment of a Confederate battery near the railroad. Prisoners taken from this division reported to their captors that Sherman's XIV and XX Corps were to march from Savannah to Charleston along the line of the railroad—a story deceptively corroborated by prisoners from other units.[48]

While waiting for the floodwaters to recede, General Sherman rode forward and joined Howard at Pocotaligo on January 24 to inspect the abandoned Confederate fortifications there. They would have been difficult to take—containing a number of well-fortified gun embrasures and a marsh immediately in front—and Howard felt lucky that the Rebels had pulled back without a fight. Even Sherman marveled at why McLaws had allowed them to take Pocotaligo so easily—unless out of fear or ignorance. "It was, to me," Sherman said, "manifest that the soldiers and people of South Carolina entertained an undue fear of our western men, and, like children, they had invented such ghost-like stories of our prowess in Georgia that they were scared by their own inventions."[49]

The Confederate show in South Carolina had been Hardee's to run ever since Beauregard had been sent to Montgomery after Christmas to restore order to Hood's shattered Army of Tennessee. Before departing, Beauregard had reminded Hardee that his goal was no longer the defense of the railroad, nor even of Charleston. He must preserve his troops at all costs, and, as with Savannah, Charleston was secondary. In the Port City the mood was one more of nervousness than of panic. From his headquarters on Morris Island, General Schimmelfennig reported that the

Confederates were prepared to take to the field. They still had a strong force there, and more troops from the Army of Northern Virginia were arriving.[50]

Everyone from Hardee to Beauregard to Governor Andrew Gordon Magrath had appealed to Richmond for reinforcements, but with Lee earnestly trying to keep Grant out of Petersburg, few troops could be spared. Finally the Confederate high command gave the okay to detach Major General Matthew C. Butler's cavalry division and General James Conner's (formerly Kershaw's) brigade from the Army of Northern Virginia to their native state. Banners and kerchiefs waved from rooftop and window, welcoming the ragged heroes as they returned to Charleston on the South Carolina Railroad. After a week of recuperation and feasting on local oysters, they boarded trains bound for the works near Salkehatchie. The new arrivals from Virginia bolstered the Rebel ranks, but Beauregard knew he needed more. He called on General Hood, who soon had men from three decimated corps of the Army of Tennessee en route to Augusta.[51] When these ten thousand troops eventually arrived, Hardee fielded a force of thirty-three thousand effectives, spread along a line from Charleston to Branchville to Augusta, against a veteran army of more than sixty thousand, and only the Union army knew Sherman's true destination.

Keeping up the ruse, Howard demonstrated on the Combahee again on January 25 for the sole purpose of rattling McLaws. Afterward General Sherman rode forward to reconnoiter the ground personally. He found the land there to be very low and swampy, cut with numerous marshes and tidal creeks. The unrelenting rains of late had swollen the Combahee River so that water filled its swamps for a mile on either side of its banks. The retreating Confederates, who held a strong fortification on the opposite side, had burned the bridges across the river behind them. They had adopted this line to protect what they thought was Sherman's next objective—Charleston. But Sherman had no designs on that city. In fact he considered it to be a "dead cock in the pit, altogether."[52]

Sherman was soon on his way back to Gregorie's Neck to speak to General Hatch about moving the Coast Division up from the Coosawhatchie and Tullifinny to Pocotaligo Station. After the division assumed its new position, its men picketed the outer works at Coosawhatchie and in back of Pocotaligo Bridge. They were also charged with continuing to probe the Confederate defenses on the Combahee near Salkehatchie. The Union strategists were confident that they could compel McLaws to retain a considerable force in the area simply by making small demonstrations along the line of the railroad from the west and south.[53]

Sherman ordered Hatch to move upon the left bank of the Combahee as if preparing to cross, clearing obstructions for his would-be skirmishers and preparing the railroad causeway for gun embrasures. He wanted Hatch's division to hold the position on the railroad near Salkehatchie for at least ten days after Howard's men left camp. At that time Hatch should cross the river and move toward Charleston, destroying the railroad as he went. Admiral John Dahlgren, despite his loss of the *Dai Ching* on the twenty-sixth at the hands of a Rebel battery on the Combahee, was told to continue to feel out the enemy defenses in the Edisto and Stono rivers. General Foster was to try to cut the railroad from the south between the Combahee and Charleston. This, Sherman expected, would cause McLaws to guard the railroad along its entire length. But he wanted there to be no mistake that the movement of Howard and Slocum to the rear of Charleston was paramount. All others were accessories and were simply to take advantage of any ground relinquished by the enemy.[54]

On January 29 the roads above Savannah were beginning to dry out, and General Slocum put his columns in motion. The XV Corps had caught up with the rest of Howard's command, and they readied themselves for the expedition into the Carolina interior. While preparations were underway, the Twenty-fifth Ohio Regiment made a reconnaissance on the Combahee, marching to the intersection of the coach road with the railroad, but did not cross the river. They exchanged shots briefly with Rebel troops in the area before returning to camp on the night of January 30. Slocum's troops were still struggling with the currents in the Savannah River at Sister's Ferry when the XV and XVII Corps broke camp on January 31. However, their real march did not begin until February 1. As Hardee had predicted, they advanced north bound for a crossing of the Salkehatchie River at River's and Beaufort's bridges. Removing the timber and crossing the swamps in their path, they continued warily toward the South Carolina Railroad at Midway. It was the same route proposed to the Union high command more than a year earlier by General Truman Seymour.[55]

While Hatch continued his feints along the Combahee line, General Foster began his operations along the coast. On February 3 General Edward E. Potter's brigade demonstrated in the South Edisto, landing the Thirty-second U.S. Colored Troops on Edisto Island. The following day, the Twenty-fifth Ohio crossed over at Combahee Ferry. Unsuccessful in its attempts to flank the Rebel works beyond the rice fields, it recrossed with only minor losses.

By this time Confederate defenses elsewhere were in dire straits. Sherman's army was pushing through the bogs and streams faster than expected, and McLaws had begun to shift troops to his right to aid in the defense of Branchville. On February 6 General Hatch reported to Sherman that the Confederates had abandoned the works at the Salkehatchie Bridge, and he intended to cross the next morning and advance toward Ashepoo. On February 7 the Fifty-fourth Massachusetts and 102nd U.S. Colored Troops under Colonel Edward Needles Hallowell marched in a driving rain, crossing the partially destroyed railroad bridge over the Combahee River. There they joined the Fourth Massachusetts Cavalry and the Third New York Artillery, and the two columns continued along the turnpike and the railroad until they encountered the enemy. A small force of Confederate cavalry with one fieldpiece contested the Union advance, dismounting at every piece of high ground to place obstructions and lob a few shells before being outflanked and withdrawing. The Federals recrossed the river and bivouacked a safe distance to the rear that night.

Sherman's XV Corps crossing the South Edisto River, engraving after a sketch by Theodore R. Davis. From *Harper's Weekly*, April 8, 1865; courtesy of South Carolina Historical Society

The next morning, they skirmished on Cuckold's Creek, destroying another portion of the railroad near Whitehall.[56]

General Sherman continued his push, preventing Hardee from concentrating his troops to defend the South Carolina Railroad. Neither General Wade Hampton's arrival at the end of January nor Beauregard's return to command produced much effect. By February 9, the line protecting the city of Columbia had been broken as the hapless Southerners took one position after another, only to be flanked easily and repeatedly by the Northern invaders. The Rebel force could not stop Sherman, and his advance stretched the Confederate lines of supply, communications, and command until they ultimately broke down. At this point Sherman decided to turn up the heat on Charleston. He ordered Hatch to advance, destroying the railroad as he went. Hatch sent the Twenty-fifth Ohio and 127th New York forward. With the support of cavalry and artillery they drove the Rebels from their works near Heyward's and Lowndes's rice plantations and did further injury to the railroad.

Also on this day General Quincy A. Gillmore returned to the command of the Department of the South. Taking over for General Foster, who left to attend to an old injury, Gillmore wasted no time in ordering an assault on James Island. In what amounted to a flanking maneuver, Potter's brigade—consisting of the Fifty-fifth Massachusetts, 144th New York, and the Thirty-second and Thirty-third U.S. Colored Troops—landed on the south end of the island on February 10, surprising and driving in the Confederate pickets there. The Yankees took numerous Confederate prisoners, among whom was Major Edward M. Manigault, former chief engineer of the Charleston & Savannah Railroad. The next day all but the Thirty-third Colored Regiment were loaded aboard transports, which set sail for Bull's Bay.[57]

At a conference on February 2, 1865, at Green's Cut, Georgia, General Beauregard had warned General Hardee that Sherman might try to turn the Confederate line north of Barnwell, and he had done exactly that, forcing McLaws to fall back upon Branchville. On February 4 Colonel M. B. McMicken asked Superintendent Haines to send an engine and car to Adams Run to take General Ambrose R. Wright to Salkehatchie. At the same time he ordered Haines to remove the stores from all depots between Edisto and Salkehatchie. The Confederates had determined to try to hold the Combahee as long as possible, but once that line was broken, Hardee was to fall back upon Charleston. If the garrison appeared to be in danger, he had to evacuate the city.[58]

The Confederates had abandoned the works at Combahee Ferry on February 11 and pulled back toward the Ashepoo. When General Hatch, whose division was busy mangling another half dozen miles of the railroad, discovered this, he ordered the Twenty-fifth Ohio across the river to take a position at Lowndes Plantation. On that same day Confederate scouts reported to Hardee the sighting of sixteen Union transport ships in Bull's Bay. With Gillmore's presence on James Island, Hatch surging in from the west, and the threat of a sizable enemy landing above the city, General Hardee had seen enough. He fired off telegrams to Beauregard in Columbia and to President Davis in Richmond apprising them of the situation. He also gave the order to send trains to transport the troops remaining along the railroad at Ashepoo back to Charleston.[59]

With Hardee's order the last remaining section of the Charleston & Savannah Railroad was left defenseless. From Salkehatchie to Savannah its bridges were burned; its cross ties were reduced to ashes; and its bent and twisted rails littered the countryside between the lowcountry hamlets that dotted its right-of-way. Thomas Drayton's "most formidable earthwork" now lay at the mercy of Gillmore, Hatch, and an army whose goal had been to break the railroad ever since the fall of Port Royal three years earlier. The abandoned railway had served its terminal cities, its state, and the Confederacy well, but it now had little to show for it. It would be quite some time—and under quite different circumstances—before locomotives would again travel its rails.

SEVENTEEN

One More River to Cross

As General Sherman's army traversed South Carolina on corduroy and pontoons, only the most stalwart secessionists could not see that time was running out on the Confederacy. As Charleston & Savannah president R. L. Singletary supervised the laying of new track from his company's Spring Street depot to the Northeastern Railroad depot across town, he paused a moment to take stock of the tall task ahead of him. In February 1864 the railroad's biggest concerns had been retiring its floating debt and finding permanent solutions to bridging the Savannah and Ashley rivers. A year later its circumstances were much graver. The city of Savannah was now in Union hands, many of the company's rails lay in ruin, and the last of the gray-clad defenders had been pulled back into Charleston. It had been more than a month since train whistles had been heard in the villages of Pocotaligo, Gopher Hill, and Hardeeville, and for the better part of three months use of the company's motive power had been almost totally at the behest of the Confederate military.

Whether from fear of the road's destruction or from patriotic loyalty, the officers of the Charleston & Savannah Railroad enjoyed a comparatively harmonious relationship with the Confederate Quartermaster Department. The same, however, could not be said for all Southern railroads. Lieutenant Colonel Frederick W. Sims, the frustrated chief of the Confederate Railroad Bureau, complained that many railroad officials felt little responsibility to the military. Any cooperation seemed to come either from a sense of moral obligation or fear of the government. Sims pointed out that the railroad men were seldom penalized for their stubbornness, and that the only railroad legislation passed by the Confederate Congress was "so full of loopholes that it [was] inoperative."[1]

The railroad bill passed in April of 1863 had authorized the impressment and redistribution of rails, rolling stock, and equipment; however, it did not provide for reimbursement for damages incurred by the companies. The railroads could withhold the resources required to run one passenger train per day, but had to run on a schedule dictated by the government. The new law assigned the supervision of the roads to Quartermaster General Alexander R. Lawton, but gave absolutely no executive authority to the Railroad Bureau. The latter point was a particular hindrance to

Sims. On the other hand, he appreciated the mutual dependence of the railroad and military, and lobbied for more protection for the carriers. Sims knew that defense could not occur without transport, nor transport without defense; if either of the two should collapse, the final goal could not be realized.[2]

As the Confederate army faltered, and its ability to supply and arm its troops became increasingly problematic, the need for public management of the railroads was underscored. The desperate times motivated the Confederate Congress to take up issues that would have been unheard of as little as one year before—the conscription of slaves into the military and government control of the railroads. South Carolina Congressman William Porcher Miles had said that the enlistment of slaves would only happen when the nation was "on the very brink of the brink of the precipice of ruin."[3] This must have been the case because the Confederate Congress passed a bill that authorized the conscription of slaves, and by a 52 to 18 vote the House passed a substantive railroad transportation bill.

The new bill built on the provisions of the 1863 railroad law but went further in giving governmental agencies power to enforce it. Not only did it threaten companies failing to cooperate, it placed employees of all carriers of military significance directly into the army. The bill authorized government aid to needy railroads, provided for reimbursement of damages incurred under the law, and set forth the role of railroad men relative to the military. When the War Department took command of any railroad, its officers, agents, and employees were considered to be part of the Confederate ground forces as long as they were engaged in government transport.[4] Only four days after its introduction, the Senate passed it unanimously on February 19, 1865. Fewer than forty-eight hours before its final reading, the city of Charleston was evacuated.

Lieutenant General Hardee had been feeling unwell, suffering from more than just the strain of the past two months. The words of President Davis's dispatch on February 14, 1865, had done nothing to improve his malaise. Only three days earlier, General Beauregard had sent word for Hardee to "execute as soon as possible the movement agreed upon the 2nd instant"[5]—namely the evacuation of Charleston. Now President Davis was ordering him to "remove whatever was not needed for defense of the city, and . . . postpone evacuation as long as prudent."[6]

General Beauregard had expected that Sherman would try to turn the Combahee line north of Barnwell but had not anticipated that he would move so rapidly. When it became evident that the Federals under Gillmore intended to move on Charleston or possibly to cut off Beauregard's communications along the Northeastern Railroad, Beauregard had sent the order to evacuate Charleston. Hardee promptly inquired as to whether he should carry it out immediately. Beauregard had decided to let Hardee decide when to carry out the movement, but stressed in his reply the next morning that there was very little time to lose.

Messages received from Richmond during this time seemed to suggest that the Confederate leadership was out of contact with reality. The War Department exhorted the commanders in the field to "fight a great battle," to "crush the enemy," and to "either destroy him or so cripple his efforts to reach Grant that reinforcements would be taken from Lee's front" at Petersburg.[7] Even President Davis's telegram on February 11 expressed the delusional hope that they could confine the enemy's activity to the coast and defeat him. Davis apparently was unaware that only a portion of Hood's reinforcements had reached the state or that recruitment efforts in South Carolina and Georgia had been so anemic. Far from being confined to the coast, Sherman's army had crossed the North Edisto and was moving in force on Orangeburg and Columbia while the fleet of ships carrying Potter's brigade appeared off the bar above Charleston at Bull's Bay. On February 12 Hardee sent trains to bring McLaws's division back to Charleston and ordered General Ambrose R. Wright's command back across the Ashepoo to the Edisto, burning bridges as they withdrew. With General Carter Littlepage Stevenson forced to fall back to Kingville, General Beauregard hoped that Hardee could at the very least check the Federal advance through the swamps and guard the crossings of the Santee River. If not, Beauregard feared the enemy might reach the Northeastern Railroad—the last avenue north—before the Charleston evacuation was complete.[8]

General Hatch's Coast Division continued to feel its way along the line of the railroad past Green Pond to reach the Ashepoo River, its march delayed only by the now bridgeless streams and rivers in their path. The Federal transports in Bull's Bay had not yet landed, but their presence was definitely being felt. As the Union forces closed in, General Hardee began to feel ill and asked General Beauregard to come to Charleston for a brief parley. Confused as to why the evacuation was not progressing more efficiently, Beauregard decided to pay Hardee a visit and prod him along. He left General Wade Hampton in command of operations in Columbia and traveled to Charleston on the morning of the fourteenth.[9]

In their discussions Beauregard got quickly to the point. He reminded Hardee of how easily Sherman had negotiated the swamps as he moved toward Orangeburg and Columbia to cut the line of retreat out of Charleston. The loss of Charleston would be a blow to morale, but it was the loss of the garrison—not of the city—that would most jeopardize South Carolina's security. In Beauregard's mind he could hold Charleston for only a few more days at most. He left little room for interpretation as he bluntly informed General Hardee that it was now "necessary to commence evacuation as soon as the necessary preparations [could] be made." Before leaving, he handed Hardee a memo detailing the exact routes to be taken by the retreating troops via Kingville, Manchester, Darlington, and Cheraw to a rendezvous point at Chester.[10]

The memo left little to chance. Beauregard's orders detailed the entire movement down to the number of days expected for each leg of the march. Hardee was satisfied. He had tried his best to hide his illness from the gallant commander, and hoped he would be up to the task. He was beginning to feel more at ease about things until he received the telegram from President Davis asking him to delay the evacuation.

President Davis had little appreciation for the severity of the situation in and around Charleston. His dispatch stymied Hardee, who wondered how he could resolve the conflicting orders of his immediate superior and the president. He set to removing nonessential military supplies from the city and hoped Beauregard and Davis would give him some latitude for the timing of the evacuation.

As Federal troops attempted to envelope the city, trains loaded with military supplies and citizens with their belongings were being sent out of Charleston. Robert L. Singletary had seen to it that the connection of the Charleston & Savannah Railroad with the Northeastern was complete, and for several weeks he and H. S. Haines had been consumed with the removal of the company's property from Charleston. They were so busy that they considered postponing the company's annual stockholders' meeting, which was scheduled for February 15, but times being as they were, they decided to proceed with it as planned. Those in attendance reelected Singletary, approved the same slate of directors as the previous year, and adjourned without conducting any additional business.[11]

Also on February 15 General Hardee wired General Samuel Cooper and President Davis in Richmond to inform them of his intention to evacuate Charleston. As the day wore on, however, his illness so racked his body and distracted his thinking that he was ultimately confined to his bed for two days. General Beauregard updated General Lee on the deteriorating circumstances in South Carolina and expressed his frustration at Hardee's procrastination. He was apparently unaware that Hardee's strength had been sapped by what was thought to be typhoid fever and that General McLaws had been asked to return to Charleston to assume command of the evacuation. President Davis did not take the news well and, in a terse response to General Beauregard, expressed bitter disappointment at the plans to abandon the city. If, however, Charleston must be given up, Davis made it clear to Beauregard that he wanted none of the Confederacy's cotton to fall into enemy hands as it had in Savannah.[12]

On the night of February 17, 1865, Confederate commanders called in troops from the various outposts around Charleston and began the evacuation of the city. Bridges on the west side of town were burned as the main force was withdrawn that evening. Smaller details of troops were left behind in the city to blow up ordnance and destroy anything else that might be of value to the enemy. At the wharf of the Charleston & Savannah Railroad, Confederate troops torched a large storehouse containing hundreds of bales of cotton and several thousand bushels of rice. Cotton burned in the streets; buildings were looted; and people ran about frantically trying to rescue their possessions. The genteel city was in a state of chaos as General McLaws abandoned it, leading his men on what had been predicted would be the "death march of the Confederation."[13]

In the end Charleston was in ruins. Its streets were littered with rubble, and its grand hotels and once-thriving businesses were deserted. A thick layer of smoke blanketed the peninsula, and the ground shook from explosions of Confederate munitions stores. On Morris Island, Union lieutenant colonel Augustus G. Bennett was scanning the horizon looking for signs that the Confederate rearguard had completely vacated the city. He entered the city that morning with a handful of soldiers from a Pennsylvania regiment, and just before midday, he ordered the Twenty-first U.S. Colored Regiment to cross over. Cheers rang out from the regiment of former slaves as they secured the city for the Union, and Old Glory once again flew over Charleston harbor.[14] As the Federal troops celebrated, the retreating Confederates received the disheartening news that General Hampton had evacuated Columbia. Both garrisons escaped and lived to fight another day; however, the morale of the state—and the Confederacy—had suffered a terrible blow.

The Northeastern Railroad ran the 102 miles from Charleston to Florence, South Carolina, connecting there with the Wilmington & Manchester and the Cheraw & Darlington lines. From its beginnings in 1857 its depot served as Charleston's gateway to the industrial North. Now it served as the point of embarkation for the retreating Confederate Army. Its Charleston depot is now remembered as the site of the greatest calamity that occurred during the city's evacuation.

On Saturday morning, February 18, 1865, the only citizens left in the city were those who did not have the means to escape. Before Federal troops entered the city, a host of these unfortunates—many of them women and children—flocked to the depot to glean anything of value that had been left behind. A large quantity of damaged gunpowder had been left in a storehouse at the depot, and some mischievous children began taking handfuls of it and tossing it into a cotton fire near the building. They unintentionally laid a trail of powder that led back to the storehouse, and when it accidentally ignited, it caused an explosion that immediately engulfed hundreds of the foragers. As it was described in the *Charleston Daily Courier* a few days later, "The spectacle which followed was horrible. In an instant the whole building was enveloped in smoke and flames. The cries of the wounded, the inability of the spectators to offer assistance to those rolling and perishing in the fire, all rendered it a scene of indescribable terror."[15] All that remained of the structure was an enormous pile of brick, mortar, and ash.

The last train out of Charleston was one belonging to the Charleston & Savannah Railroad, which was now primarily under military control. The train, reportedly carrying the assets of several of Charleston's banks, left the Northeastern depot bound for Cheraw, where a special siding awaited it on the Cheraw & Darlington Railroad. Singletary and Haines had worked tirelessly transferring the company's rolling stock, shop machinery, and other equipment to Cheraw and, just as they had done at Savannah, kept it all out of enemy hands. Since military transportation now completely occupied the trains, they were shuttled from point to point on various railroads as the army's movements dictated. Keeping track of the rolling stock required the best efforts of some of the railroad's most reliable employees. In addition to the stalwart Haines, Singletary relied on the energies of Master of Transportation R. H. Riker, conductor William Crovatt, Master of Roadway J. H. Buckhalter, and

John L. Roumillat, formerly the company's agent at Savannah. Because of the enemy's proximity, the safety of the cars was at risk on several occasions, but a combination of excellent management and good fortune minimized the company's losses of rolling stock.[16]

When he chose Cheraw, Singletary thought he had found a safe haven for his company's supplies. There was no way he that could have predicted that the quiet hamlet on the Great Pee Dee River would be directly in Sherman's path—but that is exactly what happened. General Sherman made quick work of Columbia and by February 21 he had reached Winnsboro. The Confederates under Hampton and Beauregard had fallen back to Chester, but because of Sherman's momentum, they ultimately continued up the Charlotte & South Carolina Railroad into North Carolina. At this critical juncture President Davis assigned command of the entire theater to General Joseph E. Johnston, whose goal was to consolidate the armies of Hardee, Beauregard, and Bragg and to hit Sherman somewhere between Fayetteville and the Cape Fear River.

During the final week of February, Hardee rushed his already tired troops forward in a series of forced marches to reach Cheraw and cross the Pee Dee ahead of Sherman. As he waited for the last of his troops to straggle into town, the Federals drew dangerously close. Once past Winnsboro, Sherman turned to the east, but flooded rivers again delayed his progress. General Howard's wing had been able to cross the Catawba River before the rains set in, and he was moving toward Cheraw. General Matthew Butler's cavalry rode forward and skirmished with the Federal vanguard at Chesterfield, buying precious time for their comrades. The delay gave Hardee just enough time to cross his twelve thousand men over the Pee Dee River and burn the bridge behind him. The next day Howard's column entered and occupied Cheraw. While in Cheraw, Federal troops destroyed a considerable amount of public and private property, including the shop tools and supplies of the Charleston & Savannah Railroad. On March 4, 1865, Colonel Reuben Williams led the Seventh and Ninth Illinois and the Twenty-ninth Missouri Mounted Infantry on a foray down the Cheraw & Darlington Railroad in the direction of Society Hill and Florence. The Third South Carolina Cavalry under Colonel Charles Jones Colcock were still in the area and were sent to check the Yankee raiders. After a two-hour skirmish just outside Florence, Colcock's determined regiment sent Williams's force back toward Society Hill, saving Florence's depot and preserving the city's railroad. By this time the main body of Sherman's force had crossed the river and was chasing Hardee as he retreated to the north and east.[17]

The events of the ensuing month are some of the best chronicled in United States history. Sherman continued his push into the Tarheel State, seizing Fayetteville and then setting his sights on the strategic railroad junction at Goldsboro. On his way there he defeated Hardee's corps in a brief battle at Averasboro and then survived a final desperate strike by the Confederates at Bentonville. General Johnston had decided to attack a portion of Sherman's force before they were joined by fresh troops from New Bern and Wilmington, but after a protracted battle the Confederates were beaten back. Johnston pulled his depleted army back in the direction of Raleigh, and Sherman marched into Goldsboro unopposed on March 23.

At Petersburg General Lee's lines were stretched to their breaking point. The Union army under General Ulysses S. Grant had for some time attempted to turn the Confederate right flank and finally accomplished the task with General Philip Sheridan's victory at Five Forks. On the night of April 2 Lee evacuated Petersburg and Richmond and set out to attempt to join forces with Johnston in North Carolina. Drifting westward, Grant countered Lee's every move, until the weary Lee surrendered the Army of Northern Virginia to Grant at Appomattox Courthouse, Virginia, on April 9, 1865. A week later, Johnston capitulated, surrendering to Sherman at Durham Station, North Carolina. Only the delusions of Jefferson Davis kept the last vestiges of the rebellion alive. The Confederate States of America had fallen.

Postwar Charleston was a city in shambles. Houses sat vacant; waves lapped against decaying wharves; and grass grew between the cobblestones of the streets. Union troops under General John Hatch claimed dominion over the collection of shattered, empty buildings and a population of destitute whites and freed blacks. The South had sunk all its capital into the enterprises of war, and now there was nothing left. Burned out shops, fallow fields, and a ravaged infrastructure were its legacy. As the Confederate veterans slowly wandered home and the citizens gradually returned

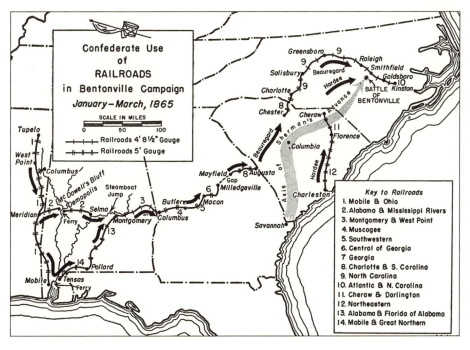

Confederate troop transportation to Bentonville, North Carolina. From *Railroads of the Confederacy*, by Robert C. Black, © 1952 by the University of North Carolina Press; renewed 1980 by Robert C. Black III. Reproduced by permission of the publisher

to the city, Charleston began to show some signs of life again. It was not normal life but at least some measure of it.

After the cessation of hostilities, the U.S. military continued to exert its control over the state. In May troops were sent to Columbia to arrest Governor Magrath and all officers of South Carolina during the rebellion. They also took possession of all public records and correspondence of prominent secessionist leaders. General Gillmore authorized General Hatch to take military control of the Northeastern Railroad, the South Carolina Railroad and all its branches, the Wilmington & Manchester Railroad, and all other railroads in the state that were required for military use. He was authorized to give whatever military aid he deemed proper, leaving the immediate control of several of the railroads to their presidents and directors as long as they were not excluded from amnesty under President Andrew Johnson's proclamation.[18]

During this period of turmoil and transition, President Singletary and Superintendent Haines returned to Charleston to take stock of the Charleston & Savannah. When Federal troops occupied the city, they counted among their spoils a considerable quantity of railroad matériel, including eight locomotives and forty-four cars of various types. None of these spoils belonged to the Charleston & Savannah Railroad. Their loss of rolling stock had been negligible compared with other roads; however, they did not survive the war unscathed. Haines counted one first-class passenger coach burned during the Federal raid at Society Hill, one platform and four boxcars destroyed in Potter's Raid on Camden in mid-April 1865, and the machinery of the locomotive *Andrew Milne*—out of service since 1860—severely damaged when Sherman burned the depot at Cheraw. Three boxcars, two stock cars, and five platform cars remained unaccounted for. Because of their use on other rail roads, the cars had been scattered from the Cape Fear River to the Chattahoochee, and Haines held out hope that they might still be found.[19]

Almost every building on the line had been destroyed by the Federals, and the bridges across the seven major rivers between Charleston and Savannah had been dismantled by the retreating Confederate army. Thirty-eight miles of track were mangled by Yankee troops, three-and-a-half miles of trestlework were destroyed; and a dense growth of weeds accelerated the decay of even the uninjured portions of the road. At Savannah 7,500 feet of track had been taken up near the junction and used by the Georgia Central Railroad, the rest of the iron between there and the Savannah River was unsalvageable. Between the Savannah River and Whitehall, a total of twelve miles of track had been taken up by the company at different points prior to the withdrawal of the rolling stock. Of that, one mile was still in Cheraw; the South Carolina Railroad was using a little more than five miles of track; and the rest lay unharmed at different points along the line.[20]

Singletary returned to Charleston on June 9, 1865, and went immediately to General Hatch's headquarters to find out what measures he need to take to resume operation of the railroad. Hatch advised Singletary that he should make a formal, written application and that John S. Ryan had made a competing bid for control of the company on behalf of a number of the railroad's creditors. The next day Singletary complied with the general's advice but puzzled over Ryan's claim on the company. He requested that the government grant him formal control of the railroad on behalf of the board of directors, and suggested a civil court might best decide whether Ryan's petition had any validity.

General Hatch submitted the letter to General Gillmore, who referred it back to Hatch, stating that the railroad could be turned back over to the board of directors if it had been purged of all disloyal men—and as long as the directors promised to repair the road without delay and satisfied Hatch of their ability to do so. After waiting two weeks, Hatch again contacted Singletary, asking him how much money would be required to repair the road, how the company planned to raise the funds necessary for the reconstruction, and if the board would pledge to commence work at once. In his reply Singletary pointed out that he had not yet examined the road and was not prepared to estimate the cost of repairs, although they would most certainly be extensive. He could speak only for himself, but his goal was most definitely to rebuild the road, and he was sure the directors would take any measures necessary to raise the required funds. Singletary knew that the rolling stock and roadbed were deteriorating and that further delays would hinder both public and private enterprise. He again asserted that he and the board of directors were the legal representatives of the corporation and that a civil court should decide any other claims on the road or its property.

Singletary was becoming impatient. The Federal government had restored the Northeastern and the South Carolina railroads to their rightful owners, but he was still being rebuffed. On July 8 he and three board members called on General Hatch to see how their situation differed from those of the other railroads. Why—they wanted to know—after an entire month had elapsed since their initial application, had the military not consented? Again, the issue came back to Ryan, who—apparently on behalf of a group of bondholders—had expressed an objection to the board's resumption of control. This was the primary reason for Hatch's hesitation.

Singletary was flabbergasted. When the company's bondholders had threatened to foreclose at the beginning of the war, the stockholders offered to let them choose half the board of directors. They had agreed to this proposal, and made their selections, all six of whom still sat on the board. Singletary felt that the board not only technically but legally represented the interests of the creditors, and that Ryan had absolutely no legal authority to represent them. In fact the creditors could not take the road or sell it because the intervention of the trustees and a Court of Equity would be necessary for either. Furthermore its sale would not return ten cents on the dollar at its current value. Singletary argued that it would benefit all concerned if he and the board of directors ran the company to the best of their ability until civil rule was reestablished, and the creditors could decide their best option at that time. If the only alternative was for the military to take and repair the road, Hatch did not think it worth the outlay. He decided to ponder the matter overnight and the next day would either return the road to Singletary and the board of directors or refer them to General Gillmore.

Singletary bristled as he read Hatch's reply the following day. The general asked him to prove that he had the confidence of the company and demanded that all disloyal members of the board be removed. If a quorum of the board asked for control of the railroad and outlined their plan for the road and its rolling stock, only then would Hatch refer their appeal to Gillmore. In response Singletary clarified a disagreement over what composed a quorum and assured Hatch that all board members were believed to have taken the Oath of Allegiance under the Amnesty Proclamation. He could not understand why the company's confidence in him was being called into question. He also did not understand why the military had concerned itself with the rights of the bondholders versus the stockholders, or why Hatch had treated the city's other railroads so much more favorably. Singletary suspected a plot by Ryan and his alliance of creditors, but he could not prove it.[21]

During this time, trade was slowly beginning to return to downtown Charleston. Shops were open all up and down King Street, and merchants began to turn their attention to the empty buildings on Meeting, Hayne, and Broad streets. Of course the Burnt District had to be rebuilt, and the economy needed a heavy infusion of money, but there were clear signs of recovery. More and more businesses were resuming operations; cities in the interior of the state were attempting to revive the cotton trade, and plans were underway for a national bank in Charleston. The missing ingredient was the railroad, but city officials expected the broken connections with Augusta and Savannah to be reestablished very soon.

While Charleston was beginning to revive, it seemed to be doing so without the Charleston & Savannah Railroad. A week after the Northeastern Railroad was transferred back to its president, an editorial in the *Charleston Daily Courier* urged the Wilmington & Manchester Railroad to complete its repairs and put Charleston back in communication with the North. The article then went on to say, "And while on the subject of railroads, we would like to know what has become of Mr. Singletary and Mr. Haines? A daily mail with Savannah is loudly called for. What has become of the energy which has hitherto characterized these public servants that, while all the roads are alive with workmen answering to the call of the public, the Savannah road lies idle?"[22] Singletary knew exactly why, but it was not his doing or Haines's. They were not trying to procrastinate—quite the opposite. They hoped to regain control soon, but Hatch continued to stall the proceedings. The time had come to take their grievances directly to General Gillmore.

In a letter dated July 10, 1865, Singletary and five members, a quorum, of the company's board of directors restated their case. They were the legal representatives of the company, its property, its credits, and its creditors; and President Singletary was officially authorized to represent the board of directors by virtue of his election by the board. General Hatch had the power to repair the railroad for military use, and if he chose not to, they felt he had no right to keep it from them as he had. They also argued that Hatch had no legal authority to decide matters of equity, which he was obviously trying to do. They petitioned General Gillmore for possession of the railroad and asked for a prompt answer so that they could finally lay the issue to rest. His reply was not prompt, but it was favorable. On August 8, 1865—just shy of two months since their initial application—Gillmore turned the railroad over to Singletary and the board.[23]

Singletary soon had a team inspecting the rail line to assess the damages and cost of repairs. Because of the heat of the "sickly season" and poor conditions in the region, Superintendent Haines was able only to perform a cursory inspection of the line, but he soon returned with his report. Based on his estimates, thirty-eight and a half miles of track needed to be relaid, not counting trestles and bridges. Half, he believed, could be straightened out and reused; however, eight and a half miles were so badly damaged that the rails had to be rerolled, and the rest were beyond repair. The company had saved twelve miles of track, which were either safely in the hands of other companies or left untouched at different points along the route.

Haines estimated the cost of repairs at somewhere between $170,000 and $200,000. This price included the construction of new water tanks, but did not include the cost of drawbridges or shops, depots, and other buildings. The largest expense in the reconstruction would be incurred on the Georgia section, where Sherman's troops had so efficiently destroyed the rails. Haines recommended carrying out the repairs in South Carolina first and using a steamboat on the Savannah River to communicate with that city until crews completed repairs on the Georgia section. Excited at the prospects of resuming operation, Haines said that he could have a portion of the necessary timber delivered before the end of the month and was of the opinion that the railroad could be opened to the Edisto River three weeks after work was begun.

The question now became one of money. With the present state of affairs, finding cash—and investors who possessed it—in the South had become nearly impossible. The Charleston & Savannah Railroad began its rise with £1,900 sterling that the company squirreled away in England during the war, likely from the sale of cotton. They also had approximately £1,000 worth of railroad supplies and equipment in Nassau, Bahamas, from the company's limited blockade-running activity.[24] Singletary also saw a great opportunity in hiring out the company's rolling stock. All 12 engines and 112 of its cars had survived the war intact, and he believed they could bring in $250 to $300 a day. Yet they needed a reliable source of cash.

To find the money needed to rebuild and retool the railroad, they decided to explore local options first. A committee of the board of directors contacted a Charleston financier, but reached an impasse. In his reply the individual—who remained nameless in all corporate records—explained to the directors the means he would use in carrying out the railroad's reconstruction and outlined conditions on which he might use them to benefit the company. Whether the potential investor was

George A. Trenholm, George W. Williams, or someone else, the guarantee he demanded for the disbursement and repayment of two hundred thousand dollars did not meet with the board's approval. These conditions being unacceptable, the board struck him from their list of prospective benefactors and began to look elsewhere.[25]

In early autumn a member of the company's board of directors, Frederick Richards, traveled north to New York City, where he met with some of the city's most influential investors. When he submitted to them a copy of the corporation's financial statement, however, it was "regarded by them as so unfavorable that any further attempt to procure the full amount needed would have been fruitless." The railroad also applied to the War Department in Washington for the purchase of enough iron to repair the portion of the road in Georgia, but they were unsuccessful. Their attempts at fund raising brought them subscriptions for a total of thirty thousand dollars, and in the end even that was unavailable because of the terms imposed by the lender.[26] Disappointed by their inability to raise capital, the board refused to give up. They had lost much valuable time, and their property was deteriorating from lack of use. They concluded that they must proceed with whatever limited means they had on hand.

Like the railroads for which they worked, most Southern railroad men had lost everything they had. Many were glad enough that the war was over and were ready to return to work. The railroads to which they returned did not resemble those they had left behind. The rights-of-way were strewn with gnarled rails, chairs, and spikes; the rail beds were overgrown with weeds. Not only did the track need relaying, but new depots had to be built, and bridges needed to be reconstructed.

As they forged ahead into the economic unknown, the Charleston & Savannah Railroad Company took steps to retain its employees by providing them some measure of financial security. On the resumption of operations the salary of the president was set at $2,000; the superintendent's was set at $1,500; and the secretary-treasurer's was set at $1,200. While these figures were numerically inferior to their previous salaries, they were several times their value in real dollars. Learning from past mistakes, they fixed the salaries of their employees on the gold standard. In a show of good faith, they made the salaries retroactive to May, later pushed back to January 1. The company expected some turnover but was surprised in October by the resignation of Secretary-Treasurer William H. Swinton. The board unanimously appointed B. M. Lee to assume the position. Fewer than two months into his term, he resigned because of health issues and was replaced by Samuel W. Fisher.[27]

Having assembled a workforce by September 1, the Charleston & Savannah brought back its cars to Charleston from Cheraw and made preparations to transport them across the Ashley River. By October 14, an engine and train were in place on the railroad and repairs were well underway on the bridge at Rantowles Creek—this time without a drawbridge. It took two months—not the three weeks Haines had predicted—but the road was finally opened to Adams Run on November 18 and to the Edisto River on November 20. Soon a thrice-weekly passenger service was in operation on the twenty-nine mile stretch between St. Andrews and the Edisto.[28]

In the latter part of December crews began work on the Edisto River Bridge. The planned structure would likely be inferior to the one it replaced, but Haines felt it would be reliable for five or six years as long as it remained free from damage. Because all available resources for obtaining timber were needed for rebuilding bridges and trestles, only one-tenth of the necessary cross ties had been placed on the road. Further complicating matters, the cost of timber and labor had risen to such an extent that Haines had to increase his original estimate of the cost of reconstruction by twenty-five thousand dollars. The inaccessibility of timber and another stretch of inclement weather delayed progress, but they hoped the bridge would be ready for passage by February.[29]

With the expectation of expanding services, Haines also began to concentrate on refurbishing his rolling stock. He had nineteen cars and two engines already in operation on the road, but the majority of the rest were on other roads. The South Carolina Railroad was using seven engines and sixty-six cars belonging to the Charleston & Savannah, and a few other cars were distributed among the Wilmington & Manchester, the Northeastern, and the Charlotte & South Carolina railroads. Almost all the Charleston & Savannah's rolling stock was in good running order; however, five of its eight-first class coaches needed extensive repairs, which had to be done in the city. In the absence of a steamer, rowboats and flats were used to ferry supplies and other material across the Ashley River, but they could not be relied on for freight or passengers, especially in bad weather. Haines needed a shop—even if just a small one—with a few machine tools in St. Andrews to reduce the amount of time and money lost in transferring the cars back and forth, but there was not enough money for it.

If Haines intended to expand the railroad's services, he would need a more reliable method for ferrying freight and passengers across the Ashley. One month after resuming passenger service, the company had netted only $965. On the other hand, receipts from car hire totaled $17,170, bringing the company's revenues up to $18,135 for the year—against expenditures of $30,823. In an effort to raise additional funds Singletary arranged to sell the company's saw mill, which was valued at around $15,000. The board also entertained contract offers from communications companies such as the American Telegraph Company and the National Express Company, but they were hesitant to accept either offer. When Samuel W. Fisher closed the books on fiscal 1865, the company owed more than $340,000 in interest on its $1.9 million debt and had little more than $1,000 cash on hand.[30]

This was all that remained of the railroad Singletary had built when no one else would and then nursed through a grueling war. It was painful for Singletary to recall the past but even more agonizing for him to stare at the present. Gone was the ample supply of slave labor that so cheaply graded the rights-of-way and laid the rails of the South. These former slaves now roamed the streets of Charleston celebrating their emancipation. Gone also were the antebellum days of ready capital and liberal credit. The railroad had tried—both at home and in the North—to raise the funds needed to keep the company afloat but had been unable to do so. Still they pressed on.

By the end of December the Charleston & Savannah Railroad proudly announced to their customers a ferry service across the Ashley River between their Mill Street depot and St. Andrews wharf. Stagecoaches were also running between the railroad's terminus on the Edisto River and Walterboro.[31] Even as the first ferry crossed the Ashley, the South Carolina legislature was in session, considering a bill that would provide some relief to the railroad. It would be timely, but would it be enough?

EIGHTEEN

Postwar Debts

Like other Southern railroads, the Charleston & Savannah had suffered greatly during the war. It had operated beyond its capacity, put its cars and employees in harm's way, and had suffered severe material losses in its aftermath. Now, along with the state that had granted its charter, it needed to begin the process of recovery.

With the close of hostilities South Carolinians began to concentrate again on railroad projects with the hope that they would somehow spur industrial development. Slavery was abolished, providing a impetus to look beyond the old agrarian system for the route to renewed prosperity. No doubt, there would still be an interest in agriculture—evidenced by the high price of cotton—but thoughts were beginning to gravitate toward an economic revival based along industrial lines. Editorials began to tout the manufacturing potential of the state and the great need of railroads toward that end. The old system had failed, and there must be a revolution of sorts, not only politically but also economically.

It is difficult for anyone not living at that particular time and at that particular place to comprehend fully the economic devastation inflicted on this part of the South during the war. In one fell swoop the monetary issues of every state and national bank, as well as the government, were rendered valueless. Men had absolutely nothing to show for years of hard work except for rolls and rolls of currency that had no more value than ornamental waste paper. Nearly every form of investment shared the same fate, and all that was left of the wealth of the region lay in the land itself. Slavery had been the driving force of the economy—and social life. The upheaval caused by emancipation left a far more lingering effect than the burning and plundering of the Union army. Physical objects could be replaced; the abolition of slavery did away with generations of habits and emotions, and filling the void left by their absence imposed lasting effects on the lives of former slaves, former masters, their families, and their future families.

During the summer of 1865 commercial activity in Charleston began to revive. Storefronts in the business district on King Street were again buzzing with activity, and the sales of cotton and rice were starting to show a profit. President Andrew Johnson reopened the port of Charleston in July 1865, and Northern capitalists flocked to the city to invest in cotton, rice, and various other business enterprises.

Historical marker at the site where the Charleston & Savannah ran along the Stono River, now part of the Links golf course at Stono Ferry. Photograph by the author

With the establishment of the Merchants' Exchange, the charter of a national bank, and the reopening of the city's railroads, Charlestonians saw themselves on the cusp of becoming once again the commercial hub of the South. Merchants were enthusiastic over their recent successes, but they could not expect the heavy volume of business they wanted until the railroads were again fully operational.

Many of their future aspirations were tied to the rail connections with the interior. Residents of country towns needed clothing, tools, and other sundries, and most felt that the railroads would be more likely to get these folks to market than any other means of transportation. Fed by pride and the desire to return the city to its former heights of prosperity, the public's expectations grew into demands, putting heavy pressure on the railroad companies. Evidence of this may be found on the pages of the *Daily Courier*, where—in specific relation to the rail companies—a writer spoke of the people's intent to "exact a rigid faithfulness to those public duties which, if neglected, retard the onward movement of this city. To look at our white population, one would say there is no idleness here; the shoulder is against the wheel, and turn it must and shall. This is the will of the people, and we expect no draw backs from any quarter."[1]

In truth, however, the people's expectations may have been too high. While the economy of Charleston was, indeed, beginning to revive, all was not yet back to normal. The military was still in control of the state, its courts, and its security. There was scarcely any money in circulation, and much of the trade was conducted on the barter system. Railroad executives found conditions chaotic. There was no longer a dependable supply of labor, no available funds, and—because of the great losses, heavy debt, and disorganized condition of the economy—no credit. Despite these obstacles, they made plans to rebuild their roads and reestablish viable freight and passenger service.[2]

It was expected that the state legislature would again come to the rescue of worthy railroad enterprises. The legislatures called under President Johnson's plan of Reconstruction began to dole out state aid for internal improvements and were almost reckless in granting credit. In the South Carolina Statehouse sympathetic members of the legislature introduced a bill proffering assistance to the Charleston & Savannah Railroad. The bill authorized the company to borrow, or raise, up to five hundred thousand dollars by the issue of bonds on a first mortgage of their property. The bonds, which would come due twenty years from their issue date, would carry a priority of lien over all other liens, bonds, mortgages, and debts of the road, including its real-estate holdings. All this was, of course, contingent on the written approval of the current holders of unendorsed bonds. In turn the legislature would also agree to postpone the existing lien of the state, making it a second lien on the property of the road. In late December 1865 the bill was passed by both houses and was submitted for the governor's signature.[3]

To discuss the passage of the relief bill, Singletary called a special meeting of the railroad's board of directors on January 12, 1866. Things were starting to move along on railroad reconstruction although work was going more slowly than projected. Before the meeting began, Haines reported on how much more expensive timber—and labor—had gotten. Before, they had been able to insulate themselves from labor expenses by reimbursing slave owners either with borrowed money or with railroad securities. Now they had to pay wages—a direct cost whose impact was more immediately felt. Fisher's report clearly showed expenses out-pacing receipts. Haines's projections were only slightly better, but hiring out the rolling stock would carry them only so far. With no credit and no cash, Singletary was starting to feel the pressure. The legislature had passed the relief bill—had opened the door to a half-million dollars—but Singletary had the unenviable task of gaining approval from the first-mortgage bondholders.

Public sentiment was turning against the railroad, and the newspapers were certainly not helping the directors' cause. The *Daily Courier* had openly criticized Singletary's and Haines's leadership, misleading readers into thinking that Haines and Singletary were to blame for the railroad's slow return to operation while General Hatch had been the true impediment from the beginning. The paper excitedly announced the return of the Northeastern and South Carolina railroads to their owners in June, but when General Gillmore finally handed the Charleston & Savannah Railroad back to Singletary and the directors in August, no mention was made in any of the city's major newspapers.

Singletary had noted the conspicuous absence of several of the company's directors during recent meetings. There were rumblings of a takeover attempt by a group of first-mortgage (second-lien) creditors. He could not say that he was surprised. He had suspected it for some time. John Ryan's relationship with the railroad had been contentious ever since the bondholders' threatened foreclosure in 1862. Singletary knew that Confederate commissary general Lucius B. Northrop had formally charged Ryan with embezzlement when Ryan served as regimental commissary under Beauregard, and although Ryan was never convicted, it was cause for Singletary to keep close tabs on him. Ryan was an opportunist and a savvy businessman; a Charleston agent for R. G. Dun & Company described him as "a sharp, shrewd, tricky man" with a bad reputation.[4] While Singletary did not know Ryan's exact motives, he was certain that they were not altruistic.[5]

For weeks the questions swirled in Singletary's mind. Was it the bondholders who were swaying public opinion? Did the bondholders control the newspapers? How is it that they held so much influence with the local military officers? He was not sure, but the movement was certainly gaining momentum. All signs pointed to Ryan as the catalyst, but this time he was not alone. He had aligned himself with some of the company's present and former directors, including James H. Taylor, Theodore D. Wagner, Moses Cohen Mordecai, and George Walton Williams.

Williams was one of the few truly wealthy individuals remaining in postwar Charleston. He had begun his professional life in Augusta, Georgia, as a protégé—and later partner—of Daniel Hand, a transplanted New Yorker. The Hand & Williams Company had become so prosperous that they had begun looking for other cities to include in their commercial domain. Charleston, with its railroad and port facilities, was the logical choice, and in 1852 Williams had started a wholesale grocery firm there with Hand's capital. George W. Williams & Company was involved in importing sugar and molasses from the West Indies, and by 1860 Williams had amassed stores, warehouses, and an industrial complex on the peninsula. A shrewd and opportunistic businessman, he had added to his fortune during the war by participating in blockade-running ventures, bringing anything from guns and medicine for the Confederate army to food and luxury items into the state.

When the Union army entered Charleston on February 18, 1865, city councilman George W. Williams had presented Union lieutenant colonel Augustus G. Bennett with Mayor Charles Macbeth's notification that the Confederate military had evacuated the city. Soon the Union command was asking Williams to lead a committee of prominent citizens charged with distributing rice, grist, and salt seized by the Union army. Having won a measure of trust from the Federal authorities, Williams was asked to travel to Washington in June with a delegation led by Judge Edward Frost, who planned to ask President Johnson to restore civil government to the city.[6]

In the midst of the destruction Williams also set about reconfiguring his financial empire. During the summer he was elected president of Charleston's Merchants' Exchange, rebuilt the Charleston Iron Works, and established a large cotton press. He and his associates reopened their business. They immediately began clearing away rubble and demolished structures along Church, Pinckney, Anson, and East Bay streets. When the city passed an ordinance offering low-interest loans to those rebuilding with brick, he replaced the ruins of the Burnt District with fifteen large warehouses for the storage of rice, cotton, and fertilizer. His company had bought sterling exchange in 1861 at $1.03 in Confederate money and sold it in 1865 at $2.25 U.S. money. With a portion of the capital obtained from this sale, he was able to establish the First National Bank of Charleston, which was organized on November 21, 1865, with Andrew Simonds as president and Williams at the helm of its board of directors.[7]

Williams had long known the benefits of railroad connections and knew that they would be essential to both his and Charleston's future prosperity. He possessed a formidable combination of local business ties, Federal influence, and wealth. He also had ties to the Charleston & Savannah Railroad. He was a holder of a significant number of the company's unendorsed bonds and had formerly sat on its board of directors. First elected in 1859, he served for three years before being voted off in 1862. One of the new directors elected in 1862 had been a surprise nominee at the company's stockholders' meeting—Singletary. Singletary was acutely aware of all these undercurrents as he prepared for the meeting. He also knew that Williams and his railroad syndicate were taking a strong interest in Charleston's railroads and their connectors.

As soon as a quorum was present, Singletary called the meeting to order. Fisher dispensed with some general housekeeping measures, and the minutes from the previous meeting were read and approved. Wasting no time, Singletary then announced the passage of the relief bill and read it in its entirety. As expected, the portion that generated the most discussion was that involving the second-lien creditors. These men might not easily accede to postponing their lien on the company. The directors wanted to make certain that these gentlemen were given all due consideration. Frederick Richards made a motion that a meeting of the second-lien bondholders be called to consider the act on February 14, 1866, and that advertisement of the meeting should begin immediately.[8] The bondholders met on February 14, but after some discussion, those in attendance decided to adjourn the meeting and reconvene on February 23, two days after the company's annual stockholders' meeting.

The stockholders met on February 21 at the hall of the Charleston Insurance and Trust Company. The meeting began with the usual formalities, and the reports of Singletary and Superintendent Haines were read and approved. In his report Singletary spoke of the difficulties of the company—its physical and financial condition. The company had begun operations on shaky ground, and it looked less and less like they would be able to pay the interest on their bonded debt. The South Carolina

General Assembly had offered assistance; however, the company's fate rested with its creditors. He had a plan that he hoped would appease them and avert a foreclosure.

Singletary's proposal offered the creditors preferred stock in the company in exchange for the creditors' unendorsed bonds. Holders of the first-mortgage (second-lien) bonds would receive preferred stock "Class Number One"; second-mortgage (third-lien) bondholders would receive preferred stock "Class Number Two"; and current stockholders would trade their certificates for new ones to be labeled as "Class Number Three." New earnings of the company would first be applied toward the interest of the state-guaranty bonds. If any surplus were left, it would be paid to the stockholders as dividends as follows: first to first-class stockholders up to 7 percent and next to second-class stockholders up to 7 percent; the remainder would be divided equally among all three classes. The plan was based on a similar offer tendered to the bondholders at a meeting on May 23, 1861, but because of the unsettled conditions caused by the start of the war, they had not been able to give it full consideration.

The next item on the agenda was the election of the president and board of directors, which brought two new members to the direction of the road. Charleston mayor Peter C. Gaillard, had been named to the board a month earlier, replacing J. K. Sass, who had died near the end of the war. Also new to the board was Andrew Simonds, president of the First National Bank of Charleston, who replaced Charles Taylor Mitchell. These gentlemen, along with Singletary and the other ten directors, were elected unanimously.

After ample time had been allowed to study Singletary's plan and the legislature's relief act, Charles M. Furman offered a resolution that the application made for renewal of the state guaranty of the company's bonds and the postponement of the state's lien be approved. He also moved to accept the provision of the relief act and to authorize the board of directors to issue bonds in place of those coming due, to exchange bonds for stock, or to make any other arrangements that would benefit the company. By acclamation they passed the resolution, and with no further business to discuss the meeting adjourned.[9]

Two days later the company's bondholders met in the same hall to consider their options. With Richard Yeadon again presiding, the act of the legislature was read, and a statement of the Charleston & Savannah Railroad's liabilities was circulated. After a brief discussion Singletary took the floor to explain the company's position. Even before the war's end, it had been apparent that meeting the seemingly insurmountable debt of $2 million would be extremely difficult. As early as May of 1861, the stockholders had felt that the road belonged to its creditors and had authorized the board of directors to offer it to them on the terms outlined earlier. This scheme, Singletary explained, was still open to them. The company had a funded debt of $1.9 million and arrears of interest totaling $342,000, making it almost impossible to affect any kind of loan. After the directors persuaded military officials to give them possession of the road, the only resources they had were from the rent of rolling stock. By passing the relief act the state agreed to postpone their lien if the second- and third-lien creditors would also postpone theirs. If the bondholders accepted the terms, the railroad could be rebuilt, but the stockholders would remain its owners and managers. The stockholders, however, preferred that the bondholders run the road—still a possibility based on the 1861 plan. If all parties agreed to this, the legislators would gladly amend the act to reflect the change, and it would not affect the offer of state aid.

Minus the expected receipts of the road, Singletary calculated that it would take $285,000 to keep down interest and rebuild the road—but without paying anything on the arrears of interest. This amount would also leave the road without a depot in Charleston and leave it dependent on the Georgia Central Railroad for its Savannah depot and for three miles of track upon which to enter the city. The Savannah River would have to be crossed via a temporary bridge. To finish the road with the aforementioned changes would take between $400,000 and $500,000. Obtaining a loan of this size in the present economic climate would be next to impossible.

As Singletary saw it, the creditors could either authorize the stockholders to borrow the necessary funds or take possession of the road by one of three ways. First, they could foreclose and sell the road under the terms of the first mortgage. This would allow payment of the state and the arrears of interest, but the remaining creditors and stockholders would lose everything. As an alternative, the creditors could take possession of the road and manage it. By this plan they would still have to borrow the money to rebuild the road, pay the interest owed to the state, apply any surplus to the liquidation of their debt, and be liable to being forced into a sale by the second-mortgage creditors. Finally the creditors could accept the offer of preferred stock by the shareholders. This would convert the lien into ownership while making no change in the security of the funded debt. This plan would give the creditors control of the company.

Singletary knew that a small coalition of bondholders could derail either of the two plans that would save the road; however, doing so would benefit neither Charleston's economy nor the bondholders themselves. After his explanation and an ensuing discussion, a resolution was offered, seconded, and approved that the bondholders accept the company's offer of preferred stock in exchange for the amount of their liens as proscribed in the stockholders' 1861 plan. The management of the company would be placed into the hands of the creditors, but the City of Charleston and the State of South Carolina would still retain a significant interest. The measure passed with no dissenting votes cast, and the proceedings of the meeting were published in the *Courier*.[10] Singletary must have wondered if he had won this small victory a little too easily. If so, his answer was soon forthcoming.

Proceedings of the meeting appeared in the newspapers in mid-March along with an advertisement in which the railroad company requested all bond creditors to turn in their bonds and asked them to acknowledge their agreement with the resolution passed at the February 23 meeting. While some began returning their bonds, the article spurred a great deal of discussion among a number of bondholders. The report of the company's finances had not been a good one. In fact it alleged the company's insolvency. The Charleston & Savannah was without any resources other than the rolling stock it owned and the road itself. The lien debt was staggering, and the stockholders had admitted—just as they had five years prior—that their company belonged in principle to its lien creditors. With both the principal and interest of the first and second mortgages unmet, several of the men were flabbergasted that Singletary was now asking for anywhere from four hundred thousand to five hundred thousand dollars to put the road back in operation. Even though his plan had been approved unanimously at the bondholders' meeting, significant questions were being raised about it.

Some of the bondholders had had misgivings about the railroad's management for years. Although they had the power to nominate half the board of directors, some felt they had not been allowed to participate fully in the railroad's operation. Several believed that it would take only two hundred thousand dollars to put the road back in running order and balked at Singletary's estimates. They also argued that the lienholders (some of whom had connections to national banks and Northern capitalists) were best equipped to provide this amount. Some believed that they would be risking a total loss with Singletary's plan and that they would have new creditors —namely the South Carolina General Assembly—substituted in their place. They argued that they were being forced to "give up their security for a contingency" and that the new debt would have precedence over the one they now held.[11]

If the bondholders decided to foreclose, they might realize a larger percentage of their claim against the company, but the stockholders would lose the ability to profit from any future success of the railroad. An alternative proposal was to farm out the railroad for a few years for an amount sufficient to repay its debts with interest and put the road back in operation. Some considered either of these alternatives preferable to Singletary's plan.

There was, however, one scheme that seemed to be gaining favor among the bondholders. It advocated that the railroad be surrendered to the bondholders because it had been recognized that the company's only asset was, in effect, their property. In this manner payment to the creditors would depend on the effectiveness of their management. They would assume control of the railroad's interests with the understanding that ownership would revert to the stockholders as soon as their liens were satisfied. To this faction of creditors, this compromise seemed to be the best alternative. The stockholders would retain possession of the road, and its management would rest with the bondholders. No matter which proposal sounded the most practical to either side, it was clear to all involved that a battle was heating up for the control of the railroad.[12]

While the bondholders and shareholders argued over the paper ownership of the company, Haines and his lieutenants busied themselves with its day-to-day operation. These were the people who put the trains in motion, laid the track, hammered the spikes, and fired the engines. Haines listened with interest to reports of aid being given to other railroads. In Georgia the United States government built temporary trestles on the Western & Atlantic and granted monetary assistance to other roads. Despite its proximity to Beaufort, however, the Charleston & Savannah had not been deemed of military importance. In South Carolina neighboring roads were able to attract wealthy Northern investors such as D. F. Appleton, R. E. Robbins, and Morris K. Jesup; however, there were no such suitors for the Charleston & Savannah.

It was unsettling for Haines to work from paycheck to paycheck, not knowing if the company would be able to pay him. Its long-term prospects did not look promising. That winter another company offered him a job, promising him a permanent position with a salary of $2,500 per year and an interest in the company. It was a good offer—one sure to provide for him and his family. Haines wanted to show his loyalty to the Charleston & Savannah, but he also had to look out for his own interests. After fretting over the decision, he took the issue before the board of directors, who had every intention of retaining their valuable superintendent and chief engineer. On a motion from Richard Yeadon, the board voted to raise Haines's salary to $3,000 per year—enough to secure his services and assure the continuation of the work at hand, which was progressing slowly but steadily.[13]

Ferry service across the Ashley River resumed in December 1866 with the use of barges, but Haines actively sought to replace the barges with a chartered steamboat. Services of the railroad were extended by the rickety stagecoaches running between its terminus and Walterboro. Haines was taking bids for the delivery of twenty thousand cross ties to be used between the Ashepoo and Salkehatchie, while Singletary negotiated terms for the purchase and sale of railroad iron owed them by the South Carolina Railroad since the war. Under the close supervision of Haines and J. W. Craig, new bridges were built across Rantowles Creek and the Edisto River. Though hindered by shortage of materials and inclement weather, construction of the road was finally completed to Anderson, a new stop between Jacksonboro and Green Pond, but storage facilities for freight had not yet been built.[14]

Without a thriving freight and passenger business, the company was still reliant on receipts from the hire of its rolling stock. To increase revenues Haines and Singletary looked to express companies in an effort to expand services. Although they had declined on a contract with the Southern Union Telegraph Company, they agreed to terms with the National Express Company, and the Southern Express—headed by Henry B. Plant—operated at all of the railroad's stations. The board of directors also considered a proposal from Mrs. William Elliott for the construction of an eating establishment on the road at Oak Lawn with the thought of eventually establishing another station there. While they assured her that they were interested in her idea, they could not commit to such an undertaking with the current financial condition of the road.

As they did when they needed cash in 1864, the company looked to their real-estate holdings. After being approached by several parties to lease the property,

Causeway originally built for tracks of the Charleston & Savannah Railroad, now occupied by a cart path next to the thirteenth fairway at the Links at Stono Ferry. Photograph by the author

Singletary decided to call in debts owed on several pieces of real estate sold before the war's end. The board authorized him to ask the planters holding the mortgages on the Whitehall, Chartrand, and Price's tracts for payment in full. If they declined, he could lease them to the highest bidder. At one point Singletary and the board also considered taking out a mortgage on the company's Spring Street property but decided against doing so when they determined that it was covered under the company's second mortgage.[15]

Work began to progress more rapidly, and on March 21 an excursion was planned that was as much for public-relations purposes as it was to inspect the line. Directors and officers of the railroad, along with local dignitaries and businessmen —such as John Henry Honour, William F. Hutson, Edward Sebring, and John E. Carew—were treated to an excursion to and dinner at the railroad's western terminus before returning to Charleston. During the meal Singletary announced to those in attendance that the road would be completed to the Ashepoo River—a total of thirty-seven miles—by March 24. The *Charleston Daily Courier* reported the following day that most of those who rode the rails that day were confident that construction would soon be pushed to its completion.[16]

In April passengers on the Charleston & Savannah Railroad departed from the Mill Street Depot every Tuesday, Thursday, and Saturday, crossed the Ashley River by steamer, and boarded the train at St. Andrews Depot. Under the direction of conductor William Crovatt and engineer H. H. Crovatt, the locomotive *Edisto* left the station at 8:00 A.M. on the short trip to the end of the line. All along the route travelers passed the charred remnants of once-stately houses, overgrown gardens, and abandoned Confederate earthworks and batteries. But planters were optimistic about the upcoming cotton crop, and the road stood to see significantly more activity as a result. Despite the tragic death of their foreman in a train accident, work continued on schedule. Crews under the direction of Haines and Craig had completed the 1,600-foot trestle across the Ashepoo, and the western limit of the road was now Whitehall Station, forty-three miles from Charleston.[17]

Just as things were beginning to look up for the railroad, a movement was started that rocked the already tenuous foundation that had been rebuilt since the war. A group of bondholders who had evidently been absent from the previous bondholders' meeting were either increasingly dissatisfied with the progress of the road or more anxious to gain control of it. This small contingent of creditors had up until this point been quietly gaining support, but they were about to take their campaign to the public in a series of articles in the *Daily Courier* that criticized the fiscal management of the company—its handling of the debt service, its extravagant payroll, and its stewardship of the company's assets. These pieces were published during the month of April and spelled the beginning of the end for Singletary and the present management of the company.

Even though the bondholders had been granted the right in 1862 to appoint half the board of directors, these creditors felt their interests were not being represented in the company's boardroom. While the directors had seemingly lost interest in its operation, the company had lost money almost every year and had now reached the point at which everyone—including Singletary—agreed it had reached insolvency. The amount of the railroad's indebtedness exceeded its assets by nearly two million dollars, and to these men, the time had arrived for the bondholders to take action. They repeatedly asserted that they were the true owners of the railroad and could see only one resolution—to fulfill the promise made to the bondholders in 1861 by surrendering the railroad to them. The central question they posed—first to the principals of the railroad and then to the citizenry of Charleston—was whether they, as the company's rightful owners, should control its operations and affairs, or whether they would continue to be denied representation in its administration.[18]

The first salvoes fired in this battle over control of the railroad were directed toward the company's president, R. L. Singletary. The disgruntled lienholders attempted to discredit his prior management of the company by claiming that, while the net earnings of the company in 1863 were enough to pay a 24 percent dividend on the company's $1 million in stock, the company had declared none. Singletary wanted the surplus—after paying expenses and interest on the funded debt—to pay for the completion of permanent bridges across the Savannah and Ashley rivers and three miles of track into the city of Savannah. While a road that is in the process of construction is not expected to pay a dividend, the creditors pointed out that no material had been purchased nor had any serious attempt been made to complete the projects. In addition they claimed that no dividend had been paid on the stock, despite the fact that the company had enough to pay three years' dividend of 7 percent on the stock. By their calculations the loss of the surplus resulting from the collapse of the Confederacy increased the real cost of the railroad to $4.3 million and caused the City of Charleston, the State of South Carolina, and the individual stockholders to absorb a loss of $1.2 million. A few more years of similar management and their bonds might not be worth the paper on which they were printed.[19]

By exploiting the company's financial predicament, the bondholders appeared to have a very short collective memory. They were most likely acting out of blatant self-interest, spinning and twisting a mixture of facts, half truths, and lies into a persuasive argument that actually started to sway public opinion. It had been no picnic for Singletary or any other railroad president during the war. The Confederate government had convinced the railroads to accept payment for their services and aid in Confederate bonds, which devalued rapidly as the government printed money almost as fast as it spent it. The runaway inflation that ensued had derailed the commercial machinery, and the bonds had deprived the railroads of cash needed to pay the inflated costs for maintenance, replacement parts, and equipment.[20]

Wartime inflation, along with increased government business, had given many Southern railroads a profit on paper, and the Charleston & Savannah had been no different. When the company's books had shown a surplus in 1863, Singletary had declared a dividend—a fact that was publicized in the *Charleston Daily Courier* but apparently lost on the bondholders. The ledger and the newspapers, however, told only part of the story. In an effort to alleviate the company's debt Singletary and the directors had arranged for funds to be available at the Bank of Charleston sufficient to meet all amounts due on their bond coupons, and they gave public notice in newspapers across the state that they were prepared to meet their obligations. Despite their opportunity many of the bondholders—men such as John S. Ryan—had not availed themselves of it. Singletary had offered third-lien bonds and later currency in exchange for the bond coupons, but several of the larger creditors had refused to come to terms. It now appeared that they were perpetrating the same scheme as the bondholders of the Alabama & Florida Railroad. They had deliberately withheld the coupons due—and payable in Confederate money—and waited until after the war to present them for payment in United States currency.[21]

The reality was that the Confederate railroads had actually been operating at a loss. No real profits could have been realized, for it would take much more than all recorded surpluses to return the roads to their prewar condition. The bondholders ignored this fact. When the Southern economy—its financial institutions and its currency—met with ruin at war's end, the company lost all the funds it had set aside for the bondholders. Since then, all available funds had been devoted to the restoration of the road and could not be spared to pay off its debt.[22]

The bondholders argued that too much of this precious cash was being paid to the company's upper-level management. Despite its enormous debt, the company was burdened with the salaries of a president, superintendent, assistant superintendent, secretary-treasurer, construction supervisor, and master machinist—among other positions. The creditors were inclined to believe that one singularly focused individual could perform the jobs of both president and superintendent and that he—along with an engineer and secretary-treasurer—could carry out the business of the road. This arrangement would save six thousand dollars per year in wages, which was an amount sufficient to pay the interest on eighty-five thousand dollars of the bonded debt.

Money had been hard to come by, and the company's poor credit had deterred investors in Charleston and New York. Even the Federal government had refused the company aid. Its only source of income came from the hire of its rolling stock, and this was expected to fund its reconstruction. The assets of the railroad were depreciating as its bonded debt accumulated at a rate of $140,000 per year. This caused the bondholders to question the board's authority to farm out the company's assets even after they had declared bankruptcy—not to mention appropriating proceeds of the hire to pursuits other than the bond and lien debt—without consulting them. There was a significant amount of deferred interest due on the state-guaranty bonds, which covered the iron of the road, and the creditors thought this should have first priority. In fact the group took Singletary to task for selling the railroad's iron to the South Carolina Railroad during the latter stages of the war, claiming that the iron was taken up without assurances or a definite time frame for its return.[23]

This last point angered Singletary, for it was an outright lie. The iron had been taken up under the direction of him and Haines to keep it out of enemy hands and to aid Confederate railroad efforts in the state. Had General Hardee had his way, the rails might have ended up scattered throughout the Confederacy and lost forever with the collapse of the Confederate government. In coming to terms with the South Carolina Railroad Singletary had kept the iron close by—on a connecting line. The terms assured the return of the iron one year after the war's conclusion, or the Charleston & Savannah would be compensated in cash.[24] One could most certainly make the case that he had acted with the stockholders' and bondholders' best interests in mind.

Had mistakes been made? Most certainly. Had there been waste? Unfortunately yes. It had taken twelve weeks longer than promised to put the road back in operation between St. Andrews and the Edisto River. At the rate their debt was accruing, this equated to thirty-four thousand dollars in interest lost. But Singletary was simply playing the poor hand he had been dealt. Couldn't these gentlemen see that the Charleston & Savannah Railroad was not much different than the other Southern railroads? The company had struggled even before the war, and the war only exacerbated the situation. Now the same men who had put such pressure on him to rebuild the road were pressuring him to give up control of the company he had worked so hard to save.

The bondholders were not concerned with Singletary's loyalty. They wanted control of the railroad. They had watched as the principal on their lien was reduced to eighteen cents on the dollar, with deferred interest of about 2 percent per year (market value). Now the management had cooked up a scheme based on a railroad-relief act that they felt was as unjust as it was impractical. These gentlemen took exception to the state's lien taking precedence over theirs and objected to the stockholders' request to be allowed to rebuild the road at their expense and without regard for past experience. They held Singletary responsible for the railroad's fiduciary loss but held equally accountable the directors who voted for the plan, as well as those who failed to voice dissent at the subsequent bondholders' meeting. Knowing their rights as outlined in the mortgage and feeling the swell of momentum behind their movement, they called for a meeting of the creditors to take action.[25]

It was now time to take their argument out of the newspaper and into the board room. Five years had elapsed since the bondholders had refrained from foreclosing, and the railroad was still unable to pay either the interest or the principle on its bonds. The company had been under different management then, but that was of no consequence to them at present. The road was important not only to South Carolina but also to Charleston's economic recovery, and heavy losses had resulted from the city's isolation from the interior. Swayed by recent editorials, holders of the first-mortgage bonds called another meeting to reevaluate their position.

On May 30, 1866, creditors representing $575,000 of the first-mortgage (second-lien) bonds—much more than the $500,000 needed to conduct business—convened at the Planters and Mechanics Bank. The gathering, chaired by J. Reid Boylston, was charged with finding an amicable solution that would secure them from further risk or loss while preserving the interests of the stockholders. Boylston appointed a committee consisting of James H. Taylor, Alexander Isaacs, Edward W. Marshall, A. R. Chisholm, and John S. Ryan to advise the creditors of the proper course of action. After some deliberation, the panel prepared a report and submitted it to the railroad's board of directors during the first week of June.

The report stuck to the themes expressed in their newspaper correspondence. The creditors reiterated their claim to true ownership of the railroad and asserted that it was only just that they have control of railroad management. The bondholders professed to have the means to put the road back into full operation. They also believed that, if they managed it appropriately, they would be able to liquidate the liens they represented. It was this fact, they reasoned, that should assure the stockholders of their chief motivation, which was the ultimate success of the road. To increase their managerial input, they proposed that the bondholders should nominate the president and a majority of the board of directors. To this end they requested that the office of president and any nine directors be vacated and filled by individuals chosen by the creditors. This would preserve the interests of the shareholders and secure the rights of the bondholders without requiring the initiation of foreclosure proceedings. After he received their communiqué, an agitated Singletary called a special board meeting, at which he assigned William F. Hutson, William C. Bee, and Henry Gourdin the task of studying the issues addressed in the memorial and reporting on it at the next board meeting.[26]

As these gentlemen studied the document, they realized its implications, both overt and covert. The names of Taylor, Marshall, and their colleagues lent a certain credibility to the proposal and commanded their immediate respect. Yet Hutson, Bee, and Gourdin also knew that somewhere behind the scenes lurked George W. Williams with his wealth and business acumen. Hutson and his committee were impressed with the bondholders' plan and hailed the "liberal offer" as a "harbinger of a new and more fortunate change in the affairs of the company." In his report back to the directors Hutson pointed out two major drawbacks to the plan. A significant number—albeit a minority—of the creditors supported the plan that was approved by the stockholders at the annual meeting. Furthermore the first-mortgage bondholders represented only one class of security, and the interests of the second-mortgage (third-lien) creditors were being ignored in the process.

The bondholders claimed to possess enough capital to rebuild the road, and this point in particular was too important not to seek the approval of the third-lien creditors. Hutson had no reason to doubt the bondholders' sincerity or their ability to fulfill their promises, but it occurred to him that their resolutions were not legally binding. Since the stockholders were in effect trustees for all classes of creditors, the proposal would have to be made through the trustees of the mortgage to enable the directors to arrive at a decision. Certainly, he thought, there could be no objection to this.

Based on Hutson's report, Singletary sent his reply back to Taylor and his committee of bondholders. Since the stockholders had declined the bondholders' offer of capital once before, Singletary thought it pointless to resubmit the offer. Instead he wanted the creditors to accede to the state's offer of monetary relief with the added stipulation that the creditors be able to select nine of the company's twelve directors. If this were done, or if their previous proposal—along with their pledge of capital—were made in means capable of legal enforcement, the directors would submit the plan to the stockholders and lobby for its acceptance. Shortly after sending his response, he had Samuel W. Fisher write a letter to Governor James Orr to inform him of recent events. In his letter Fisher assured the governor that the president and board of directors realized the gravity of the matter and that they were doing their best to look after the interests of all classes of creditors alike, be they state-guaranty, first-mortgage, or second-mortgage creditors. The company also vowed to use every exertion in preventing "one portion from reaping advantages to the injury of the other."[27]

Singletary hoped that the bondholders would view his counteroffer favorably, but they did not. They considered the conditions outlined to be a rejection of their proposal. Frustrated in their efforts, they now felt it necessary to protect their own interests and claim their rights as set forth in the mortgage—that being to foreclose. At a meeting on July 19, 1866, the creditors authorized Taylor and his consortium to set the proceedings in motion. The railroad's legal counsel, Charles A. Simonton, advised Singletary to file a bill in equity to make all classes of creditors a party in the matter. Given the sum owed to the state as well as to the first- and second-mortgage bond coupons, doing so would assure the most legally appropriate distribution of the company's property. Such would not be necessary until the bondholders took action, and until then Singletary had the authority to retain possession of the road.[28]

When Independence Day was observed in Charleston, many of the city's white inhabitants refused to celebrate it. In fact General Dan Sickles noted that as long as he had been in Charleston, he had "never seen a Carolinian raise an American flag." Despite this attitude among some, Charleston's mercantile community—led by men such as George A. Trenholm and George W. Williams—tried to lengthen the city's reach by extending railroad connections northward and attempting to reestablish Charleston as a major port. These men focused on nationwide, rather than purely Southern, endeavors, and as they looked to make Charleston a "center of Southern trade," the unfinished state of the Charleston & Savannah Railroad was truly a glaring deficiency.

Ever so slowly, however, crews were working to rebuild the road segment by segment. For whatever the railroad could find to pay them, an odd amalgam of free blacks, immigrants, and Confederate veterans laid the rails that—many hoped—would bring back prosperity to the economically convalescing Port City. With Haines and Singletary running day-to-day operations, the board of directors found themselves with little to do. Their actions in the spring, while important, had been limited to matters such as reestablishing the salaries of certain employees and officers at prewar levels, disposing of pending legal actions against the road, and allowing Governor Orr to transport provisions for the needy over the road free of charge. In June 1866 they were called on to appoint a first-mortgage trustee to replace Erastus M. Beach, who had recently resigned his position.

Beach had moved to New York City at the end of the Civil War and tendered his resignation on June 1, 1866. The wording of the mortgage deed gave the company power to substitute and appoint a replacement in the event of the death, resignation, removal, or disability of any trustee. According to the railroad's attorney, failure to accept Beach's resignation and name his replacement within ninety days would allow a majority of the bondholders to appoint a trustee of their choosing, provided he resided in South Carolina or Georgia. Heeding the advice of counsel, the board immediately accepted Beach's resignation. Three weeks later, they endorsed the selection of John E. Carew.[29]

Considered a friend of the road, Carew and the two original trustees—Edward Sebring and Isaac W. Hayne—were called on to execute the foreclosure proceedings. Sebring was a past president of the State Bank of South Carolina and a director of the Southern Express Company. Hayne had served as South Carolina attorney general during the war and had long taken an interest in the state's railroad affairs. Early in 1861 Governor Pickens had sent him to demand the surrender of Fort

Sumter from President Buchanan. Now Hayne was asked to be an agent in the capitulation of the Charleston & Savannah Railroad.

On August 16, 1866, the railroad—its rights-of-way, bridges, buildings, and property—was advertised for sale in the *Daily Courier* by the first-mortgage trustees. Just two weeks earlier, the railroad had announced that repairs had been completed on Patterson's Bridge across the Salkehatchie River and that the road was open all the way to Salkehatchie Station. It had taken an entire year to complete the fifty miles, and what little had been accomplished had been done with great difficulty. Further complicating matters, the lease on the rolling stock hired by the South Carolina Railroad expired that month, drying up the company's main source of revenue. With business stagnant, cash flow nonexistent, and the railroad now up for sale, repair work ground to a screeching halt.[30]

There were many Charlestonians who thought the sale would rejuvenate the railroad immediately. With the delays in its completion Charleston merchants found themselves in the same position relative to east-west trade that they had been in before the railroad's charter in 1853. The word was that the syndicate planning to purchase the road would take steps to complete the track to Savannah and build bridges across the Savannah and Ashley rivers at the earliest instance. Others opposed the sale, claiming the prospective buyers were nothing more than "mere speculators." These "volunteer creditors" were alleged to have bought the bonds they held with Confederate money or to have purchased quantities at twenty cents on the dollar—and now were forcing a cash sale to realize a large profit at the expense of all other classes of security but their own. The proponents of the railroad did not deny that the bondholders had legal rights but questioned their motives in the manner and timing of their exercise.[31]

Railroad spikes from tracks laid by the Charleston & Savannah Railroad and its successors, found along the causeway near the old Stono Ferry. Photograph by the author

H. S. Haines submitted his resignation to the company shortly after it was placed on the open market. Throughout his stormy six-year tenure, he had been known for his enthusiasm, dependability, and expertise, but he was afraid that might not be enough. Given his connection with the current management of the company, he was not sure that the new owners would retain him. Earlier in the year Haines had received a tempting offer but had not taken it. Now he decided it was time to look out for his own interests, and he quickly accepted the position of general superintendent of the Atlantic & Gulf Railroad.[32]

Robert L. Singletary attached his signature to an injunction to block the sale of the railroad, which ended with the words "the matters and things set forth in the above bill, so far as they came within his own knowledge, are true, and so far as he has derived them from his own information of others he believes them to be true."[33] It was the final maneuver in his battle for the railroad; a battle he had been fighting for four years. He had done his best against adverse circumstances. There was a great deal of wealth behind the railroad syndicate. Ryan, Taylor, Williams, and their associates could muster large sums of cash—much more than could the ailing railroad and its stockholders—and their campaign was gaining momentum. It now seemed only a matter of time before Singletary would have to yield. The injunction Singletary had just signed was his final stand.

The firm of Simonton and Barker filed the injunction on October 27, 1866, with the goal of preventing the sale of the railroad. Every commercial enterprise in the state was in the same financial predicament as the Charleston & Savannah; Singletary and the board of directors hoped the courts would give them some latitude. In their argument they questioned the timing, ethics, and justice of the bondholders' actions. The trustees had, according to Singletary, advertised the sale of the road without presenting the coupons of the bonds or demanding their payment. If they had done so when the company had funds available, they would have been paid. Since those funds disappeared with the collapse of the Confederate government, the company's executives wanted the bondholders to suffer the consequences of their decision.

The railroad had purchased some of the property included in the sale after the execution of the mortgage, thus Singletary felt it should not be included. He also contended that the trustees were acting ex parte without making the state legislature or the third-lien creditors a party to the sale. This was perhaps a valid point; however, his final one was not. He objected to the railroad's being sold for cash on the grounds that it restricted the number of prospective buyers. Since the railroad owned several large tracts of land in addition to its rights-of-way and given the general lack of investment capital available in the state, the bids would be limited to a select few and would result in a lower purchase price. While this was certainly true, it was no basis for a legal argument. Singletary asked that the trustees submit a statement listing the names of the creditors along with the exact amount due on their coupons and that the railroad be allowed either to pay them or sell enough of the property subject to the lien to enable their payment. In the interim he requested that the trustees halt their sale of the road and withdraw their advertisement.[34]

The bondholders had hired attorney and former Charleston & Savannah Railroad director James Butler Campbell to represent them in the matter, and he was not worried. The Rogers Locomotive and Machine Works had also filed for a restraining order to prevent the sale of nearly eighty thousand dollars worth of locomotives for which the company had not yet paid them. The Charleston & Savannah was owed a good deal of money for the hire of locomotives to the South Carolina Railroad and the Spartanburg & Union line, and the manufacturer claimed title to this money by virtue of their lien. While Campbell thought the restraining order might prevent the Charleston & Savannah from selling the locomotives, he did not think it would prohibit the bondholders from doing so. All of the contingencies set forth in the mortgage deed had been met: nonpayment of interest, lapse of sixty days after default of payment, and the written request of holders of more than half of the unpaid bonds secured by the mortgage. In Campbell's opinion the trustees had not only a right but a duty to sell the premises—or at least enough to pay the principal and interest on the unpaid bonds. This point was expressly stated in the indenture. The bondholders' case was also bolstered by the directive that no "injunction or stay of proceeding, or any process be applied for" by the railroad to preclude its sale.[35]

While the legal issues were being untangled, very little was happening on the railroad. With no source of income, repair and construction work ceased as did care of the rolling stock. The company's wharves remained in their dilapidated state, and its rolling stock sat in the railroad yard without the benefit of shelter from the elements. Its embankments had been washed away, or their earth had been removed for use in building earthworks and batteries. Culverts and ditches were filled in; bridges, trestlework, and cross ties decayed. In November, Christopher S. Gadsden replaced Haines as the railroad's superintendent, but there was little left to superintend. Very little business was being transacted at the time. In fact the company's rails and turntable in the city of Charleston were being taken up, and they had begun to lease some if their coaches to local planters for storage and residence while their houses were being rebuilt. Travelers seeking a viable mode of transportation between Charleston and Savannah were forced to choose between rickety and expensive stage lines or a coastal steamer.[36]

When all disputes were adjudicated, the road was sold at the auction house of Wardlaw and Carew on November 20, 1866. James H. Taylor, George W. Williams, and a coalition of other financiers purchased the road for the meager sum of $30,000. The acquisition was subject to liens—including the $505,000 in state-guaranty bonds—totaling $750,000. That the 102-mile railroad, connecting two of the South's largest ports, was sacrificed for approximately $7,300 per mile was testament to the sociopolitical environment of the day. The new partnership appointed Williams, along with J. Reid Boylston, John S. Ryan, W. S. Hastie, and Savannahian Dr. Francis T. Willis as trustees as they set to the task of organizing their new company.

On December 21, 1866, a new charter was issued and the company was incorporated as the Savannah & Charleston Railroad. The new owners retired the $1 million of second-lien bonds and issued in its stead an equal amount of stock. At a preliminary organizational meeting on January 27, 1867, they elected Boylston as president pro tempore. The company's five-member board of directors consisted of Taylor, Isaacs, Ryan, Hastie, and Willis. Several days later, the chancellor of the Equity Court handed down his final ruling that the property of the Charleston & Savannah Railroad should be turned over to the Savannah & Charleston Railroad. According to the directive, proceeds of the sale were to be distributed equitably as a dividend. Most of the purchase money would come back to the newly formed corporation as holders of the second-lien bonds; however, claims made by the third-lien creditors and the Cannonsborough Wharf and Mill Company were acknowledged and were reserved for judgment at a later date.[37]

NINETEEN

Terminus

The Charleston & Savannah Railroad—its assets, including rights-of-way and other property—were legally turned over to the Savannah & Charleston Railroad on January 31, 1867.[1] There was no fanfare, only a meeting of the companies' executives and attorneys in the railroad's office at 28 Broad Street. With a few signatures, the railroad changed hands. It was the end of one era and the beginning of another. Robert Singletary had done his best, but the odds had been stacked against him from the beginning. Wartime destruction and the burden of the company's debt had been insurmountable. The railroad had been unable to realize Thomas Drayton's goal of becoming part of a great seaboard rail line. Once the transfer was completed, the company that had been projected by Drayton and Charles Jones Colcock ceased to exist on paper; however, their dream of linking the Northeast with the Gulf of Mexico had not been extinguished.

Even though the route of the railroad was the same, the new company was very different. Of the officers and directors of the old Charleston & Savannah Railroad, only the secretary-treasurer, Samuel W. Fisher, remained. Some men from the old company stayed in railroading, but others did not. Few of the men who had projected the road or served it as president were ever closely associated with the new company. During the war Colonel Charles Jones Colcock had done all in his power to prevent the destruction of the railroad. A loyal Southerner to the core, he had refused to place any of his financial holdings in England—risking total loss in the event of a Northern victory rather than weakening the faith of others in the Confederacy. After the war his energies were chiefly directed toward saving the Foot Point Land Company, whose purpose had been to build and develop a new city and port where the Colleton River flowed into Port Royal Sound. It was Colcock who persuaded the U.S. government to loan military stores on credit to local planters, thus enabling many to reestablish their plantations under the prohibitive economic circumstances of the day. With his expenses increasing, the labor supply unmanageable, and cash reserves nonexistent, Colcock decided to move his family to Savannah, Georgia, where he entered the life-insurance business. He eventually made his way back to Hampton County, South Carolina, where he contentedly lived out his days planting short-staple cotton with modest success.[2]

General Thomas Drayton never returned to the railroad after he and his brigade were ordered to Virginia. In July 1862 his brigade joined General Robert E. Lee's Army of Northern Virginia and was attached to Longstreet's corps. He took part in the battles of Thoroughfare Gap, Second Manassas (Bull Run), South Mountain, and Sharpsburg (Antietam), but his leadership skills were found lacking. General Lee thought Drayton a "gentleman and a soldier in his own person" but observed that he seemed "to lack the capacity to command." His brigade was broken up, and he was detailed as a member of a military court. He returned to the field a year later, commanding a brigade in Arkansas and then a subdistrict in Texas. His last service was presiding over a court of inquiry investigating a military incursion into Missouri.

On returning to his native state, he found that his plantation had been severely damaged by Union troops, and parcels of it were sold at tax sales. He moved his family to Dooly County, Georgia, and there attempted to resume his life. With money borrowed from his brother Percival, he started a farm, but this venture was unsuccessful. Soon afterward, he took a position with the Southern Life Insurance Company and moved to Charlotte, North Carolina, in 1871. Drayton lived to see the railroad he had founded become a segment in a larger and more commercially productive network of rails—one that he had envisioned nearly thirty years before it began to take shape.[3]

Two of Drayton's successors went on to further their careers in other railroad enterprises. Bentley D. Hasell left the Charleston & Savannah in 1863, when he was hired as president of the newly organized Shelby & Broad River Railroad. After the war he went into business with Cleland Huger selling railroad supplies, primarily to Southern railroads. The firm of Huger & Hasell was backed by New York railroad financier Morris K. Jesup, who advised the men to move their business to Savannah, which he felt would soon outpace Charleston in both population and capital. By the summer of 1867 the struggling company was heavily indebted to Jesup and had to be liquidated. In 1869 Hasell was offered the presidency of the Savannah & Charleston Railroad by its new management, but their offer was not sufficient to move him. He eventually headed north to New York, where he resided in 1880, listing his occupation as "railroad manager."[4]

Perhaps one of the more successful of the Charleston & Savannah officers was William Joy Magrath. The brother of judge and former governor Andrew Gordon Magrath, William Magrath served as president of the South Carolina Railroad for sixteen years, was an officer and director of two other railroads and two steamship lines, and for a short time headed the Southwestern Railroad Bank. Known for his selflessness in both private and public life, he saw his $10,000 a year salary from the South Carolina Railroad drop to $3,500 after the war. Had he been willing to take advantage of the "wrecked railroads" as so many had, he might have emerged from it all a very wealthy man. He sacrificed heavily in his efforts to hold the South Carolina Railroad, and could not bring himself to profit at its expense. At the end of

Magrath's term the South Carolina Railroad struggled through similar circumstances to those of the Charleston & Savannah. Forced into bankruptcy, it was placed in receivership in 1878 and sold to New York capitalists three years later. Even after his retirement his knowledge of the transportation industry was respected throughout the Southeast.[5]

The last in the line of those who took up Thomas Drayton's vision was Robert L. Singletary, who—like Magrath—was no stranger to sacrifice. Eager to carry on with his life, he lived briefly in Marion, South Carolina, before moving his family to Mars Bluff, near Florence, South Carolina. There, he established a plantation, and his family became active in Hopewell Presbyterian Church. Singletary became active in politics, serving in one of Wade Hampton's Red Shirt units with some of his neighbors. When Democratic clubs were formed during the summer of 1876, he was elected president of the Jefferies Township chapter. Though not actively involved in managing any of Florence's railroads, he maintained a correspondence with his former partner, Henry M. Drane—then president of the Wilmington & Manchester—about railroad affairs elsewhere.[6]

His years of service to the Charleston & Savannah Railroad had been a test of Singletary's endurance, and he felt its failure as a bitter defeat. He worked hard, kept better records than any of his forerunners, and ushered the road through a terrible war and the first stages of Reconstruction. It was difficult for him to see the foreclosure as anything short of a personal defeat. The company's bondholders had been financed with capital placed in England before the war's end; a practice frowned on by many gentlemen of the Old South. Although the scheme was not illegal, Singletary questioned its timing. In fact he had also placed a significant portion of the railroad's assets offshore, but he could only watch as his cash reserves dwindled to nothing in the stagnant economy. In the end Singletary and the board of directors felt the brunt of much criticism, but it was the fall of the Confederacy that bankrupted the road as much as—if not more than—their management of it.

All things considered, it is difficult to find a single cause for the demise of the Charleston & Savannah Railroad. While most fingers point to the havoc wreaked by the war, the truth is that the railroad struggled almost from the time of its charter. As with many Southern railroads, the root of its failure may have been the unwillingness of Southern investors to promote industrial development. Sponsorship of regional manufacturing ventures would have provided diversification and protected the railroad from the whims of the cotton market. Instead, they preferred only those risks involved in the business of growing and selling King Cotton. For Thomas Drayton this reluctance translated to the hindrance of construction crews and astronomical prices paid for rights-of-way. The planter class demonstrated plenty of entrepreneurial initiative in enterprises they valued but seemed to shun those that would advance their ability to survive a war.

The benefits of railroads were being realized all across the United States even before the outbreak of hostilities. Thanks to railroads, the cost of transportation had been reduced by 95 percent between 1815 and 1860, and the cost of mailing a letter had dropped from fifty cents to three cents. Railroads and instant telegraph communication illuminated for the first time the value of time to American business. As conductor N. J. Bell so eloquently stated, "The building of the railroads awakened the nation, and they found that they had slept a long time, and that time was worth more than they dreamed of."[7] The iron horse reduced the cost of living by 30 percent, and transformed the United States into an industrial giant among world powers. Even before the days of the transcontinental railroad, trains helped the United States assume its place as one of the wealthiest nations in the world.[8]

Yet in the South railroads were seen as "local roads, built with local capital, to serve local needs under local management."[9] Many Southerners saw the railroads only in the context of how they could benefit agricultural endeavors. Most were not built with any larger socioeconomic purpose in mind. They were precarious roads, many built through swamplands near the coast with track laid on dirt fill or built on crude trestlework or bridges that were susceptible to fire and flood. Timber was abundant, but obtaining it was expensive, and the heat and humidity were not conducive to longevity. Poor drainage led to rapid deterioration and frequent repairs, resulting in increased expense and time of construction. In general the capital invested in track materials covered only about 25 percent of the usual construction costs. These factors, coupled with sluggish investment, inflated the cost of the Charleston & Savannah's construction by more than $1.25 million and made it necessary to take on more debt just to stay in business.[10]

The most glaring deficiency of Southern railroads was the lack of an organized system of rails. Transportation over long distances was disrupted by uncommon gauges, the lack of connections between neighboring roads, and the existence of gaps where continuous lines were needed. The Charleston & Savannah was primed for westward travel and had a much shorter reach to the east. Its tracks connected to those of the Central of Georgia three miles outside of Savannah, and its five foot gauge was compatible with tracks stretching out to Thomasville, Georgia, to the southwest and upward to Atlanta, Georgia, and Chattanooga, Tennessee. At Charleston, however, a ride aboard a trawling steamer was required to connect the Charleston & Savannah to the South Carolina and Northeastern railroads before the war. Even when the Ashley River was finally bridged, compatibly gauged tracks ran only to Charlotte and Wilmington, North Carolina.[11]

The war made unprecedented demands on the railroads, which were neither prepared nor equipped for the task. The increase in volume necessitated changes from antebellum practices—neighboring roads needed connecting; rivers needed bridging; and incompatible gauges needed correcting. The demands of the government were so heavy that little capacity remained for available civilian business. Leaders and citizens alike were confident that the war would be short and that the railroads were adequate to sustain the workload of a brief conflict. To those whose focus was on food, clothing, medicine, and the other necessities of daily life, the reliability of the railroads was taken for granted. By the time the problems with the railroads were fully appreciated, it was too late to do much more than extemporize.[12]

The railroads required constant maintenance—rails and bridge pilings needed replacing and rolling stock needed repair. Many companies relied on Northern foundries before the war—a source that became unavailable as soon as the first shots were fired on Fort Sumter. The problems that quickly ensued were the result of the South's inferior industrial capacity and its failure to make the most of the resources it actually possessed. Further exacerbating its low manufacturing capability was the loss of experienced industrial workers to the military draft. Railroads also lost valuable workers, but this was soon rectified when certain key employees were exempted from military service.[13]

The cotton embargo initially curtailed the cotton-carrying roads' chief source of revenue, and the eventual tightening of the Union blockade worked a similar effect on those roads that were dependent on more diverse sources of income as well. Procuring railroad iron and spare parts became as difficult as finding the money to pay for them. Without them maintenance was often neglected and efficiency was reduced to critical levels. Singletary, Haines, and railroad executives throughout the Confederacy became acutely aware of the fact that maintenance entails a cost whether it is performed, delayed, or left undone.[14]

Without money for rails gaps in rail lines could not be filled. Without supplies the cars could not keep rolling. Many looked to the Confederate government for assistance, but officials were slow to respond. Across the Confederacy people blamed the Confederate officers' lack of knowledge of railroading and the ignorance of Confederate quartermasters about railroad practices and the efficient distribution of rolling stock. Although things were not perfect, neither of these seemed to be a problem on the Charleston & Savannah line, where Haines and Singletary enjoyed a level of cooperation with local military personnel that appears to have been the exception when compared to other Confederate railroads.[15]

Overall the Confederate government demonstrated a failure to understand the problems facing the railroads, a lack of industrial appreciation, and an inability or unwillingness to deal with the mechanics involved in alleviating their shortcomings. The Davis administration never took charge. It overlooked certain problems while responding slowly and timidly to some and totally mismanaging others. Davis never adopted central planning. He did not give adequate authority to the Railroad Bureau, which would have enabled efficient resource allocation and distribution of manpower. To his credit the Confederate president had to placate arrogant businessmen —so commonly tolerated in peacetime—who were unaccustomed to government regulation. In a time of crisis, however, a leader must lead. Davis had been given the power to assure their cooperation, but in the end he was hesitant to use it.

As the war wore on, the cost of supplies and maintenance outstripped income. Military demands choked off the railroads' regular business and the rates paid by the Confederate government were inadequate to sustain them. Cavalier railroad

Causeway leading to Frampton Creek on the old Mackay Point Road, crossed by Federal troops advancing toward the Confederate position on October 22, 1862. Photograph by the author

managers agreed at the Montgomery Convention to accept a rate of two cents per man per mile for troops and a sliding scale for the shipment of freight. These rates covered only half the railroads' prewar operating costs. Another foolish concession was their agreement to accept payment in Confederate bonds, which quickly lost value. It was a gallant and patriotic gesture, but by this scheme Southern railroads risked almost certain financial ruin.[16]

Although the executive and legislative branches of the Confederate government failed to enact and enforce an effective transportation policy, railroad officials can not be held blameless. Some Confederate railroad managers withheld the cooperation needed to organize a coherent system of railroads. Most Southern rail executives held fast to the precepts of laissez-faire capitalism and rejected much of the government's interference. They often refused to permit their rolling stock to be used on other roads or bickered with government agencies over the priorities of civilian versus military traffic. The military even found itself in competition with railroads over iron, which was becoming as scarce as it was essential to both.

Such an attitude was more the exception than the rule on the Charleston & Savannah Railroad. Both Thomas Drayton and Robert L. Singletary served in the military and understood the soldier's point of view. Both knew the importance of interior lines of communication and of rapid mobilization. They, along with Superintendent Haines, realized that the successes of their company and the Confederate armed forces in the lowcountry were unavoidably linked. The railroad was in one sense dangerously situated, but in another sense perfectly so. Without the military, the railroad and its bridges, which were built below the head of navigation, were vulnerable to almost certain attack from the enemy troops occupying the coast. By the same token the railroad supported and was the basis for establishing the defensive line between the important cities of Charleston and Savannah. It was vital for the rapid concentration of troops at threatened points along the line and for efficient transport of provisions, munitions, and other supplies. At least in this instance, survival took precedence over profit and power.

However, Singletary and Haines could not prevent the inevitable. They guided the debt-ridden railroad through the latter stages of the war and kept it virtually unbroken until a Southern army could no longer restrain Sherman's troops. Under the circumstances one wonders if Singletary could have done better. Though the railroad faced many hardships during the war, its ultimate fate was hopelessly tied to that of the Confederate States of America. With the collapse of the Confederacy, the South lost its unpaid African American workforce and its currency, making agricultural or other commercial activity in the one-dimensional economy nearly impossible. The devastation of the railroad and decay of its track and machinery—coupled with the chaotic state of the postwar Southern economy—made initial attempts to revive the business futile. Given the economic wasteland left in Sherman's wake, it is difficult to imagine that any scheme would have been successful. Lacking manpower, materials, and the income needed to pay for them, reconstruction efforts were excruciatingly slow. With interest accruing on the company's already staggering indebtedness and the pressure exerted by the company's creditors, Singletary was forced to make a difficult decision. His scheme of taking on additional debt in order to pay the interest on existing liens proved to be his undoing.

In *Victory Rode the Rails* George E. Turner reminds us that little credit was given to "the courageous and determined railroad men who made the most of the little they had and kept the trains rolling, though ever so slowly and intermittently. Devising all manner of expedients for maintenance and repair, they kept up the struggle to stave off collapse of Southern railroad transportation."[17] This was certainly true of Singletary. He had made great personal sacrifices for the road and subsequently was made a scapegoat. He was not responsible for the debt he inherited, but he did try to find creative ways of rectifying it. Under his watch the company had realized its greatest surplus; however, because of inflation and deterioration, it was really only a phantom profit. The bondholders knew this, which is precisely why many did not redeem their coupons when given the chance to do so. In Singletary's predicament his adversaries saw opportunity and availed themselves of it. Drayton's legacy was transferred to other hands. The question was, would it be carried on?

When the newly dubbed Savannah & Charleston Railroad began operation in February of 1867, its new owners were determined to run a lean organization that operated within its means. They began by ridding their payrolls of excess employees and calling a moratorium on the free transport of freight. Payment for all shipments of freight had to be made at the Charleston depot before their departure or on their arrival.[18] The new owners' next aim was to coax investors or financial institutions into granting them a moderate loan on reasonable terms. Taylor, Boylston, and their syndicate had been dubious of their predecessors' claims as to the expense involved in rebuilding the road and the difficulty of raising funds, but they found soon enough how truthful their predecessors had been.

As the new owners pursued investment capital up and down the eastern seaboard, they found that every class of Southern security was passed over in Northern markets or made available at ruinously high interest rates. Obtaining a substantial loan was impossible as long as the company remained subject to the state's 1856 lien. At that point the board of directors decided to abandon their pursuit of a loan and to press forward with the work to the best of their ability with the means at their command.

Work crews busied themselves with bridging the Salkehatchie swamps while new superintendent John Ryan assessed the company's property and its condition. The final tally was not favorable and showed that the means available to the company were not be sufficient to complete the road. The news deflated Ryan, who had downplayed Haines earlier estimates of rebuilding costs, but Ryan kept the crews on the line until his resources were exhausted. On July 31, 1867, the project was completed to Salkehatchie and crews began work at Barnwell's Crossroads. With money from the sale of salvaged railroad iron, work was begun on August 12, and by September 1 the road was finished to Yemassee. Three weeks later the tracks reached Pocotaligo Station, and on November 2 the road was completed to Coosawhatchie, sixty-two miles from Charleston. At this point funds dried up, and the board delayed work on the line until they could find a more definite means of reaching Savannah.

When the hammers fell silent, Ryan's focus shifted to transferring the Savannah & Charleston's rolling stock from the South Carolina Railroad to his company's yard at St. Andrews. He also occupied himself inspecting bridges and wharves on the eastern half of the road and making sure his cars were retooled as efficiently as conditions allowed. While transacting little more than local business, the railroad had first-year receipts that exceeded expenditures by only $2,340. The numbers told the tale. Without a loan the railroad would remain incomplete, and without being completed, it could not yield enough revenue to pay the interest on its debt. A familiar scenario was playing itself out.[19]

Not much improved the following year as the management remained intent on staying within their means. Little work was accomplished. Not one mile of trackage was added, and expenses for the year actually exceeded receipts by nearly one hundred dollars. In two short years the railroad's financial condition had deteriorated to the point that its management went humbly back to request aid from the state legislature—a policy for which Singletary had been roundly criticized. Early in 1869 the directors of the Savannah & Charleston Railroad memorialized the assembly in the hopes of securing a lien for up to five hundred thousand dollars to rebuild the road and pay off its interest. The lien, although potentially injurious to the holders of the state bonds, would take priority over the state's statutory lien. With the support of the Charleston mercantile community, the legislature ratified the act in March, and on July 1, 1869, the company issued five hundred thousand dollars in bonds payable at 7 percent and falling due in twenty years. William Aiken, Jr., George Walton Williams, and James Robb were named as trustees.[20]

Fueled by the state's assistance, work on the road resumed. The Savannah River was bridged, and the line was restored all the way to Central Junction. A company train finally entered Savannah on March 2, 1870, over three miles of track belonging to the Central of Georgia. This arrangement continued until 1875, when a dispute with the Central sent the Savannah & Charleston elsewhere. After this trains entered the city on track rented from the Atlantic & Gulf. On the eastern end of the line, a bridge across the Ashley River was considered too expensive an undertaking, so passengers and freight were carried into the city via a steamer. The cost and inconvenience of the steamer plagued the railroad and its patrons throughout the Reconstruction Era.[21]

Elsewhere in the United States, railroad construction was booming. Between 1866 and 1873, thirty-five thousand miles of new track were laid, and the nation's first transcontinental railroad was completed in 1869. Blazing new trails and building new railroads involved large sums of money and quite a bit of risk; yet many banks and entrepreneurs became caught up in the frenzy. A group of investors planned to finance a second transcontinental road, but the banking firm of Jay Cooke & Company realized it had overextended itself and declared bankruptcy on September 18, 1873. Many other financial institutions followed suit, and the panic that ensued was devastating. Nearly one-fourth of the country's railroads declared insolvency; eighteen thousand businesses failed within a two-year span; and by 1876 unemployment was at a staggering 14 percent. Low wages and poor working conditions eroded the morale of railroad workers, and the strikes that resulted brought the entire industry to a grinding halt.

Progress on restoring South Carolina's railroads was brought to a standstill during the Panic of 1873 and the period of economic disarray that followed. No significant amount of track was laid until the end of the decade, and during the 1870s many of the state's railroads were forced into receivership. The Savannah & Charleston Railroad was in trouble even before the depression began. The company carried an outstanding bonded debt of $1.4 million, on which no interest had been paid for quite some time. On April 28, 1874, a circuit court transferred possession of the railroad to a receiver and an advisory board; its operation to accrue only to the benefit of its creditors and stockholders. In 1875 another court ordered the company sold for $1.5 million, but the South Carolina Supreme Court decided that the state should take charge of the railroad pending litigation between its many creditors. The only significant development during the years of receivership was the completion of a nine-mile connector with the Northeastern Railroad in 1878. Dubbed the Ashley River Railroad, the new line ran northward from Johns Island on the Savannah &

Charleston, crossed the Ashley further upriver, connected with the Northeastern at Ashley Junction, and entered Charleston courtesy of trackage rights obtained from that road. Despite this, the Savannah & Charleston Railroad never prospered during the Reconstruction, and with its assets subject to the discretion of the courts, it was sold at a foreclosure sale to transportation magnate Henry B. Plant and renamed the Charleston & Savannah Railway.[22]

Henry Plant made his fortune on the railroads of the South. A native of Connecticut, he was living in Augusta, Georgia, when the war broke out, working for the Adams Express Company. Soon after the first shots were fired on Fort Sumter, the company's founder determined that the portion of his company that lay within the Confederacy would be better served under Plant's direction. Alvin Adams split the company in two, and Plant took control of its holdings south of the Potomac, renaming it the Southern Express Company. He quickly won the confidence of the Southern leadership, especially the Confederate Quartermaster's Department, and the "new" company performed many essential tasks for the Confederacy, including the transport of all government funds. The communication conduits established by Plant and his assistants survived long after Lee's surrender at Appomattox. Burned out by the demands of the war and his wife's chronic illness, Plant left the Confederacy for Europe in 1863 and did not return until after the war's conclusion. In 1871 he became a principal of the Southern Railway Security Company, and several years later, with wealth accumulated during and after the war, he began purchasing insolvent railroads and assembling his transportation empire.[23]

Plant's first acquisition was the struggling Atlantic & Gulf Railroad. Like other Southern railroads of the day, the Atlantic & Gulf was deficient in equipment and in grave financial condition. Rocked by the Panic of 1873 and unable to generate any revenue, the road defaulted on its interest payments and was placed in receivership. With the cooperation of Morris K. Jesup, Plant eventually purchased the railroad in 1879 for three hundred thousand dollars. He reorganized the company, naming it the Savannah, Florida & Western Railroad. Plant placed its operation in the capable hands of General Manager Henry Stevens Haines, who had been hired as general superintendent of the Atlantic & Gulf not long after he resigned his post with the Charleston & Savannah Railroad in 1866. Haines quickly won over the ambitious new ownership with his management style and engineering expertise.

Together, Haines and Plant began refitting the Savannah, Florida & Western to become the flagship company of Plant's budding railroad system. In the spring of 1880 Haines supervised the construction of the East Florida Railway Company, the southern portion of the Waycross Short Line, which eventually provided a connection between Waycross, Georgia, and Jacksonville, Florida. Plant's next hurdle was making the railroad profitable. He had inherited a considerable debt from the old Atlantic & Gulf Railroad, and new expenses were piling up daily. The business climate was showing signs of thawing, however, and he needed to improve his connection to the markets of the North in order to capitalize on the economic recovery.

Some may have considered his goals out of reach, but Plant had a plan and the means to see it through.

In collaboration with Morris K. Jesup, W. T. Walters, and Benjamin Newcomer, Plant formed business alliances that later proved to have historic significance. Plant and Jesup joined the new Walters syndicate, which purchased controlling interest in several Atlantic coast companies. He and Jesup accumulated large holdings in the Wilmington & Weldon Railroad and in the Wilmington, Columbia & Augusta (formerly the Wilmington & Manchester). They also completed the acquisition of 7,400 shares of Northeastern Railroad stock, giving the Walters group nearly complete control of that company. The linkage of North with South was completed with Plant's purchase of the Savannah & Charleston Railroad.

With the support of the Walters alliance, Plant purchased the Savannah & Charleston Railroad on June 7, 1880, for $320,000 in cash and securities of the old company. A decree by the Court of Common Pleas of South Carolina made the sale absolute, freeing Plant from any prior obligations of the company. For this to come about, Plant had to broker a deal with the Walters syndicate and the holders of the old company's first- and second-preferred-income bonds. As a result the company's new board of directors consisted of Plant, Walters, Newcomer, Northeastern Railroad president Alfred F. Ravenel, W. H. Brawley (representing first-preferred bondholders), and Christopher G. Memminger (representing second-preferred bondholders). The newly organized Charleston & Savannah Railway immediately executed a mortgage for repairing and equipping the railroad and issued five thousand shares of stock at one hundred dollars per share. Subscribers to the bonds were to receive one-fourth of the amount paid in stock, and the company would divide the shares between the first- and second-preferred bondholders.

With the Charleston & Savannah safely in the fold, Plant and Haines concentrated their efforts toward creating an integrated system of rails. Gauges were made uniform so that the switching of trucks—the portions of railroad cars to which the wheel axles are attached—at the end of each line was not necessary. Mainline tracks were replaced with steel rails and construction was begun on a new iron bridge across the Savannah River. In addition to its Southern projects the Savannah, Florida & Western extended its westward reach to Chattahoochee, Florida, connecting there with the Pensacola & Atlantic Railroad. When the Waycross Short Line was finished, the Plant System offered uninterrupted service from Charleston to Jacksonville, as well as to New Orleans via Savannah, Waycross, Pensacola, and Mobile. Henry Plant controlled the major routes south of Charleston, and the Walters group controlled the lines to its north. New York was now linked to New Orleans and northern Florida by rail, and with Plant's investment in the Northeastern and Walters's and Newcomer's stakes in the Charleston & Savannah, the financial keystone of the network lay at Charleston.

In 1882 Plant established the Plant Investment Company to manage his holdings, and appointed Henry Haines as general manager over all railroad operations.

It seems fitting that Haines once again tended the rails between Charleston and Savannah and became the man to oversee the culmination of Thomas Drayton's dream. Under Haines's direction the Plant System expanded its reach to the south and west, reduced its expenditures, and improved its efficiency. After Plant purchased a line running from Jacksonville to Sanford, Florida, Haines began construction of the South Florida Railroad, which connected the entire system to Tampa and its steamship line to Key West and Cuba.

Haines's theory of railroad management was based on the delegation of authority as much as possible, but he stressed that responsibility should rest with the chief operating official. He was fond of the people with whom he worked and always sought to bring out their best qualities. He instituted a rigorous annual inspection of all railroads under his charge and offered cash incentives to his best supervisors and foremen to enhance efficiency and boost morale. His criticism was appropriate —never excessive—and he was always glad to acknowledge the efficiency and success of his subordinates. In 1892, the year of Thomas Drayton's death, Haines was promoted to vice president of the Savannah, Florida & Western Railroad, while he continued to serve as general manager of the entire system. He remained in this capacity until 1895, when he severed his ties with the Plant System.

Henry S. Haines became perhaps the most renowned of the officers of the old Charleston & Savannah Railroad. After leaving the Plant System, he served as commissioner of the Southern States Freight Association and briefly as vice president of the Atlantic & Danville Railroad. He served as president of the American Society of Mechanical Engineers and the American Railway Association, and he was one of the founders of the American Railway Guild. His reputation in engineering and railroading spread worldwide. Foreign governments consulted him on issues ranging from railroad gauges to modes of increasing water supply, and his writings on railroad legislation, railroad management, and the railroad's role in public service were standard texts for many years. Today, the municipality of Haines City, Florida, on the old South Florida line, still bears his name.[24]

Though there were still improvements to be made, Henry Plant organized a constellation of previously defunct railroads in the Southeast, opening the region to agriculture, industry, and tourism. He had the determination and wherewithal to continue building at a time when others gave up (or were taken over), and his success paved the way for the future development of both the Plant System and the Atlantic Coast Line. The Charleston & Savannah Railway, operating as part of the Plant System, was absorbed into the Savannah, Florida & Western on May 10, 1901. Because of a large, potentially uncooperative minority, the Charleston & Savannah received only nonvoting preferred stock to distribute to its owners. In the spring and summer of 1902 the Plant System and the Atlantic Coast Line finalized negotiations for the roads of the Savannah, Florida & Western to be consolidated with and become property of the Atlantic Coast Line Railroad.[25]

Twentieth century technology spurred a remarkable transformation of the railroad running between Charleston and Savannah. Wooden trestles were razed in favor of iron bridges. The once single-tracked beds crossing the South Carolina lowcountry became double-tracked arteries, facilitating bidirectional travel. Sturdier steel rails weighing 115 to 132 pounds per yard replaced unreliable, wrought-iron rails weighing between 35 and 68 pounds per yard to accommodate a heavier workload. Bigger and better rolling stock accommodated larger payloads, and Pullman cars delivered more passengers than ever to destinations up and down the eastern seaboard. Gone were the old 4–4–0 "American" locomotives bearing names such as *Southward Ho* and *Mayor Macbeth*. They were replaced by superheated locomotives distinguished only by number. Later more advanced models of steam engines were employed, including the Baldwin Pacific, the Mikado, and the Santa Fe. The last variety of steam engine used on the line—the 4–8–4 Class R-1—could achieve speeds of up to eighty or ninety miles per hour. These remained in service until 1945, when the diesel engines became the mainstay and the Champion held reign over the rails from New York to Miami, hurtling over the line where rusty wood-burning engines once ran the gauntlet of enemy artillery.

Through the years the railroads and the companies that ran them continued their evolution. On July 1, 1967, the Atlantic Coast Line and its chief competitor, the Seaboard Air Line, merged to form the Seaboard Coast Line Railroad Company. At the end of 1982 the Seaboard Coast Line united with the Louisville & Nashville, the Clinchfield, the Atlantic & West Point, and the Georgia railroads to create the Seaboard System. In 1985 the Seaboard System and the Chessie System consolidated their railroad operations under the umbrella of the CSX Corporation, later CSX Transportation. It had taken more time and more money than anyone ever dreamed would be necessary, but a system of rails linking the major port cities from Boston to New Orleans had been completed, and the cities of Charleston and Savannah were firmly ensconced upon it.

From its beginnings the Charleston & Savannah Railroad had become intractably intertwined with the affairs of the state that granted its original charter. The railroad could in a sense be viewed as a microcosm of the South Carolina economy. It was organized and managed by Southern gentlemen, most of whom were staunch advocates of John C. Calhoun's states' rights philosophy. Built to secure regional commercial interests, the railroad was financed largely by state and local governments. Any form of centralized control or Federal government assistance was stoutly resisted for fear of forfeiting even the smallest bit autonomy. The founders took particular pride in their railroad's being a Southern enterprise despite their dependence on the manufacturing might of the North for much of their capital equipment. The Charleston & Savannah was built by the muscle of slave labor to service primarily agrarian pursuits; it ran through a state in which most voting citizens opposed industrialization with every fiber of their being.

Present-day CSX Transportation railroad tracks at Coosawhatchie, not far from the point at which a Charleston & Savannah Railroad train loaded with Confederate troops was ambushed by the Forty-eighth New York on October 22, 1862. Photograph by the author

The railroad served both its home state and the Confederacy heroically during the Civil War as a conduit for troops and supplies. It became the cornerstone of the Confederate defensive strategy along the South Carolina coast, and its destruction was the objective of no fewer than eight major and minor enemy offensives. The railroad participated materially in its own defense and remained intact until late in the war, broken only after it had exhausted all resources and manpower and the half-starved Confederate army was overwhelmed from front and flank by General Sherman, making its lines untenable. As predicted by both Union and Confederate strategists, the city of Charleston and the remainder of the South Carolina defenses did not last long afterward. The railroad faltered during the postwar period and was finally done in by the chaos and corruption of the Reconstruction era. Slowly, however, its remnant emerged from the ashes, was rebuilt and nursed back to health by laissez-faire capitalists, and became an integral part of a larger, unified system that served the common good of all Americans—Northern or Southern, black or white.

The Charleston & Savannah Railroad—once hailed by Thomas Drayton as "the most formidable earthwork that could have been devised"—is now a part of a railroad system that revolutionized commercial transportation on the Atlantic coast. The Charleston & Savannah began as an imperfect scheme, originated by imperfect men during a volatile period of America's past; but—despite its flaws—it eventually won vindication for all who had labored for its success. History is made by people such as these, who faced challenges and met them. By examining how the Charleston & Savannah and military leaders identified, addressed, and solved their problems—how their prejudices affected their decisions and the consequences of their actions—we enhance our understanding of Southern railroads, their early management, and their role in the American Civil War.

We also learn the stories of people not very different from those we know today, who found themselves players in an extraordinary drama that continued long after their passing. As Robert Black reminds us, "White or black, officer or employee, free citizen or slave, the typical railroader of the Confederate South worked hard, achieved varying rewards, and left few personal traces." Many railroaders are better remembered for their political careers than for their exploits on the iron rails, while the names of Drayton, Singletary, and Haines survive on pages of history too infrequently turned. The memory of a great many others—the Buckhalters and the Stroheckers—has evaporated like the smoke of some long-forgotten freight train. The names of the myriad African American slaves, whose long and honorable service to the American railroad has been too often neglected, will regrettably never be recorded. The essence of being American is to understand the lessons of America's past—both the good and the bad. Each generation holds onto a new set of common beliefs, sharing the inheritance of ideals, no matter our ancestry.

As today's modern diesel engines streak past Salkehatchie, across the Combahee River and through Yemassee on their way to Savannah, they pay homage to the wheezing steam locomotives with rusty wheels and dented boilers that long-ago served so faithfully in the face of adversity. The weather-worn earthworks at Honey Hill, Pocotaligo, and Coosawhatchie conjure up sounds from the past—a distant train whistle, the crackle of musketry, the report of cannon. One by one the scenes flash before us: of the proud regiments of black troops rushing forward behind the Stars and Stripes to break the rails that many of them laid with their own hands; of smartly dressed Citadel cadets fighting alongside grizzled veterans in defense of the railroad and their beloved Palmetto State; and finally, of burning stacks of cross ties—premonitory funeral pyres for the railroad, the Confederacy, and a society that would no longer be the same.

The story of the railroad is the story of the Union soldier and the Confederate, the planter and the slave, the railroad capitalist, manager, and workmen. The actions of the men who built it, defended it, captured and destroyed it placed as much of their mark on the road as those who reclaimed it, bought and sold it, and further developed it. To leave their story untold would be to diminish their efforts, their ambitions, and their sacrifices. The rails were their common link—a memorial to their shared past, a foreshadowing of their common destiny.

NOTES

INTRODUCTION

1. Freehling, *Prelude to Civil War*, 177; *Charleston Courier*, March 5, 1829, and December 31, 1829.
2. Journal of the House of Representatives of the State of South Carolina, 1829, 65; South Carolina Department of Archives and History (hereafter SCDAH); *Columbia Telescope*, December 1, 1829, and *Southern Patriot*, December 3, 1829, cited in Freehling, *Prelude to Civil War*, 178.
3. Freehling, *Prelude to Civil War*, 11.
4. Freehling, *Road to Disunion*, 216.
5. Freehling, *Prelude to Civil War*, 178.
6. *Charleston Mercury*, February 2, 1830, and February 11, 1830; *Columbia Telescope*, February 5, 1830; Freehling, *Prelude to Civil War*, 178–79.
7. Fraser, *Charleston! Charleston!*, 220.
8. Coleman, ed., *A History of Georgia*, 165.
9. Freehling, *Prelude to Civil War*, 304; Fraser, *Charleston! Charleston!*, 223.

CHAPTER 1: BIRTH OF A RAILROAD

1. Ford, "The Changing Geographic Pattern of South Carolina's Railroad System, 1860–1902," 18–37.
2. Charleston & Savannah Railroad. *Report of the Chief Engineer*, 20–22.
3. Charleston & Savannah Railroad, *Annual Report*, 1857, 9
4. Hemphill, ed., *Men of Mark in South Carolina*, 3:87; Courtney, "Charles Jones Colcock," 33–39; Salley, "Captain John Colcock and Some of His Descendants," 231.
5. *Charleston Mercury*, June 23, 1853.
6. Ibid., June 23, 1853, and July 4, 1853.
7. Evans, ed. *Confederate Military History*, 6:387–89.
8. T. F. Drayton to W. B. Hodgson, August 8, 1853, Thomas Fenwick Drayton Papers, South Caroliniana Library.
9. Thomas A. Bryson, "Willian Brown Hodgson," *Dictionary of Georgia Biography*, 1:464–65.
10. *Savannah Daily Morning News*, July 1, 1853; *Charleston Mercury*, October 13, 1853.
11. *Charleston Mercury*, July 4, 1853; Drayton to Hodgson, August 8, 1853.
12. *Charleston Daily Courier*, November 10, 1853.
13. Ibid.

14. Ibid.
15. *Savannah Daily Morning News*, December 3, 1853.
16. Ibid., December 14, 1853.
17. South Carolina Acts and Joint Resolutions, 1853, no. 4142, 271–80, SCDAH; *Charleston Daily Courier*, February 23, 1854.
18. *Charleston Daily Courier*, March 1, 1854.
19. Ibid., February 23, 1854.
20. *Savannah Daily Morning News*, April 13, 1854.
21. Charleston & Savannah Railroad, *Report of the Chief Engineer*, 6.
22. Ibid., 7–8.
23. Ibid., 7–8, 10–11.
24. Ibid., 5.
25. Ibid., 9.
26. Charleston & Savannah Railroad, *Annual Report*, 1857, 7–8.
27. *Charleston Daily Courier*, July 14, 1854.
28. Drayton to Hodgson, August 8, 1853; *Charleston Daily Courier*, July 14, 1854.
29. Ibid.
30. Charleston & Savannah Railroad, *Annual Report*, 1855, 13.
31. Charleston & Savannah Railroad, *Report of the Chief Engineer*, 2.
32. *Charleston Daily Courier*, August 13, 1855.
33. Charleston & Savannah Railroad, *Report of the Chief Engineer*, 5–7.
34. *Charleston Daily Courier*, August 24, 1855.
35. Charleston & Savannah Railroad, *Annual Report*, 1856.

CHAPTER 2: CONSTRUCTION, CONNECTION, AND THE CONVENTION

1. Charleston & Savannah Railroad, *Annual Report*, 1855, 6.
2. *Charleston Daily Courier*, March 18, 1858.
3. Charleston & Savannah Railroad, *Annual Report*, 1858, 4.
4. Ibid., 1855, 7.
5. Edward L. Parker to the Reverend I. H. Cornish, October 5, 1859, Special Collections, Hill Memorial Library, Louisiana State University Libraries.
6. Charleston & Savannah Railroad, *Annual Report*, 1855, 7; Nelson, *Iron Confederacies*, 17.
7. *Charleston Daily Courier*, June 23, 1853; *Savannah Daily Morning News*, July 1, 1853; *Charleston Mercury*, July 4, 1853, and October 13, 1853.
8. Charleston & Savannah Railroad, *Annual Report*, 1859, 11.
9. Ibid., 1857, 6.
10. Perry, *Moving Finger of Jasper County*, 35.
11. *Daniel Blake v. Charleston & Savannah Railroad* [pamphlet] (N.p., 1869), Daniel Blake Papers, South Caroliniana Library; Charleston & Savannah Railroad, Minutes, May 11, 1864.
12. Charleston & Savannah Railroad, *Annual Report*, 1856; ibid., 1857, 3; ibid., 1858, 19; ibid., 1859, 15.
13. *Charleston Mercury*, November 5, 1860; Charleston & Savannah Railroad, *In Equity. Bill for Injunction, etc. the Charleston & Savannah Railroad Company vs Isaac W. Hayne, John E. Carew, et al.*

14. In the Matter of the Account between the Charleston & Savannah Railroad and John S. Ryan—Statement, James Butler Campbell Papers, South Carolina Historical Society; Charleston & Savannah Railroad, *Annual Reports*, 1858 and 1859.

15. Charleston & Savannah Railroad, *Annual Report*, 1855, 11–13.

16. Ibid., 1857, 13; DeSaussure, "Quinine as a Prophylactic of Intermittent and Remittent Fevers."

17. Charleston & Savannah Railroad, *Annual Report*, 1858, 5; ibid., 1859, 15.

18. Ibid., 1858, 6; T. F. Drayton to Jefferson Davis, April 9, 1858, in Rowland, ed., *Jefferson Davis—Constitutionalist*, 216–17.

19. In the Matter of the Account between the Charleston & Savannah Railroad and John S. Ryan—Statement, James Butler Campbell Papers, South Carolina Historical Society.

20. T. F. Drayton, Report to Governor M. L. Bonham, November 29, 1858, Governor Milledge L. Bonham Papers, SCDAH; *American Railroad Journal*, June 5, 1858, 363; Charleston & Savannah Railroad, *Annual Report*, 1859, 7.

21. Heyward, *Seed from Madagascar*, 107–9.

22. *Charleston Daily Courier*, June 23, 1853; *Savannah Daily Morning News*, April 16 and April 17, 1860.

23. *American Railroad Journal*, April 14, 1860, 315; *Charleston Daily Courier*, April 21, 1860; Charleston & Savannah Railroad, *Annual Report*, 1861, 4–5.

24. *Charleston Mercury*, April 23, 1860.

25. *Charleston Daily Courier*, April 23, 1860; *Charleston Mercury*, April 23, 1860.

26. Heyward, *Seed from Madagascar*, 109–11.

27. *Charleston Daily Courier*, April 21, 1860.

28. Beringer, Hattaway, Jones, and Still, eds., *Why the South Lost the Civil War*, 66.

29. *Charleston Mercury*, April 21, 1860.

30. Catton and Catton, *Two Roads to Sumter*, 202.

31. Catton, *The Coming Fury*, 27.

32. Ibid., 30.

33. Fraser, *Charleston! Charleston!*, 242.

34. Ripley, *Siege Train*, vi–vii.

35. Charleston & Savannah Railroad, *Annual Report*, 1861, 5; *Charleston Mercury*, January 20, 1860.

36. *Charleston Daily Courier*, August 8, 1860.

37. Charleston & Savannah Railroad, *Annual Report*, 1861, 5.

38. Ibid., 5–7.

39. Charles Heyward and Charleston & Savannah Railroad, indenture for land, August 11, 1858, Heyward Family Papers, South Caroliniana Library; Charleston & Savannah Railroad, *Annual Report*, 1861.

40. J. H. Steinmeyer to Charleston & Savannah Railroad, deed, January 16, 1856, Charleston County, S.C., Register of Mesne Conveyances, book S18, p. 253; Charleston & Savannah Railroad, *Annual Report*, 1861, 12–13.

41. *Savannah Daily Evening News*, November 21, 1859; Charleston & Savannah Railroad, *Annual Report*, 1861, 21; *American Railroad Journal*, October 27, 1860, 962; *Illustrated London News*, 40 (January 4, 1862): 30; Smith, *Civil War Savannah*, 101.

42. Charleston & Savannah Railroad, *Annual Report*, 1861, 14–15.

CHAPTER 3: A CONFEDERATE RAILROAD

1. *Charleston Daily Courier*, November 5, 1860.
2. *Charleston Mercury*, November 5, 1860.
3. Ibid.; *Charleston Daily Courier*, November 5, 1860.
4. *Charleston Mercury*, November 5, 1860.
5. Ibid., November 8 and 9, 1860; *Charleston Daily Courier*, November 9, 1860.
6. Nicolay and Hay, *Abraham Lincoln*, 319–24.
7. Cauthen, *South Carolina Goes to War*, 55.
8. *Charleston Mercury*, November 10, 1860.
9. *Charleston Daily Courier*, November 10, 1860; Cauthen, *South Carolina Goes to War*, 43.
10. *Charleston Mercury*, November 10, 1860.
11. Ibid.
12. Ibid.; *Charleston Daily Courier*, November 10, 1860.
13. *Charleston Daily Courier*, November 10, 1860.
14. *Charleston Mercury*, September 26, 1861; Cauthen, *South Carolina Goes to War*, 58.
15. Cauthen, *South Carolina Goes to War*, 58; *Charleston Daily Courier*, November 10, 1860; *Keowee Courier*, November 17, 1860.
16. *Charleston Daily Courier*, November 12, 1860.
17. Cauthen, *South Carolina Goes to War*, 60.
18. Edgar, *South Carolina: A History*, 351–52; Riecke, *Recollections of a Confederate Soldier*, 6–7; Freehling, *The South vs. the South*, 40; *Charleston Mercury*, December 21, 1860.
19. Black, *Railroads of the Confederacy*, 5.
20. Ibid., 37.
21. Ibid., 38.
22. Nelson, *Iron Confederacies*, 17.
23. Black, *Railroads of the Confederacy*, 42.
24. Charleston & Savannah Railroad, *Annual Report*, 1859, 15; ibid., 1861, 19.
25. Black, *Railroads of the Confederacy*, 4.
26. Charleston & Savannah Railroad, *Annual Report*, 1859, 13.
27. Black, *Railroads of the Confederacy*, 13; *Charleston Daily Courier*, March 18, 1857.
28. Charleston & Savannah Railroad. *Report of the Chief Engineer*, 2; Black, *Railroads of the Confederacy*, 13.
29. Charleston & Savannah Railroad, *Annual Report*, 1861, 24–25.
30. Drayton, Annual Report to the Legislature, SCDAH; *Charleston Mercury*, January 19, 1860; Charleston & Savannah Railroad, *Annual Report*, 1861, 24–25; Black, *Railroads of the Confederacy*, 21.
31. Charleston & Savannah Railroad, *Annual Report*, 1861; Linder, *Historical Atlas of the Rice Plantations of the ACE River Basin—1860*, 326–27, 485–86, 611–13.
32. Charleston & Savannah Railroad, *Annual Report*, 1861, 24–25.
33. Black, *Railroads of the Confederacy*, 23; Lander, "Charleston: Manufacturing Center of the Old South," 333.
34. Catton, *The Civil War*, 8; Eaton, *A History of the Southern Confederacy*, 61.
35. Woodman, *King Cotton and His Retainers*, 207–8; Chadwick, "Diary of Samuel Edward Burges, 1860–1862," 158; Eaton, *A History of the Southern Confederacy*, 237.

36. Woodman, *King Cotton and His Retainers*, 207; Catton, *The Civil War*, 38; McPherson, *Battle Cry of Freedom*, 438.

37. Woodman, *King Cotton and His Retainers*, 208–9; Eaton, *A History of the Southern Confederacy*, 240–41; Black, *Railroads of the Confederacy*, 36, 79.

38. *Charleston Daily Courier*, November 17, 1860.

39. Black, *Railroads of the Confederacy*, 39.

40. Charleston & Savannah Railroad, *Annual Report*, 1861, 10.

CHAPTER 4: OPENING GUNS

1. Klein, *Days of Defiance*, 133.
2. Burton, *The Siege of Charleston*, 17.
3. Catton, *The Civil War*, 23.
4. Burton, *The Siege of Charleston*, 26.
5. Ibid., 33–34.
6. Chesnut, *A Diary from Dixie*, 36.
7. Black, *Railroads of the Confederacy*, 53.
8. Ibid., 54.
9. Ibid., 55.
10. *Charleston Mercury*, July 8, 1861; *Charleston Daily Courier*, June 1, 1861.
11. Russell, *My Diary North and South*, 74.
12. Chadwick, "Diary of Samuel Edward Burges, 1860–1862," 157.
13. Charleston & Savannah Railroad, *Annual Report*, 1861, 4.
14. Ibid., 16.
15. Ibid.
16. *Charleston Daily Courier*, April 11, 1861; Charleston & Savannah Railroad, *In Equity. Bill for Injunction*.
17. *Charleston Daily Courier*, May 23, 1861.
18. Service records of Capt. Robert L. Singletary, Eighth S.C. Infantry, Compiled Military Service Records of Confederate Soldiers, National Archives, microcopy 267, roll 233; *Charleston Daily Courier*, April 7, 1863.
19. Charles O. Haines, "Memoir of Henry Stephens Haines," 1695–97; Johnson, "Plant's Lieutenants," 387–88.
20. Crist, ed., *The Papers of Jefferson Davis*, 7:93.
21. *Charleston Mercury*, September 25, 1861.
22. Jones, *Port Royal Under Six Flags*, 229.
23. Evans, *Confederate Military History*, 4:31–33.
24. Ibid.
25. Jones, *Port Royal Under Six Flags*, 231–32; U.S. War Department, *War of the Rebellion: Official Records of the Union and Confederate Navies in the War of the Rebellion* (hereafter *O.R.N.*), 12:302.
26. Major A. M. Huger, quoted in Evans, *Confederate Military History*, 4:33.
27. Jones, *Port Royal Under Six Flags*, 231–32; *O.R.N.*, 12:302.
28. Burton, *The Siege at Charleston*, 72–74; Carse, *Department of the South*, 12; Johnson and Buel, *Battles and Leaders of the Civil War*, 1: 686–89; *O.R.N.*, 12:264.
29. Lee, *Recollections and Letters of General Robert E. Lee*, 55.

30. U.S. War Department, *War of the Rebellion: Official Records of the Union and Confederate Armies in the War of the Rebellion* (hereafter *O.R.* and series 1 unless otherwise indicated), 6:357.
31. *Savannah Daily Morning News*, November 11, 1861.
32. *Charleston Daily Courier*, October 11, 1861.
33. Ibid., September 14, 1861.
34. *Tri-Weekly Watchman*, October 29, 1862.
35. Hoyt, ed., "To Coosawhatchie in December 1861," 6–7.
36. *O.R.*, 6:325.
37. R. E. Lee, quoted in Burton, *The Siege of Charleston*, 79; *O.R.*, 6:322.
38. Freeman, *R. E. Lee*, 617.
39. Smith, *Civil War Savannah*, 50; Freeman, *R. E. Lee*, 615, 644–47.
40. Lee, *Recollections and Letters of General Robert E. Lee*, 58–60.
41. *O.R.*, 6:149, 203.
42. Ibid., 6:198.
43. Ibid., 6:144.
44. Ibid., 6:145.
45. Ibid., 6:45–53, 66–71.

CHAPTER 5: THE MOST FORMIDABLE EARTHWORK

1. *Charleston Mercury*, January 9, 1862.
2. Wise, *Gate of Hell*, 10.
3. Halliburton, ed., *Saddle Soldiers*, 17.
4. *O.R.*, 6:390.
5. Heyward, *Seed from Madagascar*, 131–32.
6. Dowdey, *The Wartime Papers of R. E. Lee*, 121–23.
7. *O.R.*, 6:391–93.
8. Ibid., 6:395–96.
9. *Charleston Mercury*, January 8, 1862.
10. *Charleston Daily Courier*, March 4, 1862.
11. Ibid., August 17, 1861.
12. Black, *Railroads of the Confederacy*, 91.
13. J. C. Pemberton to T. A. Washington, January 29, 1862, and Military Order, General J. C. Pemberton to General Maxcy Gregg, January, 24, 1862, C.S.A. Fourth Military District Papers, South Caroliniana Library.
14. R. E. Lee to J. C. Pemberton, February 2, 1862, C.S.A. Fourth Military District Papers, South Caroliniana Library.
15. *Charleston Daily Courier*, February 11, 1862.
16. Ibid.
17. Ibid., October 17, 1861.
18. Marie G. Magrath, biographical notes, William J. Magrath Papers, South Carolina Historical Society.
19. *Charleston Mercury*, February 11, 1862.
20. *Charleston Daily Courier*, September 13, 1862.
21. Black, *Railroads of the Confederacy*, 90.

22. *Charleston Daily Courier*, September 13, 1862.

23. Ibid., November 21, 1861; J. C. Pemberton to I. W. Hayne, C.S.A. Fourth Military District Papers, South Caroliniana Library.

24. Burton, *The Siege of Charleston*, 93.

25. *O.R.*, 6:101–10.

26. Ibid., 14:496.

27. Ibid., 14:21.

28. *Charleston Daily Courier*, May 31, 1862.

29. *O.R.*, 14:525; Halliburton, ed., *Saddle Soldiers*, 30; Camp, ed., "The War for Independence, North and South: The Diary of Samuel Catawba Lowry," 191; Company records, Seventeenth S.C. Infantry, Compiled Military Service Records of Confederate Soldiers, National Archives, microcopy 267, roll 291; Baxley, *No Prouder Fate*, 48.

30. *O.R.*, 14:23, 25; *Charleston Daily Courier*, May 31, 1862.

31. *O.R.*, 14:23.

32. Ibid.; *Charleston Daily Courier*, May 31, 1862; Taylor, Matthews, and Power, eds., *The Leverett Letters*, 137.

33. *O.R.*, 14:25; company records of the Eleventh S.C. Infantry, Compiled Military Service Records of Confederate Soldiers, National Archives, microcopy 267, roll 246.

34. Taylor, Matthews, and Power, eds., *The Leverett Letters*, 137.

35. Halliburton, ed., *Saddle Soldiers*, 31.

36. *O.R.*, 14:23–26.

37. Ibid., 14:26; *Charleston Daily Courier*, May 31, 1862.

38. *O.R.*, 14:24, 26.

39. Ibid., 14:21.

40. Ibid., 14:527.

CHAPTER 6: SECESSIONVILLE TO POCOTALIGO

1. *O.R.*, 14:27–28, 86.

2. Halliburton, ed., *Saddle Soldiers*, 33; *O.R.*, 14:536–37.

3. *O.R.*, 14:554.

4. Morrill, *The Civil War in the Carolinas*, 224.

5. Ibid., 223–24.

6. Ibid., 224–29; Burton, *The Siege of Charleston*, 106.

7. Burton, *The Siege of Charleston*, 108.

8. *O.R.*, 14:52.

9. Burton, *The Siege of Charleston*, 98.

10. Ibid., 111.

11. Silverman, Thomas, and Evans, *Shanks: The Life and Wars of General Nathan George Evans, C.S.A.*, 103; *O.R.*, 14:586–87.

12. *Charleston Daily Courier*, September 13, 1862.

13. Gardner, *Gardner's New Orleans Directory for 1861*, ix; *O.R.*, 14:526, 554, 573.

14. *Charleston Daily Courier*, September 13, 1862.

15. Black, *Railroads of the Confederacy*, 104; *Charleston Daily Courier*, August 29, 1862.

16. *Charleston Mercury*, September 8, 1862.

17. Ibid.
18. Ibid.
19. Black, *Railroads of the Confederacy*, 105; *Charleston Daily Courier*, October 27, 1862.
20. Hagood, *Memoirs of the War of Secession*, 98–109.
21. Ibid., 98–102.
22. Camp, "The War for Independence, North and South," 182–97.
23. Hagood, *Memoirs of the War of Secession*, 106–12.
24. Camp, "The War for Independence, North and South," 182–97.
25. Company records, Eleventh S.C. Infantry, Compiled Military Service Records of Confederate Soldiers, National Archives, microcopy 267, roll 246; *Charleston Daily Courier*, July 11, 1862; Halliburton, ed., *Saddle Soldiers*, 37–38.
26. Camp, "The War for Independence, North and South," 182–97.
27. Hagood, *Memoirs of the War of Secession*, 104–5.
28. Ibid., 182–97; ibid., 106–7.
29. Ibid., 106–7.
30. Western Carolina Historical Research, *The Pocotaligo Expedition*, 4.
31. Ibid., 6–7.
32. *O.R.*, 14:144–48.
33. *New South*, October 25, 1862; *O.R.*, 14:144–48, 150–53.
34. *O.R.*, 14:180–84.
35. Ibid., 14:651; company records, Twenty-sixth S.C. Infantry, Compiled Military Service Records of Confederate Soldiers, National Archives, microcopy 267, roll 350.
36. *O.R.*, 14:115–19; company records, Eleventh S.C. Infantry, Compiled Military Service Records of Confederate Soldiers, National Archives, microcopy 267, roll 246.
37. Clement and Wise, *Mapping the Defense of the Charleston & Savannah Railroad*, 26.
38. *O.R.*, 14:150–53, 180–84.
39. Ibid.
40. Schmidt, *The Battle of Pocotaligo*, 180–81; *O.R.*, 14:180–84.
41. *O.R.*, 14:150–53, 180–84.
42. Ibid., 14:175–77.
43. Ibid., 14:185–87.
44. Schmidt, *The Battle of Pocotaligo*, 63; *O.R.*, 14:185–87.
45. *O.R.*, 14:175–77.
46. Ibid., 14:175–77, 184, 188. Reports vary as to whether it was the engineer or the fireman who was killed, but Colonel Wilson's report seems to be the most accurate regarding all other events of the day, and his account was chosen for the description of the battle.
47. *O.R.*, 14:175–77, 185–87.
48. Schmidt, *The Battle of Pocotaligo*, 72; *O.R.*, 14:188; *Charleston Mercury*, October 23, 1862.
49. *O.R.*, 14:148, 180.

CHAPTER 7: THE BUSINESS OF WAR

1. *Charleston Daily Courier*, October 23, 1862, and October 31, 1862.
2. Ibid., September 26, 1862, and September 29, 1862.
3. Burton, *The Siege of Charleston*, 119; *O.R.*, 14:692.
4. *O.R.*, 14:741–42.

5. Camp, "The War for Independence, North and South," 182–97; Mackintosh, ed. *"Dear Martha . . ."—The Confederate War Letters of a South Carolina Soldier*, 92–93; *O.R.*, 14:655.

6. *Charleston Daily Courier*, November 10, 1862, and November 28, 1862; Charleston & Savannah Railroad, Minutes, December 16, 1863.

7. *Savannah News*, December 4, 1862; Confederate Post Office Department, Letters Sent, Inspection Office, February 2, 1863 to February 3, 1864, RG 109, chap. 11, 46:324, Records of the Confederate Post Office, National Archives.

8. *Charleston Daily Courier*, September 22, 1862; *Charleston Mercury* September 23, 1862.

9. Smith, *Civil War Savannah*, 101; Lieutenant George Mercer, in Lane, *Times That Prove People's Principles*, 69.

10. Hasell to Wadley, December 24, 1862, Railroad Bureau Archives, Valentine Museum.

11. Hasell to Hayne, December 12, 1862, Governor Milledge L. Bonham Papers, SCDAH; Hasell to Wadley, December 24, 1862.

12. Hasell to Wadley, December 24, 1862; *Charleston Daily Courier*, December 12, 1862; Black, *Railroads of the Confederacy*, 125.

13. Hasell to Hayne, December 12, 1862, Governor Milledge L. Bonham Papers, SCDAH.

14. Black, *Railroads of the Confederacy*, 111.

15. *O.R.*, ser. 4, 2:275.

16. Ibid.

17. Black, *Railroads of the Confederacy*, 112.

18. Hasell to Wadley, December 24, 1862.

19. *O.R.*, ser. 4, 2:270–72.

20. Silverman and Murphy, eds., "Our Separation is Like Years: The Civil War Letters of Deopold Daniel Louis," 141–47.

21. Company records, Eleventh S.C. Infantry, Compiled Military Service Records of Confederate Soldiers, National Archives, microcopy 267, roll 246.

22. *Charleston Daily Courier*, November 2, 1861.

23. Ibid., September 2, 1861, Garrison, "Administrative Problems of the Confederate Post Office Department," 232–250.

24. Report of the Postmaster General of the Confederate States of America, November 27, 1861, cited in Garrison, "Administrative Problems of the Confederate Post Office Department."

25. Ibid.

26. Confederate Post Office Department, Letters Sent, Inspection Office, February 2, 1863 to February 3, 1864, RG 109, chap. 11, 32:382, 444; 33:121, 177, Records of the Confederate Post Office Department, National Archives.

27. Ibid., chap. 11, 46:324.

28. *Savannah Republican*, October 5, 1861, and December 31, 1861; *Charleston Daily Courier*, September 8, 1862.

29. *Charleston Daily Courier*, December 2, 1862.

30. B. D. Hasell to M. L. Bonham, Governor Milledge L. Bonham Papers, SCDAH; Report of Gov. M. L. Bonham, Governor Milledge L. Bonham Papers, SCDAH.

31. A. H. Brown to R. L. Singletary, March 14, 1863, Charleston & Savannah Railroad, Minutes, March 15, 1863.

CHAPTER 8: SINGLETARY'S INHERITANCE

1. *Charleston Mercury*, February 19, 1863.
2. *Charleston Daily Courier*, March 7, 1863, and April 21, 1863.
3. Charleston & Savannah Railroad, Minutes, May 25, 1863.
4. Ibid.
5. Swinton to Washington, July 17, 1863, Confederate Papers Relating to Citizens or Business Firms, microfilm, roll 1002, Mfm 346, National Archives; Fuller to Swinton, July 28, 1863, Confederate Post Office Department, Letters Sent, Inspection Office, February 2, 1863 to February 3, 1864, RG 109, chap. 11, 46:324; Myers to Lee, September 18, 1861, Hutson Lee Papers, South Carolina Historical Society.
6. *Charleston Daily Courier*, April 30, 1863.
7. A. H. Brown to R. L. Singletary, March 14, 1863, Charleston & Savannah Railroad, Minutes.
8. Hasell to Wadley, December 24, 1862; *Charleston Daily Courier*, April 7, 1863; Charleston & Savannah Railroad, *Annual Report*, 1864, 11.
9. *O.R.*, 14:799; Clement and Wise, *Mapping the Defense of the Charleston & Savannah Railroad*, 27, 28, 34; Jones, *Enlisted for the War*, 74.
10. Burton, *The Siege of Charleston*, 136.
11. Ibid., 141–43.
12. Black, *Railroads of the Confederacy*, 119–21; *O.R.*, ser. 4, 2:508–10.
13. Black, *Railroads of the Confederacy*, 122.
14. Ibid., 134–35.
15. Charleston & Savannah Railroad, Minutes, May 25, 1863.
16. Ibid.
17. Ibid.
18. *Charleston Daily Courier*, May 27, 1863; Charleston & Savannah Railroad, *Annual Report*, 1864, 4–6; Charleston & Savannah Railroad, Minutes, December 16, 1863.
19. Charleston & Savannah Railroad, Minutes, May 29, 1863.
20. Charleston & Savannah Railroad, *Annual Report*, 1864, 11.

CHAPTER 9: UNBROKEN LINES

1. *O.R.*, 14:290, 308.
2. *Charleston Daily Courier*, January 17, 1856.
3. Higginson, *Army Life in a Black Regiment*, 164.
4. Ibid., 165–66.
5. Ibid., 166.
6. *O.R.*, 28, pt. 1:194; Higginson, *Army Life in a Black Regiment*, 165–66.
7. Ibid., 28, pt. 1:194–95, 198.
8. Ibid., 28, pt. 1:196, 199; Higginson, *Army Life in a Black Regiment*, 171–73.
9. *O.R.*, 28, pt. 1:195, 198; Higginson, *Army Life in a Black Regiment*, 174.
10. *O.R.*, 28, pt. 1: 195–98.
11. Burton, *The Siege of Charleston*, p. 160; *O.R.*, 28, pt. 1:68–69.
12. Burton, *The Siege of Charleston*, 154–60.
13. *O.R.*, 28, pt. 2:73, 193.

14. Morrill, *The Civil War in the Carolinas*, 336.
15. *O.R.*, 28, pt. 2: 192.
16. Clark, *Railroads in the Civil War*, 82.
17. Ibid., 84–86.
18. Ibid., 96–97.
19. Ibid., 92, 97.
20. *O.R.*, 29, pt. 2:706, 708, 713.
21. *Charleston Daily Courier*, May 2, 1861.
22. Halliburton, ed., *Saddle Soldiers*, 111.
23. *Charleston Daily Courier*, September 14, 1863; Taylor, *Reminiscences of My Life in Camp*, 28.
24. *Charleston Mercury*, September 14, 1863; *O.R.*, 28, pt. 1: 729–30.
25. Halliburton, ed., *Saddle Soldiers*, 116–18.
26. Spears, *British North Americans (Canadians) Biography. Canadians in the Civil War Web Site*; Greeley, "The Military Telegraph Service."
27. Coxe, "Chickamauga," 291, and Bowen to Simms (telegram), September 14, 1863, Railroad Bureau Archives. Valentine Museum.
28. Dickert, *History of Kershaw's Brigade*, 263–24; Coxe, "Chickamauga," 291; Clark, *Railroads in the Civil War*, 100.
29. Coxe, "Chickamauga," 291; Dubose Eggleston, quoted in Wyckoff, *A History of the 2nd South Carolina Infantry*, 89.
30. Clark, *Railroads in the Civil War*, 124.
31. Peake to Sims (telegram), September 16, 1863, Railroad Bureau Archives, Valentine Museum.
32. Turner, *Victory Rode the Rails*, 285.

CHAPTER 10: THE SQUEAKY WHEEL

1. *O.R.*, 28, pt. 1:113.
2. *O.R.*, 28, pt. 1:111.
3. *O.R.*, 28, pt. 1: 432.
4. Smith, *Civil War Savannah*, 124–26.
5. *Charleston Mercury*, November 3, 1863; *Charleston Daily Courier*, November 3, 1863.
6. *Charleston Mercury*, November 3, 1863.
7. Black, *Railroads of the Confederacy*, 118; Charleston & Savannah Railroad, *Annual Report*, 1864, 10, 31.
8. *Journal of the Congress of the Confederate States of America*, 5:90, 95; *O.R.*, ser. 4, 1:1081; Black, *Railroads of the Confederacy*, 129–30.
9. Knudsen, *The Official Records of Company D*, 331, 509, 651; *O.R.*, ser. 4, 2:792; *Journal of the Congress of the Confederate States of America*, 6:529; Confederate Post Office Department, Letters Sent, Contract Bureau, chap. 40, 34:459, Records of the Confederate Post Office Department, National Archives; Black, *Railroads of the Confederacy*, 173, 215.
10. Extended Draft Act of February 17, 1864, para. 5, sec. 10, in *O.R.*, ser. 4, 3:180.
11. *Charleston Daily Courier*, April 16, 1861, April 21, 1863, and November 3, 1863; Charleston & Savannah Railroad, *Annual Report*, 1864, 31; Black, *Railroads of the Confederacy*, 125, 193.

12. *Charleston Mercury*, April 24, 1863; Charleston & Savannah Railroad, *Annual Report*, 1864, 11, 27–28.

13. Charleston & Savannah Railroad, *Annual Report*, 1864, 10.

14. *O.R.*, 28, pt. 2:545.

15. Wise, *Lifeline of the Confederacy*, 120; Lawton to Winnemore, August 11, 1863, Records of the Confederate Quartermaster Department, National Archives.

16. Lawton to Winnemore, August 11, 1863, Lawton to Cameron, August 24, 1863, Lawton to Adams, August 31, 1863, and Lawton to Cuyler, September 2, 1863, Records of the Confederate Quartermaster Department, National Archives; Solomons to Sims (telegram), September 12, 1863, Railroad Bureau Archives, Valentine Museum.

17. Lawton to Seixas, October 13, 1863, Records of the Confederate Quartermaster Department, National Archives; Sharp to Sims, October 21, 1863, Peake to Sims, October 22, 1863, Foss to Bayne (telegram), October 10, 1863, Railroad Bureau Archives, Valentine Museum.

18. Black, *Railroads of the Confederacy*, 172; *Charleston Mercury*, December 1, 1863; Wise, *Lifeline of the Confederacy*, 244–45.

19. Adams to Sims (telegram), December 11, 1863, Railroad Bureau Archives, Valentine Museum; Lawton to Cameron, December 17, 1863, Records of the Confederate Quartermaster Department, National Archives.

20. Singletary to Sims (telegram), December 22, 1863, Ravenel to Sims (telegram), December 22, 1863, and Lee to Sims (telegram), December 22, 1863, Railroad Bureau Archives, Valentine Museum.

21. Lee to Sims, December 22, 1863, Lawton to Lee, December 23, 1863, Lawton to Cothran, December 25, 1863, and Alexander to Cameron, January 2, 1864, Records of the Confederate Quartermaster Department, National Archives; Haines to Sims (telegram), December 22, 1863, Railroad Bureau Archives, Valentine Museum.

22. Black, *Railroads of the Confederacy*, 172; G. W. Adams to Maj. F. W. Sims, telegram, December 11, 1863, and Maj. Hutson Lee to Maj. F. W. Sims, telegram, December 22, 1863, Railroad Bureau Archives, Valentine Museum.

23. *Charleston Daily Courier*, February 8, 1861.

24. Charleston & Savannah Railroad, *Annual Report*, 1864, 18.

25. *Charleston Daily Courier*, March 5, 1863; Charleston & Savannah Railroad, *Annual Report*, 1864, 23.

26. Payments made to the Charleston & Savannah Railroad, recorded in Maj. Hutson Lee's quartermaster's ledger, Confederate Papers Relating to Citizens or Business Firms, microfilm, roll 1002, Mfm 346, National Archives.

27. DeCredico, *Patriotism for Profit*, 100–3.

28. Charleston & Savannah Railroad, Minutes, December 16, 1863; *Charleston Daily Courier*, November 10, 1862, and *Savannah Daily Morning News*, August 6, 1863.

29. *Charleston Daily Courier*, February 27, 1863, May 26, 1863, December 19, 1863; Charleston & Savannah Railroad, *Annual Report*, 1864, 4.

30. *O.R.*, 28, pt, 1:112.

31. Ibid., 28, pt. 1:112–13.

32. Ibid., 28, pt. 1:110; ibid., 28, pt. 2:579.

33. Ibid., 28, pt. 2:567.

34. Ibid., 28, pt. 2:580, 594; Boineau to Heyward December 28, 1863, Heyward Family Papers, South Caroliniana Library.

35. *O.R.*, 28, pt. 2:580–81.

CHAPTER 11: UNREALIZED GAINS

1. *Charleston Mercury*, December 16, 1863.

2. Ibid.

3. Clement and Wise, *Mapping the Defense of the Charleston & Savannah Railroad*, 35.

4. Wise, *Lifeline of the Confederacy*, 63.

5. Charleston & Savannah Railroad, *Annual Report*, 1864, 8; Clark, *Railroads in the Civil War*, 69.

6. Records of Cotton Shipped through the Confederate Customhouse at Charleston, South Carolina, 1861–1865, RG 366, stack 450, row 80, compartment 11, shelf 03, entry 748, National Archives.

7. Clark, *Railroads in the Civil War*, 70; Charleston & Savannah Railroad, Minutes, June 8, 1864; Turner, *Victory Rode the Rails*, 69.

8. *O.R.*, 53:96.

9. Ibid., 53:95.

10. Ibid., 35, pt. 1:321–22.

11. Ibid., 35, pt. 1:30–31, 106–7.

12. Ibid., 35, pt. 1:107–8; Hyde, *History of the One Hundred and Twelfth Regiment of New York Volunteers*, 63–65; Baker, *Cadets in Gray*, 76.

13. *O.R.*, 35, pt. 1:107–8; Baker, *Cadets in Gray*, 54, 76.

14. *O.R.*, 35, pt. 1:144; Hyde, *History of the One Hundred and Twelfth Regiment*, 63–65; Baker, *Cadets in Gray*, 77–79.

15. *O.R.*, 35, pt. 1:144–45; Evans, *Confederate Military History*, 4:299.

16. *O.R.*, 35, pt. 1:145.

17. Ibid.

18. Ibid., 35, pt. 1:145–46.

19. Ibid., 35, pt. 1:147; Hyde, *History of the One Hundred and Twelfth Regiment*, 63–65.

20. *O.R.*, 35, pt. 1:622; Gregorie, ed., "Diary of Captain Joseph Julius Wescoat, 1863–1865," 19; Halliburton, ed., *Saddle Soldiers*, 128.

21. *O.R.*, 35, pt. 1, 322–23.

22. Ibid., 35, pt. 1, 493–94.

23. *Charleston Daily Courier*, February 13, 1864, February 27, 1864; Charleston & Savannah Railroad, *Annual Report*, 1864, 7.

24. Charleston & Savannah Railroad, *Annual Report*, 1864, 5.

25. In the Matter of the Account between the Charleston & Savannah Railroad and John S. Ryan—Statement, James Butler Campbell Papers, South Carolina Historical Society.

26. Ibid.

27. Charleston & Savannah Railroad, Minutes, February 17, 1864.

28. Ibid., February 18, 1864.

29. Ibid., May 9, 1864; J. B. Campbell to W. C. Bee, May 9, 1864, James Butler Campbell Papers, South Carolina Historical Society.

30. J. B. Campbell to W. C. Bee, May 9, 1864, James Butler Campbell Papers, South Carolina Historical Society.

31. Charleston & Savannah Railroad, Minutes, May 11, 1864, and June 8, 1864.

32. Ibid., July 16, 1864.

CHAPTER 12: FOSTER TRIES CHARLESTON

1. Black, *Railroads of the Confederacy*, 220; *Memorial to the Honorable the Senate and House of Representatives of the Congress of the Confederate States of America*, 1864.

2. *Memorial to the Honorable the Senate and House of Representatives*.

3. Black, *Railroads of the Confederacy*, 220; *Memorial to the Honorable the Senate and House of Representatives*.

4. Charleston & Savannah Railroad, Minutes, June 8, 1864.

5. *O.R.*, 35, pt. 2:423.

6. Ibid., 35, pt. 2:470–71; *Charleston Daily Courier*, April 27, 1864.

7. Phelps, *The Bombardment of Charleston*, 92.

8. *O.R.*, 35, pt. 1:7–9, 400, 514; Moore, ed., *The Rebellion Record*, 10:200–202.

9. Moore, ed., *The Rebellion Record*, 10:200–202.

10. *O.R.*, 35, pt. 1:400–401; Moore, ed., *The Rebellion Record*, 10:200–202.

11. *O.R.*, 35, pt. 2:526.

12. Higginson, *Army Life in a Black Regiment*, 249; *O.R.*, 35, pt. 1, 78–79, 158, 166; Ripley, *Siege Train*, 193.

13. *O.R.*, 35, pt. 1:166; Ripley, *Siege Train*, 193–94.

14. Ripley, *Siege Train*, 196–97.

15. *O.R.*, 35, pt. 1:15, 40, 124, 167.

16. Ibid., 35, pt.1:14.

17. Ibid., 35, pt. 1:409; accounts of the skirmish near White Point are from ibid., 35, pt. 1:14, 50–51, 125, 408.

18. Ibid., 35, pt. 1:14–17.

19. Ibid., 35, pt. 2:561–66.

20. Andrews, *Footprints of a Regiment*, 138.

21. Accounts from *O.R.*, 35, pt. 1: 15–16, 85, 267; Burton, *The Siege of Charleston*, 292–93; Andrews, *Footprints of a Regiment*, 139–41.

22. *O.R.*, 35, pt. 1:142.

23. Ibid., 35, pt. 1:16, 85, 142, 255, 268; Andrews, *Footprints of a Regiment*, 144.

24. *O.R.*, 35, pt. 1:142, 255, 268; Andrews, *Footprints of a Regiment*, 144–45.

25. *O.R.*, 35, pt. 2:574–75.

26. Burton, *The Siege of Charleston*, 295.

CHAPTER 13: UNDER SIEGE

1. Ward, ed., *Southern Railroad Man*, 143.

2. Black, *Railroads of the Confederacy*, 29; Charleston & Savannah Railroad, Minutes, May 29, 1863.

3. Varhola, *Every Day Life During the Civil War*, 58, 88; Charleston & Savannah Railroad, Minutes, June 4, 1864.

4. *Charleston Mercury*, May 12, 1864, and June 7, 1864; Cameron to Lawton, April 23, 1864, and Corley to Lawton, May 2, 1864, Records of the Confederate Quartermaster Department, National Archives.

5. Charleston & Savannah Railroad, Minutes, June 8, 1864.

6. *Charleston Mercury*, June 7, 1864; Charleston & Savannah Railroad, Minutes, August 24, 1864.

7. *Charleston Daily Courier*, April 27, 1864.

8. Black, *Railroads of the Confederacy*, 30.

9. Ripley, *Seige Train*, 235.

10. Phelps, *The Bombardment of Charleston*, 101.

11. *Charleston Daily Courier*, July 30, 1864; Adams, *Reminiscences of the Nineteenth Massachusetts Regiment*, chap. 13; Baird, letter in *Scioto Gazette*, September 24, 1864.

12. McElroy, *Andersonville*, 513–15.

13. Abbott, *Prison Life in the South*, 103.

14. Ford, *Life in the Confederate Army*, 33–34.

15. *O.R.*, 35, pt. 2:637.

16. Ibid., 35, pt. 2:635, 637; Evans, ed., *Confederate Military History*, 1:665; Hughes, *General William Hardee*, 252.

17. Hughes, *General. William Hardee*, 253.

18. Charleston & Savannah Railroad, Minutes, September 28, and October 26, 1864.

19. Ibid., November 25, 1863.

20. Ibid., March 16, 1864, and April 20, 1864.

21. Ibid., June 29, 1864.

22. Ibid.

23. Ibid., July 27, 1864, and August 24, 1864.

24. Ibid., August 24, 1864.

25. Ibid.

26. Ibid., October 26, 1864; Singletary to Campbell, September 26, 1864, and Campbell to Singletary, October 19, 1864, James Butler Campbell Papers, South Carolina Historical Society.

27. Charleston & Savannah Railroad, Minutes, November 22, 1864.

CHAPTER 14: HONEY HILL

1. Smith, *Civil War Savannah*, 37–38.

2. *O.R.*, 39, pt. 3:162; Smith, *Civil War Savannah*, 146–51.

3. Hughes, *General William Hardee*, 250.

4. Smith, *Civil War Savannah*, 156; *O.R.*, 44:871; Hughes, *General William Hardee*, 254.

5. Lawrence, *A Present for Mr. Lincoln*, 168–69; Hughes, *General William Hardee*, 254; Smith, *Civil War Savannah*, 156.

6. Lawrence, *A Present for Mr. Lincoln*, 169; Hudson, "A Confederate Victory at Grahamville: Fighting at Honey Hill," 9–21.

7. *O.R.*, 44:414–15.

8. Ibid., 44:906, 415.

9. Hughes, *General William Hardee*, 256.

10. Hudson, "A Confederate Victory at Grahamville," 22.

11. *O.R.*, 39, pt. 3:740; ibid., 44:505, 525.

12. Courtney, "Fragments of War History Relating to the Coast Defense of South Carolina, 1861–65 and the Hasty Preparations for the Battle of Honey Hill," 69.

13. *O.R.*, 44:422.

14. Emilio, *A Brave Black Regiment*, 248; Cozzens and Girardi, eds., *The New Annals of the Civil War*, 448.

15. Cozzens and Girardi, eds., *The New Annals of the Civil War*, 448–49.

16. Emilio, *A Brave Black Regiment*, 248–49.

17. Trudeau, *Like Men of War*, 319.

18. Ibid., 319–21; *O.R.*, 44:422.

19. Jones, "The Battle of Honey Hill," 363; Courtney, "Fragments of War History," 74–75.

20. Hudson, "A Confederate Victory at Grahamville," 23; Jones, "The Battle of Honey Hill," 363; Trudeau, *Like Men of War*, 322.

21. Sneden, Bryan, and Lankford, eds. *Eye of the Storm*, 279.

22. Hudson, "A Confederate Victory at Grahamville," 23; *O.R.*, 44:415.

23. Jones, "The Battle of Honey Hill," 363.

24. Trudeau, *Like Men of War*, 322–23.

25. De la Cova, *Cuban Confederate Colonel*, 236.

26. Hudson, "A Confederate Victory at Grahamville," 22; *Charleston Mercury*, December 5, 1864; *O.R.*, 44:423.

27. *O.R.*, 44:422–26.

28. Emilio, *A Brave Black Regiment*, 248–52; Hudson, "A Confederate Victory at Grahamville"; *O.R.*, 44:426.

29. Emilio, *A Brave Black Regiment*, 255.

30. Trudeau, *Like Men of War*, 326.

31. Courtney, "Fragments of War History," 82.

32. *Savannah Republican*, December 3, 1864.

33. *O.R.*, 44:906, 909, 911, 914.

34. Ibid., 44:430.

35. Emilio, *A Brave Black Regiment*, 254; Cozzens and Girardi, eds., *The New Annals of the Civil War*, 462; Trudeau, *Like Men of War*, 324–26, 328–29; *O.R.*, 44:430.

36. Emilio, *A Brave Black Regiment*, 257–58.

37. *O.R.*, 44:433–34.

38. Emilio, *A Brave Black Regiment*, 259; *O.R.*, 44:423.

39. Emilio, *A Brave Black Regiment*, 260–61; *O.R.*, 44:423–25.

40. *O.R.*, 44:416; *Charleston Daily Courier*, December 6, 1864.

CHAPTER 15: SHERMAN'S NECKTIES

1. Baker, *Cadets in Gray*, 136.

2. Ibid., 136.

3. *O.R.*, 44:920.

4. Ibid., 44:416–17, 920, 923.

5. Federal Writers' Project, "Interview with William Rose, Ex-slave of Edisto Island, in *Slave Narratives, South Carolina*, 14, pt. 4:48–50.

6. *O.R.*, 44:929; Baker, *Cadets in Gray*, 134.

7. *O.R.*, 44:931.

8. Ibid., 44:443.

9. Ibid., 44:443, 930.

10. Ibid., 44:931; Evans, ed., *Confederate Military History*, 6:356.

11. *O.R.*, 44:439, 444.

12. Baker, *Cadets in Gray*, 137–38; *O.R.*, 44:439, 935.

13. Baker, *Cadets in Gray*, 138–40; *O.R.N.*, 16:87; *O.R.*, 44:444.

14. Baker, *Cadets in Gray*, 139–40.

15. *O.R.*, 44:935.

16. Ibid., 44:936.

17. Ibid., 44:447; Baker, *Cadets in Gray*, 141–42.

18. *O.R.N.*, 16:87.

19. *O.R.*, 44:447; Baker, *Cadets in Gray*, 143.

20. Baker, *Cadets in Gray*, 143.

21. Ibid., 144.

22. Bond, *The Story of the Citadel*, 77; Ravenel, "The Boy Brigade of South Carolina," 417.

23. Emilio, *A Brave Black Regiment*, 269.

24. Boineau to Heyward, December 7, 1864, Heyward Family Papers, South Caroliniana Library.

25. *Charleston Mercury*, December 8, 1864.

26. *O.R.*, 44:941–42.

27. *O.R.N.*, 16:89.

28. Herriot, "Fighting in South Carolina," 415.

29. Ibid.

30. *O.R.*, 44:441; Emilio, *A Brave Black Regiment*, 270–71; Herriot, "Fighting in South Carolina," 415; *O.R.N.*, 16:89.

31. Emilio, *A Brave Black Regiment*, 270–71.

32. Smith, *Civil War Savannah*, 162.

33. Hughes, "Hardee's Defense of Savannah," 48; Jones, "The Siege and Evacuation of Savannah," 68–69; Lawrence, *A Present for Mr. Lincoln*, 173.

34. Jones, "The Siege and Evacuation of Savannah," 50–51.

35. Ibid.

36. *O.R.*, 44:49, 218, 276.

37. Ibid., 44:208, 218, 235, 257, 276; Lawrence, *A Present for Mr. Lincoln*, 175.

38. *O.R.*, 44: 49, 208.

39. Ibid., 44: 257; Lawrence, *A Present for Mr. Lincoln*, 175.

40. *O.R.*, 44: 218, 257; Lawrence, *A Present for Mr. Lincoln*, 175; Jones, "The Siege and Evacuation of Savannah," 51.

41. Jones, "The Siege and Evacuation of Savannah," 51.

42. Roman, *The Military Operations of General Beauregard*, 312–14; Jones, "The Siege and Evacuation of Savannah," 49.

43. Black, *Railroads of the Confederacy*, 258–60.

44. *Charleston Mercury*, December 7, 1864; *O.R.*, 44:676; Black, *Railroads of the Confederacy*, 258–60.

45. *O.R.*, 44:258, 266.

46. De la Cova, *Cuban Confederate Colonel*, 173.

47. Ibid., 232, 241; Colonel A. J. Gonzales to Colonel T. B. Roy, November 11, 1864, Ambrosio Gonzales Papers, South Caroliniana Library; Bowman and Irvin, *Sherman and His Campaigns*, 297; Hughes, *General William Hardee*, 271.

48. *O.R.*, 44:676, 681, 689.

49. *O.R.N.*, 16:489–90.

50. Ibid., 16:486, 488–90; *O.R.*, 44:208; Evans, ed., *Confederate Military History*, 6:367–68; Smith, *Civil War Savannah*, 165; Lawrence, *A Present for Lincoln*, 177.

51. *Charleston Mercury*, December 12, 1864.

52. *Charleston Daily Courier*, December 12, 1864.

CHAPTER 16: RUNNING THE GAUNTLET

1. Ravenel, "Fighting in South Carolina," 417; Herriott, "Fighting to the End," 167.

2. *O.R.*, 44:723–24, 750; Herriott, "Fighting to the End," 167.

3. Charleston & Savannah Railroad, *Annual Report*, 1866, 6.

4. Heyward Family Papers; Charleston & Savannah Railroad, Minutes, December 28, 1864, and *Annual Report*, 1866, 5.

5. *O.R.*, 44:953–54.

6. Ibid., 44:958–60.

7. Ibid., 44:960.

8. Hughes, "Hardee's Defense of Savannah," 56.

9. Jones, "The Siege and Evacuation of Savannah," 79.

10. *O.R.*, 44:959–60; Roman, *The Military Operations of General Beauregard*, 315–17.

11. Smith, *Civil War Savannah*, 182.

12. Ibid., 185.

13. *O.R.*, 44:750.

14. Ibid., 44:450–51; Emilio, *A Brave Black Regiment*, 273–74.

15. *O.R.*, 44:719–20.

16. Sherman, *Memoirs of General W. T. Sherman*, 2:204, 216.

17. *O.R.*, 44:966.

18. Hughes, "Hardee's Defense of Savannah," 58.

19. Smith, *Civil War Savannah*, 190–91; *O.R.*, 44:967; Scaife, *Joe Brown's Pets*, 150.

20. Hughes, "Hardee's Defense of Savannah," 59; Andrews, *Footprints of a Regiment*, 154.

21. Scaife, *Joe Brown's Pets*, 150.

22. Ibid.; Andrews, *Footprints of a Regiment*, 154; Smith, *Civil War Savannah*, 198–99.

23. Haines to Singletary, December 23, 1864, Hutson Lee Papers, South Carolina Historical Society; Roman, *The Military Operations of General Beauregard*, 320.

24. Thomas, *Memoirs of a Southerner*.

25. Smith, *Civil War Savannah*, 198–99.

26. *Recollections and Reminiscences*, 396; Herriott, "Fighting in South Carolina," 415; Ravenel, "The Boy Brigade of South Carolina," 417; Baker, *Cadets in Gray*, 149.

27. *O.R.*, 44:974–76.

28. Hughes, *General William Hardee*, 272; *O.R.*, 44:975.

29. *O.R.*, 44:985; Andrews, *Footprints of a Regiment*, 155.

30. Andrews, *Footprints of a Regiment*, 155.

31. Emilio, *A Brave Black Regiment*, 275.

32. *O.R.*, 44:443–45.

33. Ibid., ser. 4, 3:968–70.

34. Special Order No. 2, General William Hardee, January 3, 1865, Charleston & Savannah Railroad, Minutes, January 7, 1865.

35. Ibid.

36. Ibid., January 13, 1865.

37. Hughes, "Hardee's Defense of Savannah," 59.

38. *O.R.*, 44, pt. 1:191.

39. Emilio, *A Brave Black Regiment*, 275; *O.R.*, 44: p. 824.

40. *O.R.*, 47, pt. 2:1116–17.

41. Brooks, ed., *Stories of the Confederacy*, 228–29.

42. Ibid.; *O.R.*, 47, pt. 1:191–93.

43. Andrews, *Footprints of a Regiment*, 158; *O.R.*, 47, pt. 1:191–93.

44. Andrews, *Footprints of a Regiment*, 158; *O.R.*, 47, pt. 1:1068.

45. Emilio, *A Brave Black Regiment*, 277.

46. Ibid., 278.

47. *O.R.*, 47, pt. 1:193; Emilio, A *Brave Black Regiment*, 278.

48. *O.R.*, 47, pt. 2: 1020, 1029–30, 1034–36.

49. Sherman, *Memoirs of General W. T. Sherman*, 2:254–57.

50. *O.R.*, 47, pt. 2:152.

51. Dickert, *History of Kershaw's Brigade*, 504–7; Evans, ed., *Confederate Military History*, 4:359.

52. Sherman, *Memoirs of General W. T. Sherman*, 2; 254–57.

53. *O.R.*, 47, pt. 1:18; ibid., 47, pt. 2:150–51.

54. Ibid., 47, pt. 2:163–64; ibid., 47, pt. 1:19; Emilio, *A Brave Black Regiment*, 281.

55. *O.R.*, 47, pt. 1:18–19, 1069; ibid., 53:96.

56. Ibid., 47, pt. 2: 325; Emilio, *A Brave Black Regiment*, 284.

57. Emilio, *A Brave Black Regiment*, 285; *Charleston Daily Courier*, February 11, 1865.

58. Evans, ed., *Confederate Military History*, 4:360–61; McMicken to Haines, February 4, 1865, Hutson Lee Papers, South Carolina Historical Society.

59. *O.R.*, 47, pt. 2:402; Emilio, *A Brave Black Regiment*, 285–86; *O.R.*, 47, pt. 2:1167.

CHAPTER 17: ONE MORE RIVER TO CROSS

1. *O.R.*, ser. 4, 3:1086.

2. Black, *Railroads of the Confederacy*, 280.

3. Dickert, *History of Kershaw's Brigade*, 508.

4. Black, *Railroads of the Confederacy*, 280.

5. *O.R.*, 47, pt. 2:1157.

6. Ibid., 47, pt. 2:1181.

7. Dickert, *History of Kershaw's Brigade*, 510.

8. *O.R.*, 47, pt. 2:1157, 1167, 1181.

9. Emilio, *A Brave Black Regiment*, 285–86; Roman, *The Military Operations of General Beauregard*, 348; *O.R.*, 47, pt. 2:1167.

10. *O.R.*, 47, pt. 2:1179–80.

11. Emilio, *A Brave Black Regiment*, 293; *Recollections and Reminiscences*, 412; Charleston & Savannah Railroad, Minutes, February 10, 1865.

12. *O.R.*, 47, pt. 1:1071; ibid., 47, pt. 2:1201, 1204; Roman, *The Military Operations of General Beauregard*, 639.

13. Burton, *The Siege of Charleston*, 318; *Charleston Daily Courier*, February 20, 1865; Cauthen, *South Carolina Goes to War*, 294.

14. Trudeau, *Like Men of War*, 356–57.

15. *Charleston Daily Courier*, February 20, 1865.

16. Charleston & Savannah Railroad, *Annual Report*, 1866, 25–26.

17. Evans, *Confederate Military History*, 4:365–67; Dickert, *History of Kershaw's Brigade*, 519–21; *O.R.*, 47, pt. 1:254; Brooks, *Stories of the Confederacy*, 229–30.

18. *O.R.*, 47, pt. 3:560–61.

19. Charleston & Savannah Railroad, *Annual Report*, 1866, 29; Black, *Railroads of the Confederacy*, 287.

20. Charleston & Savannah Railroad, *Annual Report*, 1866, 27.

21. Ibid., 8, 13–24.

22. *Charleston Daily Courier*, June 16, 1865.

23. Charleston & Savannah Railroad, *Annual Report*, 1866, 24, and Minutes, August 8, 1865.

24. Charleston & Savannah Railroad, Minutes, July 25, 1865.

25. Charleston & Savannah Railroad, *Annual Report*, 1866, 8; *Charleston Daily Courier*, April 23, 1866.

26. Charleston & Savannah Railroad, *Annual Report*, 1866, 8–9.

27. Charleston & Savannah Railroad, Minutes, December 2, 1865.

28. Charleston & Savannah Railroad, *Annual Report*, 1866, 26.

29. Ibid., 27.

30. Ibid., 28.

31. *Charleston Daily Courier*, December 20, 1865.

CHAPTER 18: POSTWAR DEBTS

1. *Charleston Daily Courier*, June 16, 1865.

2. Ibid., June 2, 1865; Derrick, *Centennial History of the South Carolina Railroad*, 234.

3. Simkins and Woody, *South Carolina during Reconstruction*, 187–88; Charleston & Savannah Railroad, Minutes, January 12, 1866.

4. *Charleston Daily Courier*, August 4, 1868; Strickland, "Ethnicity and Race in the Urban South," 63.

5. Charleston & Savannah Railroad, Minutes, July 18, 1866; Moore, *Confederate Commissary General—Lucius Bellinger Northrop and the Subsistence Bureau of the Southern Army*, 178.

6. Fraser, *Charleston! Charleston!*, 269–70; Coulter, *George Walton Williams*, 27; Simkins and Woody, *South Carolina during Reconstruction*, 269.

7. Simkins and Woody, *South Carolina during Reconstruction*, 269; Fraser, *Charleston! Charleston!*, 282; *Charleston Daily Courier*, June 9 and November 22, 1865.

8. Charleston & Savannah Railroad, Minutes, January 12, 1866; *Charleston Daily Courier*, January 15, 1866.

9. Charleston & Savannah Railroad, Minutes, January 12 and 29, 1866; Charleston & Savannah Railroad, *Annual Report*, 1866, 9.

10. *Charleston Daily Courier*, March 15, 1866; and Charleston & Savannah Railroad, Minutes, March 13, 1866.

11. *Charleston Daily Courier*, March 17 and March 26, 1866.

12. Charleston & Savannah Railroad, Minutes, March 13, 1866; Charleston & Savannah Railroad, *Annual Report*, 1866, 11–12; *Charleston Daily Courier*, March 17 and March 26, 1866.

13. Coulter, *George Walton Williams*, 16; Charleston & Savannah Railroad, Minutes, February 15, 1866; *Charleston Daily Courier*, March 15 and March 19, 1866.

14. *Charleston Daily Courier*, February 2 and March 8, 1866; Charleston & Savannah Railroad, Minutes, January 12 and March 13, 1866.

15. Charleston & Savannah Railroad, Minutes, April 26, 1866.

16. *Charleston Daily Courier*, March 22, 1866.

17. Ibid., April 19 and April 25, 1866.

18. Ibid., April 3 and April 23, 1866.

19. Ibid., April 3, 1866.

20. Clark, *Railroads in the Civil War*, 39–40.

21. *Charleston Daily Courier*, April 30, 1863; Charleston & Savannah Railroad, *In Equity. Bill for Injunction*, 9; Charleston & Savannah Railroad, Minutes, March 13, 1866; Black, *Railroads of the Confederacy*, 220.

22. Black, *Railroads of the Confederacy*, 219; Charleston & Savannah Railroad, *In Equity. Bill for Injunction*, 9.

23. *Charleston Daily Courier*, April 23, 1866.

24. Charleston & Savannah Railroad, Minutes, January 12, 1866.

25. *Charleston Daily Courier*, April 3 and April 26, 1866.

26. Charleston & Savannah Railroad, Minutes, June 8, 1866.

27. Ibid., June 20 and June 22, 1866.

28. Ibid., July 25, 1866.

29. Fraser, *Charleston! Charleston!*, 281–82; E. M. Beach to C&SRR, June 1, 1866, in Charleston & Savannah Railroad, Minutes, June 8, 1866; Simonton to C&SRR, June 6, 1866, in Charleston & Savannah Railroad, Minutes, June 20, 1866.

30. *Charleston Daily Courier*, August 2 and August 16, 1866; Simkins and Woody, *South Carolina during Reconstruction*, 194.

31. *Charleston Daily Courier*, August 18 and August 21, 1866.

32. Charleston & Savannah Railroad, Minutes, August 29, 1866; Haines, "Memoir of Henry Stephens Haines," 1695–97.

33. Charleston & Savannah Railroad, *In Equity. Bill for Injunction*.

34. Ibid.

35. J. B. Campbell, legal opinion, October 1, 1866, Railroad Litigation Papers, 1863–1881, James Butler Campbell Legal Papers, South Carolina Historical Society; *Charleston Daily Courier*, August 16, 1866.

36. Savannah & Charleston Railroad, *Annual Report*, 1868, 2, and Minutes, November 15, 1866; Stover, "The Ruined Railroads of the Confederacy," 382–83.

37. Savannah & Charleston Railroad, Records, 1866–1871; *Charleston Daily Courier*, December 27, 1866; Simkins and Woody, *South Carolina during Reconstruction*, 192.

CHAPTER 19: TERMINUS

1. Savannah & Charleston Railroad, Records, 1866–1871, February 1, 1867.

2. Courtney, "Charles Jones Colcock," 34–39.

3. Evans, ed., *Confederate Military History*, 4:387–89; James H. Easterby, "Thomas Fenwick Drayton," 446–47.

4. *Hill and Swayze's Confederate States Rail Road and Steam Boat Guide*, 62; partnership agreement, Huger & Hasell, January 10, 1865; Jesup to Huger & Hasell, March 15, 1866; and Jesup & Co. to Huger, July 18, 1867; Cleland Huger Papers, South Caroliniana Library; Hasell, "Genealogical Chart Showing the Descent and Alliance from the Colonial Period of Cruger, Son of Henry Walton" (1897), Bentley D. Hasell Files, New York Genealogical Society.

5. Marie G. Magrath, biographical notes, William J. Magrath Family Papers, South Carolina Historical Society; Derrick, *Centennial History of the South Carolina Railroad*, 253–57.

6. *Recollections and Reminiscences*, 412; H. M. Drane to R. L. Singletary, April 9, 12, and 14, 1873, Robert L. Singletary Papers, South Caroliniana Library; King, *Rise Up So Early*, 74, 84.

7. Ward, ed., *Southern Railroad Man*, 141.

8. Clark, *Railroads in the Civil War*, 10, 221–23.

9. Ibid., 10.

10. Black, *Railroads of the Confederacy*, 8–13.

11. See "The Railroads of the Confederate States" on pp. 48–49.

12. Clark, *Railroads in the Civil War*, 19–21; Turner, *Victory Rode the Rails*, 104.

13. Clark, *Railroads in the Civil War*, 47–50, 55–56.

14. Ibid., 68–69; Turner, *Victory Rode the Rails*, 104, 107.

15. Clark, *Railroads in the Civil War*, 58; Turner, *Victory Rode the Rails*, 107.

16. Clark, *Railroads in the Civil War*, 39–41, 55–56; Turner, *Victory Rode the Rails*, 104.

17. Turner, *Victory Rode the Rails*, 243.

18. Savannah & Charleston Railroad, Records, February 1, 1867.

19. Savannah & Charleston Railroad, *Annual Report*, 1868, 2–5; Savannah & Charleston Railroad, Records, 1866–1871; Simkins and Woody, *South Carolina during Reconstruction*, 193–94.

20. Simkins and Woody, *South Carolina during Reconstruction*, 194–95; indenture, George Walton Williams Papers, South Carolina Historical Society.

21. Simkins and Woody, *South Carolina during Reconstruction*, 193–95.

22. Collins, "Charleston and the Railroods," 49; Simkins and Woody, *South Carolina during Reconstruction*, 194–95; Prince, *Seaboard Air Line Railway*, 26, and *Atlantic Coast Line Railroad*, 56.

23. Nelson, *Iron Confederacies*, 42–43; Johnson, "Plant's Lieutenants," 381–83; Hoffman, *Building a Great Railroad*, 129.

24. Johnson, "Plant's Lieutenants," 388–392; Haines, "Memoir of Henry Stephens Haines"; Hoffman, *Building a Great Railroad*, 54–59, 122–24.

25. Hoffman, *Building a Great Railroad*, 129–33.

BIBLIOGRAPHY

MANUSCRIPT COLLECTIONS

Louisiana State University Libraries
    Special Collections, Hill Memorial Library
South Carolina Department of Archives and History, Columbia, South Carolina
    Governor Milledge L. Bonham Papers
    South Carolina General Assembly, Miscellaneous Communications
South Caroliniana Library, University of South Carolina, Columbia, South Carolina
    Daniel Blake Papers
    C.S.A. Fourth Military District Papers
    Thomas Fenwick Drayton Papers
    Ambrosio Gonzales Papers
    Maxcy Gregg Papers
    Heyward Family Papers
    Cleland Huger Papers
    Robert L. Singletary Papers
South Carolina Historical Society, Charleston, South Carolina
    James Butler Campbell Papers, 1814–1897
    Hutson Lee Papers, 1858–1865
    Magrath Family Papers, 1865–1956
    George Walton Williams Papers, 1845–1907
Davis Library, University of North Carolina, Chapel Hill
    Documenting the American South Collection
Hargrett Library, University of Georgia, Athens
    Cornelius C. Platter Civil War Diary, 1864–1865
New York Genealogical and Biographical Society
    Cruger Family / Bentley D. Hasell File

COMPANY AND PUBLIC RECORDS

Act to Charter the Charleston & Savannah Railroad. South Carolina Acts and Joint Resolutions, no. 4142 (1853): 271–80. South Carolina Department of Archives and History.
Charleston & Savannah Railroad. Records, 1863–1867. Robert Scott Small Library, Special Collections, College of Charleston.

Charleston County, Book of Mesne Conveyances. South Carolina Department of Archives and History.

Confederate Railroad Bureau Archives. Valentine Museum, Richmond, Virginia.

Confederate Records Relating to Private Citizens and Businesses. National Archives and Records Administration, Washington, D.C.

Drayton, Thomas F. Annual Report to the Legislature on the Conditions and Prospects of the Charleston & Savannah Railroad, November 29, 1858. General Assembly, Misc. Communications, South Carolina Department of Archives and History.

Journal of the House of Representatives of the State of South Carolina, 1829. South Carolina Department of Archives and History.

Reports and Resolutions, State of South Carolina. Abstract Report of the Charleston & Savannah Railroad, 1859, 1860; Report of the Charleston & Savannah Railroad, 1862, 1863. South Carolina Department of Archives and History.

Savannah & Charleston Railroad. Memorial and Petition of the Savannah and Charleston Railroad Company to the Legislature of the State of South Carolina. Charleston Library Society.

———. Minutes, 1869–1878. South Carolina Historical Society.

———. Records, 1866–1871. South Carolina Historical Society.

Records of the Confederate Post Office Department, 1861–1865. National Archives and Records Administration, Washington, D.C.

Records of the Confederate Quartermaster Department. National Archives and Records Administration, Washington, D.C.

Records of Cotton Shipped through the Confederate Customhouse at Charleston, South Carolina. National Archives and Records Administration, Washington, D.C.

South Carolina Acts and Joint Resolutions, 1853. South Carolina Department of Archives and History.

U.S. War Department. Compiled Service Records of Confederate Soldiers Who Served in Organizations from the State of South Carolina (microcopy no. M267), 1861–65. National Archives and Records Administration, Washington, D.C.

U.S. War Department. Compiled Service Records of Confederate General and Staff Officers and Nonregimental Enlisted Men (microcopy no. M331), 1861–65. National Archives and Records Administration, Washington, D.C.

NEWSPAPERS AND PERIODICALS

American Railroad Journal
Charleston Daily Courier
Charleston Mercury
Columbia Telescope
Illustrated London News
Keowee Courier
New South (Beaufort, S.C.)
Savannah Daily Evening News
Savannah Daily Morning News
Savannah News
Savannah Republican

Scioto Gazette
Tri-Weekly Watchman (Sumter, S.C.)

BOOKS, ARTICLES, THESES, INTERNET SITES

Abbott, A. O. *Prison Life in the South during the Years 1864 and 1865.* New York: Harper, 1866.

Adams, John G. B. *Reminiscences of the Nineteenth Massachusetts Regiment.* Boston: Wright-Potter Printing, 1899.

Andrews, William Hill. *Footprints of a Regiment: A Recollection of the First Georgia Regulars, 1861–1865.* Atlanta: Longstreet Press, 1992.

Atlas to Accompany the Official Records of the Union and Confederate Armies. 2 vols. Washington, D.C.: U.S. Government Printing Office, 1891, 1895.

Baker, Gary R. *Cadets in Gray: The Story of the Cadets of the South Carolina Military Academy and the Cadet Rangers in the Civil War.* Columbia, S.C.: Palmetto Bookworks, 1989.

Baxley, Neil. *No Prouder Fate: The Story of the 11th South Carolina Volunteer Infantry.* Bloomington, Ind.: Authorhouse, 2005.

Beringer, Richard E., Herman Hattaway, Archer Jones, and William Still, Jr., eds. *Why the South Lost the Civil War.* Athens: University of Georgia Press, 1986.

Black, Robert C. *Railroads of the Confederacy.* Chapel Hill: University of North Carolina Press, 1952.

Bond, Oliver James. *The Story of the Citadel.* Greenville, S.C.: Southern Historical Press, 1989.

Bowman, Samuel M., and Richard B. Irwin. *Sherman and His Campaigns: A Military Biography.* New York: Richardson, 1865.

Brooks, U. R., ed. *Stories of the Confederacy.* Columbia, S.C.: State Co., 1912.

Burton, E. Milby. *The Siege of Charleston, 1861–1865.* Columbia, S.C.: University of South Carolina Press, 1970.

Camp, Vaughn, Jr., ed. "The War for Independence, North and South: The Diary of Samuel Catawba Lowry." *South Carolina Historical Magazine* 79 (July 1978): 182–97.

Carse, Robert. *Department of the South: Hilton Head Island and the Civil War.* Columbia, S.C.: State Printing Co., 1961.

Catton, Bruce. *The Civil War.* Boston: Houghton Mifflin, 1987.

———. *The Coming Fury.* Garden City, N.Y.: Doubleday, 1961.

Catton, William, and Bruce Catton. *Two Roads to Sumter.* New York: McGraw-Hill, 1963.

Cauthen, Charles Edward. *South Carolina Goes to War, 1860–1865.* Chapel Hill: University of North Carolina Press, 1950.

Chadwick, Thomas W. "Diary of Samuel Edward Burges, 1860–1862." *South Carolina Historical and Genealogical Magazine* 48 (July 1947): 141–63.

Charleston & Savannah Railroad, *Annual Report.* 8 vols. Charleston, 1855–1859, 1861, 1864, 1866.

———. *By-laws of the Charleston & Savannah Railroad Company: Adopted at a Meeting of the Stockholders, 26th July, 1854.* Charleston, S.C.: Walker & Evans, 1854.

———. *In Equity. Bill for Injunction, etc. the Charleston & Savannah Railroad vs Isaac W. Hayne, John Carew, et al.* Charleston, S.C.: Courier Job Press, 1866.

———. *Report. Engineer's Office, Charleston & Savannah Railroad to the President and Directors, Charleston & Savannah Railroad.* Charleston, S.C.: Walker & Evans, 1855.

———. *Report of the Chief Engineer on the Preliminary Survey of the Charleston & Savannah Railroad.* Charleston, S.C.: Walker & Evans, 1854.

Chesnut, Mary Boykin. *A Diary from Dixie.* Edited by Isabella D. Martin and Myrta Lockett Avary. New York: Appleton, 1905.

Clark, John Elwood. *Railroads in the Civil War: The Impact of Management on Victory and Defeat.* Baton Rouge: Louisiana State University Press, 2001.

Clement, Christopher Ohm, Stephen R. Wise, et al. *Mapping the Defense of the Charleston & Savannah Railroad—Civil War Earthworks in Beaufort and Jasper Counties, South Carolina.* Report prepared for the American Battlefields Protection Program Grant. Columbia: University of South Carolina Department of Archaeology, 2000.

Coleman, Kenneth, ed. *A History of Georgia.* 2nd ed. Athens: University of Georgia Press, 1991.

Coleman, Kenneth, and Charles Stephen Garr, ed. *Dictionary of Georgia Biography.* 2 vols. Athens: University of Georgia Press, 1983.

Collins, Frederick Burtrumn. "Charleston and the Railroads: A Geographic Study of a South Atlantic Port and Its Strategies for Developing a Railroad System, 1820–1860." M.S. thesis, University of South Carolina, 1977.

Coulter, E. Merton. *George Walton Williams: The Life of a Southern Merchant and Banker, 1820–1903.* Athens, Ga.: Hibriten Press, 1976.

Courtney, William A. "Charles Jones Colcock: A Typical Citizen and Soldier of the Old Regime." *Southern Historical Society Papers* 26 (January–December 1898): 33–39.

———. "Fragments of War History Relating to the Coast Defense of South Carolina, 1861–65 and the Hasty Preparations for the Battle of Honey Hill." *Southern Historical Society Papers* 26 (January–December 1898): 62–87.

Coxe, John. "Chickamauga." *Confederate Veteran* 30 (August 1922): 291–94.

Cozzens, Peter, and Robert I. Girardi, eds. *The New Annals of the Civil War.* Mechanicsburg, Pa.: Stackpole Books, 2004.

Crist, Lynda L., ed. *The Papers of Jefferson Davis.* Vols. 7 and 8. Baton Rouge: Louisiana State University Press, 1992, 1995.

DeCredico, Mary A. *Patriotism for Profit: Georgia's Urban Entrepreneurs and the Confederate War Effort.* Chapel Hill: University of North Carolina Press, 1990.

De la Cova, Antonio Rafael. *Cuban Confederate Colonel: The Life of Ambrosio José Gonzales.* Columbia: University of South Carolina Press, 2003.

Derrick, Samuel Melanchthon. *Centennial History of South Carolina Railroad.* Columbia: State Co., 1930.

DeSaussure, Henry W. "Quinine as a Prophylactic of Intermittent and Remittent Fevers." *Charleston Medical Journal and Review* 15 (July 1860): 433–41.

Dickert, David Augustus. *History of Kershaw's Brigade.* Wilmington, N.C.: Broadfoot Publishing, 1990.

Dowdey, Clifford, ed. *The Wartime Papers of R. E. Lee.* New York: Bramhall House, 1961.

Eaton, Clement. *A History of the Southern Confederacy.* New York: Macmillan, 1954.

Edgar, Walter B. *South Carolina: A History.* Columbia: University of South Carolina Press, 1998.

Emilio, Luis F. *A Brave Black Regiment: The History of the Fifty-Fourth Regiment of Massachusetts Volunteer Infantry, 1863–1865.* New York: Da Capo, 1995.

Evans, Clement A., ed. *Confederate Military History*. 13 vols. Atlanta: Blue and Gray Press, 1970.

Federal Writers' Project. "Interview with William Rose, Ex-slave of Edisto Island." In *Slave Narratives, South Carolina*, 4:48–50. N.p., 2000?

Ford, Arthur P., and Marion J. Ford. *Life in the Confederate Army—Being the Personal Experiences of a Private Soldier in the Confederate Army* [and] *Some Experiences and Sketches of Southern Life*. New York & Washington, D.C.: Neale, 1905.

Ford, Ralph Watson. "The Changing Geographic Pattern of South Carolina's Railroad System, 1860–1902." Master's thesis, University of South Carolina, 1973.

Fraser, Walter J. *Charleston! Charleston!: The History of a Southern City*. Columbia: University of South Carolina Press, 1991.

Freehling, William W. *Prelude to Civil War: The Nullification Controversy in South Carolina, 1816–1836*. New York: Oxford University Press, 1965.

———. *Road to Disunion: Secessionists at Bay, 1776–1854*. New York: Oxford University Press, 1990.

———. *The South vs. the South: How Anti-Confederate Southerners Shaped the Course of the Civil War*. New York: Oxford University Press, 2001.

Freeman, Douglas Southall. *R. E. Lee: A Biography*. 4 vols. New York & London: Scribners, 1942.

Gardner, Charles. *Gardner's New Orleans Directory for 1861*. New Orleans: Charles Gardner, 1861.

Garrison, L. R. "Administrative Problems of the Confederate Post Office Department." *Southwestern Historical Quarterly* 19 (January 1916): 232–50.

Golay, Michael. *A Ruined Land: The End of the Civil War*. New York: Wiley, 1999.

Greeley, A. W. "The Military Telegraph Service." In *The Photographic History of the Civil War*. 10 vols. 8:342–63. New York: Review of Reviews, 1911.

Gregorie, Anne King, ed. "Diary of Captain Joseph Julius Wescoat, 1863–1865." *South Carolina Historical Magazine* 59 (January 1958): 11–23; (April 1958): 84–95.

Hagood, Johnson. *Memoirs of the War of Secession*. Columbia, S.C.: State Printing Co., 1910.

Haines, Charles O. "Memoir of Henry Stephens Haines." *Transactions of the American Society of Civil Engineers* 98 (1928):1695–97.

Halliburton, Lloyd, ed. *Saddle Soldiers: The Civil War Correspondence of General William Stokes of the Fourth South Carolina Cavalry*. Orangeburg, S.C.: Sandlapper Publishing, 1993.

Hemphill, J. C., ed. *Men of Mark in South Carolina*. 3 vols. Washington, D.C.: Men of Mark Publishing, 1907–9.

Herriott, Robert. "Fighting in South Carolina." *Confederate Veteran* 30 (November 1922): 415.

———. "Fighting to the End." *Confederate Veteran* 31 (May 1923): 167, 198.

———. "On the South Carolina Coast, 1864–65." *Confederate Veteran* 32 (May 1924): 169, 195.

Heyward, Duncan Clinch. *Seed from Madagascar*. Chapel Hill: University of North Carolina Press, 1937.

Higginson, Thomas Wentworth. *Army Life in a Black Regiment*. 1870. Reprint, Alexandria, Va.: Time-Life Books, 1982.

Hill and Swayze's Confederate States Rail-Road & Steam-Boat Guide. Griffin, Ga.: Hill & Swayze, 1862.

Hoffman, Glenn. *Building a Great Railroad: A History of the Atlantic Coast Line Railroad.* United States: CSX Transportation Corporate Communications and Public Affairs, 1998.

Hoyt, William D., Jr., ed. "To Coosawhatchie in December, 1861." *South Carolina Historical Magazine* 53 (January 1952): 6–12.

Hudson, Leonne M. "A Confederate Victory at Grahamville: Fighting at Honey Hill." *South Carolina Historical Magazine* 94 (January 1993): 19–33.

Hughes, Nathaniel Cheairs. *General William J. Hardee—Old Reliable.* Baton Rouge: Louisiana State University Press, 1965.

———. "Hardee's Defense of Savannah." *Georgia Historical Quarterly* 47 (March 1963): 43–67.

Hyde, William Lyman. *History of the One Hundred and Twelfth Regiment of New York Volunteers.* Fredonia, N.Y.: McKinstry, 1866.

Johnson, Dudley S. "Plant's Lieutenants." *Florida Historical Quarterly* 49 (April 1970): 381–92.

Johnson, Robert U., and Clarence C. Buel, eds. *Battles and Leaders of the Civil War.* New York: Yoseloff, 1956.

Jones, Charles C., Jr. "The Battle of Honey Hill." *Southern Historical Society Papers* 13 (January–December 1885): 355–67.

———. "The Siege and Evacuation of Savannah, Georgia, in December, 1864." *Southern Historical Society Papers* 17 (January–December 1889): 60–85.

Jones, Katherine M. *Port Royal under Six Flags.* Indianapolis & New York: Bobbs-Merrill, 1960.

Journal of the Congress of the Confederate States of America, 1861–1865. 7 vols. Washington, D.C.: U.S. Government Printing Office, 1904–05.

King, G. Wayne. *Rise Up So Early: A History of Florence County, South Carolina.* Spartanburg, S.C.: Reprint Co., 1981.

Klein, Maury. *Days of Defiance.* New York: Knopf, 1997.

Knudsen, Lewis F. *The Official Records of Company D (South Carolina Rangers), 5th South Carolina Cavalry Regiment, 1861–1865.* Columbia, S.C.: L. F. Knudsen, 1990.

Lander, Ernest M., Jr. "Charleston: Manufacturing Center of the Old South." *Journal of Southern History* 26 (August 1960): 330–51.

Lane, Mills, ed. *Times That Prove People's Principles: Civil War in Georgia, A Documentary History.* Savannah: Library of Georgia, 1993.

Lawrence, Alexander A. *A Present for Mister Lincoln: The Story of Savannah from Secession to Sherman.* Macon, Ga.: Ardivan Press, 1961.

Lee, Captain Robert E. *Recollections and Letters of General Robert E. Lee, by His Son.* New York: Doubleday, Page, 1924.

Linder, Suzanne Cameron. *Historical Atlas of the Rice Plantations of the ACE River Basin—1860.* Columbia: South Carolina Department of Archives and History, 1995.

Longacre, Edward G., ed. "We Left a Black Trail in South Carolina: Letters of Corporal Eli S. Ricker, 1865." *South Carolina Historical Magazine* 82 (July 1981): 210–24.

Mackintosh, Robert Harley, Jr., ed. *"Dear Martha . . ."—The Confederate War Letters of a South Carolina Soldier, Alexander Faulkner Fewell.* Columbia, S.C.: R. L. Bryan, 1976.

Easterby, James H. "Thomas Fenwick Drayton," in *Dictionary of American Biography*, edited by Allen Johnson and Dumas Malone, 5: 446–47. New York: Scribners, 1930.

Massey, Mary Elizabeth. *Ersatz in the Confederacy.* Columbia: University of South Carolina Press, 1952.

McElroy, John. *Andersonville: A Story of Rebel Military Prisons.* Toledo, Ohio: D. R. Locke, 1879. Project Gutenberg Ebook, http://www.gutenberg.org/ebooks/3072 (accessed May 19, 2004).

Memorial to the Honorable the Senate and House of Representatives of the Congress of the Confederate States of America. Columbia, S.C., 1864.

Moore, Frank, ed. *The Rebellion Record—A Diary of American Events.* Vol. 10. New York: Putnam, 1867.

Moore, J. N. *Confederate Commissary General—Lucius Bellinger Northrop and the Subsistence Bureau of the Southern Army.* Shippensburg, Pa.: White Mane Publishing, 1996.

Morrill, Dan. *The Civil War in the Carolinas.* Charleston, S.C.: Nautical and Aviation Publishing, 2002.

Nelson, Scott Reynolds. *Iron Confederacies: Southern Railways, Klan Violence, and Reconstruction.* Chapel Hill: University of North Carolina Press, 1999.

Nicolay, John G., and John Hay. *Abraham Lincoln: A History.* Edited by Paul M. Angle. Chicago: University of Chicago Press, 1966.

Olmstead, Charles H. *An Address Delivered Before the Georgia Historical Society, March 3, 1879.* Savannah: J. H. Estill, 1879.

Perry, Grace Fox. *Moving Finger of Jasper County.* Ridgeland?, S.C.: Jasper County Confederate Centennial Commission, 1962.

Phelps, W. Chris. *The Bombardment of Charleston, 1863–1865.* Gretna, La.: Pelican Publishing, 2002.

Prince, Richard E. *Atlantic Coast Line Railroad: Steam Locomotives, Ships, and History.* Bloomington: Indiana University Press, 2000.

———. *Seaboard Air Line Railway: Steam Boats, Locomotives, and History.* Bloomington: Indiana University Press, 2000.

Ravenel, Samuel W. "The Boy Brigade of South Carolina." *Confederate Veteran* 29 (November–December 1921): 417–18.

Recollections and Reminiscences, 1861–65 through World War I. Vol. 1. South Carolina: United Daughters of the Confederacy, 1990.

Richardson, James D., ed. *Messages and Papers of the Confederacy.* Vol. 1. Nashville, Tenn.: U.S. Publishing, 1905.

Riecke, Anthony W. *Recollections of a Confederate Soldier of the Struggle for the Lost Cause.* Charleston, S.C.: A. W. Riecke, 1879.

Ripley, Warren, ed. *Siege Train: The Journal of a Confederate Artillery Man in the Defense of Charleston.* Columbia: University of South Carolina Press, 1986.

Robinson, Joseph M. "The Defense of the Charleston & Savannah Railroad, 1861–1865." Master's thesis, University of South Carolina, 1950.

Roman, Alfred. *The Military Operations of General Beauregard in the War Between the States, 1861 to 1865.* New York: Harper, 1884.

Rowland, Dunbar, ed. *Jefferson Davis—Constitutionalist.* Jackson: Mississippi Department of Archives and History, 1923.

Rowland, Lawrence S., Alexander Moore, and George C. Rogers Jr. *History of Beaufort County*, Vol. 1. Columbia: University of South Carolina Press, 1996.

Russell, William Howard. *My Diary North and South.* Edited by Fletcher Pratt. Gloucester, Mass.: P. Smith, 1969.

Salley, A. S. "Captain John Colcock and Some of His Descendants." *South Carolina Historical and Genealogical Magazine* 3 (October 1902): 216–41.

Savannah & Charleston Railroad. *Annual Report.* Charleston: Courier Job Press, 1868.

Scaife, William. *Joe Brown's Pets: The Georgia Militia, 1861–1865.* Macon, Ga.: Mercer University Press, 2004.

Schmidt, Lewis G. *The Battle of Pocotaligo, October 22, 1862.* Allentown, Pa.: L. G. Schmidt, 1993.

Sherman, William T. *Memoirs of General W. T. Sherman.* 2 vols. 1875. Reprint, New York: Library of America, 1990.

Silverman, Jason H., Samuel N. Thomas Jr., and Beverly D. Evans IV. *Shanks: The Life and Wars of General Nathan George Evans, C.S.A.* Cambridge, Mass.: Da Capo Press, 2002.

Silverman, Jason H., and S. R. Murphy, eds. "Our Separation is Like Years: The Civil War Letters of Deopold Daniel Lewis." *South Carolina Historical Magazine* 87 (July 1986): 141–47.

Simkins, Francis Butler, and Robert Hilliard Woody. *South Carolina during Reconstruction.* Chapel Hill: University of North Carolina Press, 1932.

Simms, Mary C., and Mary C. Simms Oliphant Furman. *South Carolina from the Mountains to the Sea.* Columbia: State Printing, 1964.

Smith, Derek. *Civil War Savannah.* Savannah.: Frederic C. Biel, 1997.

Sneden, Robert Knox, Charles F. Bryan Jr., and Nelson D. Lankford, eds. *Eye of the Storm.* New York: Free Press, 2000.

Spears, Linda. *British North Americans (Canadians) Biography. Canadians in the Civil War Web Site,* http://members.tripod.com/PvtChurch/bios/forsterbio.html

Stover, John F. *The Railroads of the South, 1865–1900: A Study in Finance and Control.* Chapel Hill: University of North Carolina Press, 1955.

———. "The Ruined Railroads of the Confederacy." *Georgia Historical Quarterly* 42 (December 1958): 376–88.

Strickland, Jeffery. "Ethnicity and Race in the Urban South: German immigrants and African-Americans in Charleston, South Carolina during the Reconstruction." Ph.D. dissertation, Florida State University, 2003.

Talliaferro, William B. "Defense of Charleston from July 1st to July 10th, 1864." *Southern Historical Society Papers* 2 (October 1876): 192–204.

Taylor, Frances Wallace, Catherine Taylor Matthews, and J. Tracy Power, eds., *The Leverett Letters: Correspondence of a South Carolina Family, 1851–1868.* Columbia: University of South Carolina Press, 2000.

Taylor, Susie King. *Reminiscences of My Life in Camp with the Thirty-third United States Colored Troops, Late First South Carolina Volunteers.* 1902. Reprint, New York: Markus Wiener, 1988.

Thomas, Edward J. *Memoirs of a Southerner, 1840–1923.* Savannah, 1923.

Trudeau, Noah Andre. *Like Men of War: Black Troops in the Civil War, 1862–1865.* Edison, N.J.: Castle Books, 2002.

Turner, George Edgar. *Victory Rode the Rails.* Lincoln: University of Nebraska Press, 1953.

U.S. War Department. *War of the Rebellion: Official Records of the Union and Confederate Armies in the War of the Rebellion.* 128 vols. Washington, D.C.: U.S. Government Printing Office, 1902.

U.S. War Department. *War of the Rebellion: Official Records of the Union and Confederate Navies in the War of the Rebellion.* 31 vols. Washington, D.C.: U.S. Government Printing Office, 1901.

Van Deusen, John G. *Economic Basis of Disunion in South Carolina.* New York: Columbia University Press, 1928.

Varhola, Michael J. *Every Day Life During the Civil War.* Cincinnati, Ohio: Writer's Digest Books, 1999.

Ward, James A., ed. *Southern Railroad Man: Conductor N. J. Bell's Recollections of the Civil War Era.* Dekalb: Northern Illinois University Press, 1994.

Western Carolina Historical Research. *Civil War in Southwestern South Carolina.* Montmorenci, S.C.: Western Carolina Historical Research, 1997.

———. *The Pocotaligo Expedition, Southwestern South Carolina, October 21–23, 1862.* Montmorenci, S.C.: Western Carolina Historical Research, 1997.

Wise, Stephen R. *Lifeline of the Confederacy: Blockade Running during the Civil War.* Columbia: University of South Carolina Press, 1988.

———. *Gate of Hell: Campaign for Charleston Harbor, 1863.* Columbia: University of South Carolina Press, 1994.

Woodman, Harold D. *King Cotton and His Retainers.* Lexington: University of Kentucky Press, 1968.

Wyckoff, Mac. *A History of the 2nd South Carolina Infantry, 1861–1865.* Fredericksburg, Va.: Sergeant Kirkland's Museum and Historical Society, 1994.

INDEX

Italic page numbers refer to illustrations. Charleston & Savannah Railroad is abbreviated as "C & S Railroad."

Abel's battery, 238
Abney, Joseph, 117
abolitionism, 3, 143
Adams, Alvin, 310
Adams, George W., 164, 166
Adams, John, 203
Adams Express Company, 34, 310
Adams Run, S.C.: and Birney, 192, 194; and Davis, 161; and Hagood, 108, 109–10; and Ripley, 190; and Robertson, 197, 198
Aiken, Hugh Kerr, 146, 147, 151, 178
Aiken, William, Jr., 161, 309
Aiken, William, Sr., 1
Alabama & Florida Railroad, 292
Aldrich, Alfred P., 13
Allston, Joseph Blythe, 114
Allston, Robert F. W., 15, 41
American Railway Association, 312
American Railway Guild, 312
American Society of Mechanical Engineers, 312
American Telegraph Company, 279
Ames, Adelbert, 177, 179
Anaconda Plan, 69
Anderson, E. C., 255
Anderson, George T., 152–53, 156
Anderson, Robert, 59, 60, 61, 191
Andersonville prison, 155, 203, 204
Andrew Milne (locomotive), 31, 52, 274
Andrews, W. H., 253, 256, 260
Appleby, W. P., 154
Appleton, D. F., 289
Army of Northern Virginia, 128, 138, 151, 175–76, 263, 273
Army of Tennessee, 128, 156, 175
Army of the Cumberland, 151, 156

Army of the Potomac, 138, 152, 172
Arsenal Academy, 226
Arsenal Cadets, 231, 232, 234
Ashepoo (locomotive), 52
Ashepoo River, 21, 153, 190, 194, 269
Ashley (locomotive), 29, 52, 53
Ashley River: and C & S Railroad steamer service, 46, 124, 207, 208, 209, 279, 289; lack of railroad bridge across, 17, 18, 22, 23, 27, 32, 38, 73, 89, 124, 136, 207; and Savannah & Charleston Railroad steamer service, 309; telegraph line over, 153
Ashley River Bridge: and Cannonsborough Wharf and Mill Company, 207–8; construction of, 105–6, 120, 124–25, 126, 131–32, 133, 135–36; and New Bridge Ferry, 73; ownership of, 131, 135–36, 141; safety of, 208, 209; and Singletary, 133, 135–36, 181, 207–9
Ashley River Railroad, 309–10
Atlanta, Ga.: C & S Railroad as alternative route east of, 46; and Confederate strategy, 151, 212; and Confederate troop transport, 152, 155; and W. T. Sherman, 185, 204, 206, 209, 210
Atlanta Rolling Mill, 54
Atlantic & Danville Railroad, 312
Atlantic & Gulf Railroad: and C & S Railroad, 309; and Haines, 298; and Hardee, 237, 240; and Plant, 310; and Savannah, 210, 240, 257; and troop transport, 180, 213, 216
Atlantic Coast Line Railroad, 312
Augusta, Ga., 18, 81, 212
Augusta & Savannah Railroad, 210
Augusta Battalion, 235

Bachman, William K., 236, 246
Bacon, Edwin H., 231
Baird, Absalom, 242–43, 244
Baird, G. W., 217
Baker, Lawrence S., 221, 228, 238, 255, 256
Bank of Charleston, 23, 140, 292
Bank of South Carolina, 134
Bank of the State of Georgia, 11
Banks, Hugh R., 133
Baring, Charles, 21
Barker, Samuel, 21
Barnard, George N., photograph of Sherman's troops, *241*
Barnwell, Robert W., 55
Barton, William B., 116, 117, 118, 119
Bartow, Francis Stebbins, 41, 44, 225
Battery Gregg, 137
Battery Island, 99
Battery Pringle, 192, 195, 196, 197, 198
Battery Simkins, 193
Battery Tynes, 196
Battery Wagner, 144, 149, 150–51, 158, 175, 223
Battery Wright, 192
Bayley, Thomas, 189
Beach, Erastus M., 28, 296
Beaufort, S.C., 69, 104, 176, 177, 259, 289
Beaufort Volunteer Artillery, 68, 96, 112, 114, 116, 118, 218, 243
Beauregard, Pierre G. T.: in Charleston, 240, 269; and Davis, 149, 161, 171, 252; and defense of Charleston, 156, 158, 169, 171; and Du Pont, 121, 137; and evacuation of Savannah, 254, 255; and Finnegan, 179, 180; and Fort Sumter, 60, 61; and Gregorie's Neck, 234; and Hagood, 111; and Hardee, 240, 248, 250, 256, 262, 266, 268, 269, 270; and Hood, 262; and James Island, 148–49; and Johns Island, 180; and Samuel Jones, 188, 237, 240, 248; and Lee, 187, 270; and Manassas, 62, 63; and Pemberton's regiments, 92; photograph of, *121*; and Port Royal, 68; relationship with C & S railroad, 128; Richmond assignment, 138; in Savannah, 237, 238, 239, 249; and Seddon, 160; and W. T. Sherman, 211, 268, 272; and troop transport, 122–23, 149, 150, 172, 177; and William Walker, 112, 158, 169, 171; and Western Coalition, 151
Bee, William C., 15, 44–45, 86, 134, 184–85, 258, 294–95
Beecher, James C., 219
Bee's Creek earthworks, 227, 229

Behling, Luder F., 27
Bell, N. J., 200, 304
Ben De Ford (naval vessel),112
Benham, Henry W., 99, 101–2, 103, 104
Benjamin, Judah, 81
Bennett, Augustus G., 271, 284
Bennett, Isaac S. K., 2, 65
Bennett, William T., 219, 223
Bennett's Point, 189–90
Bentonville, N.C., map of Confederate troop transport to, *273*
Berry, J. W. R., 154
Berry, Thomas J., 68
Birney, William, 189, 190, 191, 192, 194–95, 196
Black, Alexander, 1
Black, Robert, 315
Black, Samuel L., 239, 240
Blair, Frank, 259, 260–61
Blake, Daniel, 21, 27
Blake, Francis D., 192
Bluffton Boys, 58
Boineau, L. H., 172, 234
blockade, Federal: and cotton, 55, 56, 174, 305; Davis's policy toward, 174–75; and *Planter*, 92; and Southern railroads, 83, 84, 164, 165; tightening of, 72; and George Williams, 284
Bonaud, A., 193, 195, 197, 198
Bonham, Milledge L., 131, 209, 226
Boston (naval vessel), 189–91
Boyce, Robert, 105
Boyd's Landing, 73, 215, 216, 228, 229, 234
Boyd's Neck, 228, 229, 234
Boylston, J. Reid, 294, 299, 300, 308
Bradley, John, 18
Bradley, Richard W., 27, 86
Bragg, Braxton, 151, 156, 157, 160, 176, 212, 221
Branchville, S.C., and railroad development, 12, 14
Branchville & Savannah Railroad, 15, 17, 19, 20, 36
Brannan, John M., 112, 114–16, 118
Brawley, W. H., 311
Breckinridge, John, 36
Brenholtz, Thomas S., 95
Broad River: and defense of C & S Railroad, 80, 83; defense of headwaters, 81; and Foster, 259; and Honey Hill, 214–16, 224; map of, *77*; and Mitchel, 111; and Isaac Stevens, 76, 78–79; and William Walker, 173
Brooks, J. H., 116

Broun, Joseph, 75
Broun, Thomas L., 75
Brown, Alex H., 131, 135–36
Brown, John, 143
Brown, Joseph E., 43, 212
Bryan, Goode, 152–53
Buchanan, James, 42, 59, 297
Buckhalter, J. H., 117, 118–19, 153–54, 271, 315
Buckner, B. F., 19
Buist, Henry, 45, 65, 86
"Bureau" Battalion, 193
Burges, Samuel, 64
Burnside, Ambrose E., 128, 152
Butler, Matthew C., 263, 272

Cadet Battalion, 226, 230–33, 235, 255
Cadet Rangers, 178–79
Cairo & Fulton Railroad, 20
Caldwell, John, 120
Calhoun, John C., 2, 313
California, 35
Cameron, Simon, 60
Camp Allen, 129
Campbell, James Butler: on board of C & S Railroad, 18, 27; and bondholders of C & S Railroad, 86, 299; and Cannonsborough Wharf and Mill Company, 208, 209; and Ryan, 182, 183, 184–85
Camp Croft, 108
Camp Elzey, 129
Camp Heyward, 129
Camp Jones, 129
Camp Lee, 108
camp life, 108–9, 110, 128–29
Camp Pillow, 108
Camp Pocotaligo, 129
Camp Simons, 108
Cannonsborough Wharf and Mill Company, 207–8, 300
Canonicus (naval vessel), 215
Carew, John E., 290, 296
Carman, Ezra, 239, 251, 252
Carolina (steamer), 43
Carr, C. D., 86
Central America (steamship), 27
Castle Pinckney, 59, 137
Central Railroad and Banking Company of Georgia, 30, 50, 134
Chamberlain, Charles V., 44, 184
Chaplin, Benjamin, 21
Chaplin, J. C., 190
Chapman's Fort, 190

Charles's battery, 180, 196
Charleston, S.C.: and absentee planters, 1–2; agrarian orientation, 2–3, 5; attitudes toward C & S Railroad, 34, 287; commercialism and industrialism, 4, 2, 3, 4, 5, 26, 281; and completion of C & S Railroad, 41; and construction of C & S Railroad, 26; Davis's visit to, 161; defenses, 70, 80, 81, 82, 89–90, 93, 99, *100*, 101, 104, 105, 106, 108, 110, 112, 133, 137, 152–53, 156, 158, *159*, 161, 169, 173, 191–92; economy, 4, 281–83, 287, 294, 296; evacuation of, 266, 268, 269–71; firewood shortages in, 123; and Foster, 191–99, 202, 203, 209; Gillmore's attack on, 151; harbor defenses, 80, 161, 192; Independence Day in, 296; and interest in C & S Railroad, 287; and Lincoln's election, 42; municipal aid to C & S Railroad, 16, 39, 50, 64; postwar conditions, 273–74, 276; prisoners of war in, 203–4, 205; and railroad development, 4, 7, 11, 12–15, 26, 282; and Seymour, 176; and W. T. Sherman, 261, 262–63, 265; and shipping capability, 74; shortages and sacrifices of, 202, 203; siege of, 158, 161, 162, 167, 186, 202, 203; as site of 1853 railroad convention, 11, 13–14; as site of 1860 Democratic National Convention, 6, 33, 34–35; as site of secession convention, 45–46; stock in C & S Railroad, 30, 86–87, 88; and terminus of C & S Railroad, 21–23; and trade, 6, 7, 18, 22, 60, 283, 296; Union forces outside, 156, *159*; U.S. navy attack on harbor, 136, 137–38, 139
Charleston Battalion, 104
Charleston Bridge Company, 131, 135–36, 141, 207
Charleston Daily Courier: on Battle of Honey Hill, 225; on bondholders of C & S Railroad, 140; on Charleston fire, 271; on C & S Railroad, 15, 40, 135, 244, 283, 287, 290, 291, 292, 294; on mail service, 131; and postwar expectations, 282; and sale of C & S Railroad, 297; on slave labor, 143; on Wilmington & Manchester Railroad, 276
Charleston Free Market, 203
Charleston Insurance and Trust Company, 15, 285
Charleston Iron Works, 285
Charleston Light Dragoons, 73, 116
Charleston Mercury, 34, 41, 43, 44, 80, 244
Charleston & Savannah Railroad: accidents, 120–21, 163; adversaries, 26, 37; advertisement for stock, *20*; as alternative route east

Charleston & Savannah Railroad (*continued*) of Atlanta, 46; and black regiments of U.S. Army, 142; and blockade-running, 174; board of directors, 18, 27, 32, 36, 38, 39, 58, 64–65, 86–88, 128, 133, 181, 184, 206–7, 247, 258–59, 270, 275–78, 279, 283, 285, 286, 289, 291, 293–94, 298; bondholders, 86–87, 88, 285–88, 291–95, 297, 298–99, 307; bonds, 27–28, 30, 31, 36, 37, 39, 50, 58, 64–65, 133, 134, 135, 139–40, 169, 181–85, 207, 258, 283, 285, 286; causes of failure, 303; causeway for tracks of, *290;* ceremonial spikes for final rail, 32; charter, 15; civilian freight, 83–84, 88, 201, 203; and Confederate railroad convention in Columbia, 106–7; commercial identity, 72; completion, 6, 32–33, 37, 57; concessions to landowners, 38; connections to other lines, 81, 105, 126, 131, 136, 304; construction estimates, 16–17, 37; construction of, 15, 24–32, 142–43, 304; cooperation with military, 85, 91, 97–98, 120, 122, 123–24, 128, 130, 135, 168–69, 173, 174, 201, 247, 267, 305, 314; and cotton shipments, 165–66, 168; and danger of military assault, 88; day-to-day operation, 36, 63, 288–89, 296; defense of, 18, 90–91, 93–97, 99, 107–8, 111, 112, 114–19, 120, 128–29, 158, 160, 161–62, 169, 171, 214, 225, 226, 266; and defense of Charleston, 70–72, 80, 101, 104, 137, 169, 314; and defense of Savannah, 70–72, 237, 238, 239, 240, 243, 244–45; development of, 7, 11–13; engraving from *Harper's Weekly, 25;* and evacuation of Charleston, 270, 271–72; and evacuation of Savannah, 257; and evacuation of slaves, 82; fares, 33–34, 57, 107, 134, 167, 207; as feeder line, 46; finances, 36, 39, 50, 57, 58, 64, 65, 84, 86, 107, 125–26, 133–35, 139–40, 168–69, 181–84, 201, 209, 258, 267, 277–80, 283, 285, 286–87, 288, 313; first through trips, 32–33; and Florida, 180; and Foster, 192, 196, 214–15, 246, 264; and freight tariffs, 168, 207; and freight transport, 34, 56, 63, 72, 107, 123, 124, 168, 202; fuel sources, 53; gala for, 40–41, 43–44; and Georgia Central Railroad, 15, 16, 17, 20, 22, 32, 40, 47, 72, 167, 169, 242, 274, 304; grading, 51; and Hardee, 206, 240; and Hatch, 189, 226, 234, 236, 246, 247, 250, 265, 266, 274–77, 283; hiring of slave labor, 202; historical marker, *282;* and iron supply, 124, 257; leadership, 85–88, 120, 133; and Lee, 80, 81, 82–83, 85, 90; locomotives, 52–54, 163, 205, 237, 243, 274, 277; and mail service, 123–24, 129–31, 201; maintenance, 52, 53–54, 58, 133–34, 163, 164, 174, 205; and Northeastern Railroad, 16, 17, 21, 22, 23, 46, 105, 126, 131, 136, 270, 304; notice of fare increases, *134;* notice of rice-storage rates, *125;* notice to civilian shippers, *83;* notice to passengers crossing Ashley River, *123;* operating expenses, 162; passenger service, 33, 63, 85, 123–24, 134, 141, 167, 201–2, 203, 234, 279, 290–91; and Pon Pon River raid, 145–48; postwar assessment of, 274; postwar control of, 275, 288–89; postwar reconstruction of, 278–79, 283, 289–90, 291, 296, 297; postwar recovery of, 281; postwar repair costs, 277, 279; and prisoners of war, 203–5, *204;* proposed defenses of area crossed by, *170;* railroad spikes from, *297;* and railroad workers, 162–63, 188, 200–201, 278; real-estate holdings, 134, 201, 283, 289–90, 298; right-of-way costs, 19, 26–27, 29, 30, 37–38, 298; and Ripley, 82–83; rolling stock, 52–53, 72, 84, 88–89, 98, 105, 125–26, *162, 163, 164,* 167, 205, 247, 254, 258, 277, 288, 289, 293, 297, 299; route, *8–9,* 13–21; sale of, 297–300, 301; schedule, *57;* and Schimmelfennig, 177; and Seymour, 176; and Thomas Sherman, 76; and W. T. Sherman, 242, 250, 272; and Siege of Charleston, 167; speed of, 33–34; state guaranty bond, *28;* state support of, 131; steamer service of, 124, 207, 208, 209, 289; stock, 15, 18, 19–20, 28, 29–30, 37, 50, 64, 84, 128; stockholders' resolutions notice, *65;* storage rates, 125; storage space, 164; taking up rails of, 257, 259, 260; and taxation, 187; and telegraph lines, 153, 160, 250; timberland owned by, 53; and trade, 7, 60; and troop transport, 63, 65, 71–72, 85, 88, 91, 115–16, 122, 123, 125, 149, 150, 155–57, 168, 169, 177, 187, 192, 198, 199, 214, 216–17, 224, 225, 226–27, 228, 231, 233, 254, 266; turnouts, 89; Union and Confederate positions adjacent to, *249;* Union troops as threat to, 65, 93, 104, 105, 122, 138; Union attacks on, 110–12, 114–18, 121, 145–49, *148,* 194, 229–37, 246–47, 251, 314

Charleston & Savannah Railway, 310, 311, 312

Charlotte & South Carolina Railroad, 165, 272, 279

Chatfield's brigade, 114

Cheraw & Darlington Railroad, 271, 272
Chesnut, James, 61, 221, 227–28, 229, 231, 255
Chesnut, Mary Boykin, 61
Chesnut Light Artillery, 145, 147
Chessie System, 313
Chickamauga, Battle of, 156, 160
Chisholm, A. R., 117, 294
Chisholm's Island, 78
Christ, Benjamin C., 93, 95, 96
Citadel Academy, 43, 60, 178, 194, 226, 315
Citadel Cadet Battalion, 226, 230–33, 235, 255
Citadel Cadet Rangers, 178–79
Civil War, role of railroads in, 52, 61. *See also specific battles and generals*
Clarke, J. G., 248, 253–54
Clay, Henry, 2, 50
Clingman, Thomas L., 73, 218
Coast Division, 214, 224, 251, 261, 263, 269
Coffin, George M., 234
Cohen, Solomon, 13, 14
Colcock, Agnes Bostick, 216
Colcock, Charles Jones: and Blair, 260; bloodhounds of, 154; and construction of C & S Railroad, 25; and Davis, 161; and development of C & S railroad, 7, 10–13, 15, 301; and evacuation of Savannah, 255–56; and Florence, 272; and Grahamville, 161, 171, 216, 217–18; and Honey Hill, 221, 223; illness of, 117; and Johns Island, 195; and military assignments, 65–66, 73, 301; portrait of, *66*
Colcock, Judge Charles Jones, 7
Colcock, William Ferguson: on board of C & S Railroad, 18, 27, 32; as customs collector for Charleston, 65; and gala for C & S Railroad, 41, 42; and development of C & S railroad, 13; resignation as Charleston customs collector, 41; and secession, 44, 45, 58
Coleman, Kenneth, 3
Coles Island, 99, 193
Colquitt, Alfred H., 179–80, 188
Columbia, S.C.: evacuation of, 271; railroad conventions, 106–7, 124, 186; and W. T. Sherman, 265, 269, 272; as site of secession convention, 45
Combahee (locomotive), 32–33, 52
Combahee Plantation, 234
Combahee River: Confederate defensive of, 256, 257, 263, 264, 266; and Union forces, 263, 264–65
commercial goods, prices of, 4

Compromise of 1850, 35
Confederate army: and C & S Railroad, 85; conditions, 129; desertions from, 129; and former slaves as Federal soldiers, 142; and iron from Savannah River Bridge construction, 38; troop strength of, 74, 75, 76, 90, 228; and Wassaw Island battery, 210. *See also specific units*
Confederate Bureau of Conscription, 162, 163
Confederate Conscription Act (1862), 162
Confederate currency: and cost of living, 140, 183, 200, 201, 258; and cotton, 55; devaluation of, 186, 187, 209; and postwar economy, 281, 285
Confederate navy: blockade-running policy, 174; capacity, 70; gunboats, 240; ironclad-ship program, 84; and iron supply, 107, 124
Confederate Ordnance Bureau, 164
Confederate Post Office Department, 130–31, 163
Confederate prison camps, 203
Confederate Railroad Bureau, 138, 152, 163, 164, 166–67, 174, 267, 305
Confederate States of America: conscription of railroad workers, 162–63, 188, 268, 305; coordination of railroads, 61–62, 65, 83, 85, 107, 122–23, 125, 126–28, 138–39, 164, 167, 174, 206, 267–68, 304–5; and cotton production, 54–55, 56; economy, 3, 55; fall, 273; fares paid to railroads, 173–74; funds for war effort, 55; and interline cooperation, 47; and iron supply, 258, 306; major railroads, 46, 47, *48–49*, 50, 51; manufacturing capacity, 54; and produce loans, 55–56; railroad legislation, 138–39, 267–68; soldiers' sacrifices for, 129; and Southern nationalism, 34; tax legislation, 186, 187; treasury bonds, 55, 135, 173, 292, 306
Connecticut artillery, 93
Conner, Henry W., 3
Conner, James, 41, 42, 45, 263
Constitutional Democratic Convention, 35
Constitutional Democratic Party, 35–36
Convention of Southern States, 60
Cooper, Samuel, 128, 237, 270
Coosawhatchie (locomotive), 52
Coosawhatchie, Battle of, *115*, 116–19, 121
Coosawhatchie River, 73
corn production, 56
cost of living, 200, 201
cotton embargo, 55, 84, 305
cotton exports: demand for, 54–55; and Federal blockade, 55, 56, 174, 305

cotton factors: and Federal blockade, 56; and secession crisis, 55
cotton plantations: and Federal raiding parties, 81, 82; and postwar recovery, 281; removal of slaves from, 81–82; self-sufficiency, 4
Cotton State alliance, 35, 42
Crafts, George J., 21
Craig, J. W., 289, 291
Crane, W. T., engraving of Forty-eighth New York, *118*
Crovatt, H. H., 290
Crovatt, William, 271, 290
CSX Transportation, 313, *314*
Cuyler, R. R., 106, 123, 164–65, 168

Dahlgren, John, 149, 158, 196, 199, 250, 252, 263–64
Dai Ching (naval vessel), 190, 191, 263
Daniel, Charles P., 230, 232
Davant, Robert J., 11
Davis, Charles O., 21
Davis, Isaac B., 168
Davis, J. C., 242
Davis, Jefferson: and Beauregard, 149, 161, 171, 252; blockade-running policy, 174–75; and Bragg, 151, 160; and cotton embargo, 55; and defense of Charleston, 173; delusional hopes, 269, 273; and Drayton, 11, 30, 67; elected president of the Confederacy, 60; and Fort Sumter, 61; and Hardee, 205, 266, 268, 269–70; and Lee, 69, 151; and Pemberton's forces, 105; and prisoners of war, 203; and railroad operations, 107, 164, 173, 174, 305; visit to Charleston, 161; visit to Savannah, 160; and Wadley, 126
Davis, Theodore R.: sketch of Sherman's troops in Pocotaligo, *262*; sketch of Sherman's XV Corps, *265*
Davis, W. H. H., 195
DeCredico, Mary, 168
Deer Creek Swamp, 21
DeFoe, Robert, 154
Delaware, Lackawanna & Western Railroad, 53
DeLorme, Thomas M., 192
Democratic National Convention of 1860, 6, 33, 34–35
Democratic Party: 1860 platform of, 35; and sectionalism, 34; and slavery, 6, 34–35; split in, 35–36, 40
Department of Refuge. *See* Department of South Carolina, Georgia, and East Florida
Department of South Carolina, Georgia, and East Florida, 67, 69, 80, 89, 138, 188, 211

Department of Southern Virginia and North Carolina, 187
Department of the South, 144
DeSaussure, Henry W., 29
DeSaussure, William Davie, 73
DeSaussure, Wilmot G., 228
desertions from Confederate army, 129
DeTreville, Richard, 27
DeVeaux's (Gregorie's) Neck, 232, *233*, 234–36, 251, 256–57, 259, 263
disease, 129. *See also* malaria; yellow fever
District of Georgia, 171
District of South Carolina, 205
Donelson, Daniel S., 79, 90, 92
Douglas, Stephen A., 34, 35
Drake, Jeremiah, 177
Drane, Henry M., 31–32, 66, 133, 136, 140, 182, 184, 303
Drayton, Elizabeth Pope, 30
Drayton, Percival, 67, 102, 302
Drayton, Thomas Fenwick: and adversaries of C & S Railroad, 26, 36; and alternative railroad connection, 93; arms deal with U.S. War Department, 42; and Ashley River crossing, 207; background, 11; at Bluffton, 73; and construction of C & S Railroad, 23, 24, 25, 26, 131–32, 303; and defenses north of Savannah, 81; and earthworks, 266; and Federal forces on May River, 90; and finances of C & S Railroad, 57, 64, 135; and first trips of C & S Railroad, 33; and Fort Walker, 67, 68–69, 94; and gala for C & S Railroad, 6, 40–41, 42; at Hardeeville, 72, 88, 101; and interstate railroad cooperation, 47; legacy, 307; and meeting of secessionist railroads, 62; military duties, 87, 302, 306; photograph of, *10*; and Pocotaligo, 94; and Port Royal Railroad, 36–37; and Confederate Post Office Department, 130; as president of C & S Railroad, 18, 27, 64, 65, 67, 87, 88; and Provisional Army, 67; and railroad development, 7, 11, 12–13, 15, 301, 312, 314, 315; and railroad politics, 85; and C & S Railroad's role in war effort, 70–71; Richmond assignment, 105, 107; and right-of-way, 30; and roadbed grading, 29; and Ryan, 182; Singletary compared to, 133; and stock sales, 19–20, 29–31; and use of slave labor, 24, 143
Drayton, William, 11
Dunkin, A. B., 86
Dunovant, Robert G. M., 68, 69, 73–74
Du Pont, Samuel F., 67, 69, 121, 137–38

8th Michigan, 76, 78, 93, 102, 103, 104
8th North Carolina Volunteers, 73
8th South Carolina Infantry, 66, 133
8th Tennessee, 92
11th South Carolina: and Florida, 180; and Green Pond, 142; and James Island, 158; and Morris Island, 158; near Coosawhatchie, 112, 114, 129; and Petersburg, 188; and Pocotaligo, 96, 97, 109, 117, 129; support of Evans, 101; and Union army tapping telegraph, 154
18th Georgia Battalion, 150
18th South Carolina, 105, 180
82nd Ohio, 242
Earle, William E., 190–91
earthwork batteries: and attacks on Charleston, 103–4; and defense of C & S Railroad, 151, 160, 232–33; in District of the South, 111; and Drayton, 266; extensive construction of, 102; and Frampton's plantation, 114; and Hartridge, 238; and Honey Hill, 218, 224, 227; and Lee, 76; and Pemberton, 99; and slave labor, 216–17. *See also specific batteries*
Eason, James M., 22–23
East Florida Railway Company, 310
Eaton, Samuel C., 118
Edisto (locomotive), 52, 290
Edisto Island, 93, 99, 264
Edisto River, 160, 171
Edisto River Bridge, 279, 289
Edwards, A. C., 231, 232, 235
Edwin Lewis (naval vessel), 189–90, 191
Eggleston, Dubose, 155
1860 Association, 41–42, 44
Ellen (naval vessel), 78
Elliott, George Parsons, 25
Elliott, Stephen, 68–69, 78, 96, *96*, 97, 114, 116
Elliott, Thomas, 95
Elliott, William, 21
Elliott, Mrs. William, 289
Elmore Insurance Agency, 86
Emilio, Luis, 261
Enoch Dean (naval vessel), 145, 146, 147
Evans, Nathan G. "Shanks," 82, 101, 105, 107, 133, 187

1st Battalion South Carolina Sharpshooters, 117
1st Connecticut Artillery, 145, 147
1st Georgia Infantry, 150, 198
1st Georgia Regulars, 195, 196, 197–98, 253, 256, 260
1st Georgia Reserves, 230, 236
1st Massachusetts Cavalry, 93, 112, 214
1st Michigan Engineers and Mechanics, 242
1st New York Light Artillery, 244
1st South Carolina Artillery, 65, 101, 194
1st South Carolina Cavalry, 93, 187, 190, 193, 231–32
1st South Carolina Mounted Militia Regiment, 65
1st South Carolina Sharpshooters, 114, 117
1st South Carolina Volunteer Regiment, 144, 145, 147, 153, 154, 173
1st U.S. Artillery, 112, 115
4th Georgia Cavalry, 197, 198
4th Louisiana Regiment, 101, 104
4th Massachusetts Cavalry, 195, 264
4th New Hampshire, 111, 114
4th South Carolina Cavalry, 81, 94, 153–54, 161, 187
4th Virginia Regiment, 179
5th Georgia Infantry, 105, 193, 198, 229, 230, 231, 232, 236
5th Pennsylvania, 76, 78, 93, 95
5th South Carolina Cavalry, 187
XIV Corps, 242, 262
14th South Carolina, 73–74, 78, 90
XV Corps, 259, 260, 264, *265*
15th South Carolina Infantry, 68, 73, 105
41st New York, 177
47th Georgia Regiment: and Battery Pringle, 198; and Edwards, 231, 232, 235; and Fort Johnson, 193; and Honey Hill, 216, 218, 220–21; and James Island, 101; and Johns Island, 197; and Samuel Jones, 229, 230; and U.S. Army attack on railroad tracks, 118
47th Pennsylvania, 111, 114, 115
48th New York, 78, 111–12, 116, *118*, 150
51st Georgia Infantry, 105
52nd Pennsylvania Volunteers, 193
54th Georgia Volunteers, 171
54th Massachusetts: advance to South Carolina, 259, 261, 264; and C & S Railroad, 236, 256; and earthworks, 228; and Gregorie's Neck, 251; and Honey Hill, 214, 215, 216, 219, 223, 224; and James Island, 149, 150
54th New York, 177
55th Massachusetts: attacks on C & S Railroad, 229; and Honey Hill, 214, 216, 219, 220, 221, 222, 223, 224; and James Island, 192, 265
55th Pennsylvania, 111

56th New York Infantry, 214, 218, 230, 235
57th Massachusetts Regiment, 143
57th New York Infantry, 214
59th Virginia, 178, 179, 180
Farmers and Merchants Bank, 134
Federal blockade: and cotton, 55, 56, 174, 305; Davis's policy toward, 174–75; and *Planter*, 92; and Southern railroads, 83, 84, 164, 165; tightening of, 72; and George Williams, 284
feeder lines, 46
Ferebee, Wilson, 19
Ferguson, Samuel, 248, 252
Fernandina & Cedar Keys Railroad, 31, 133
Fingal (blockade-runner), 74
Finnegan, Joseph, 177, 179, 180–81
First Military District of South Carolina, 150
First National Bank of Charleston, 285
Fiser, J. C., 256, 260
Fisher, Samuel W., 278, 279, 283, 285, 295, 301
Fitzpatrick, Benjamin, 35
Florence Stockade, 205
Florida, 176–77, 179, 180–81
Floyd, John B., 42
Folly Island, 149, 150, 159, 177, 180, 188
food shortages, 56
Foot Point Land Company, 301
Forest City Foundry, 54
Forster, William, 153, 154–55
Fort Beauregard, 68–69, 70
Fort Donelson, 82
Fort Fisher, 261
Fort Henry, 82
Fort Johnson, 61, 104, 192, 193, 199
Fort McAllister, 210, 248
Fort Moultrie: and Confederate torpedoes, 137; as Federal installation, 59; and Foster, 191; South Carolina troops' occupation of, 59, 60
Fort Pemberton, 99
Fort Pulaski, 70, 81, 91, 144, 210
Fort Ripley, 137
Fort Sumter: and Anderson, 59, 60, 61, 191; and Confederate torpedoes, 137; as Federal installation, 59; and Foster, 199; and Gillmore, 144; U.S. reinforcements to, 59–61
Fort Walker, 67, 68–69, 70
Foster, John Gray: and Charleston, 191–99, 202, 203, 209; and C & S Railroad, 192, 196, 214–15, 246, 264; and Goldsboro, 122; and Hatch, 225; injury of, 265; portrait of, *191*; repositioning of troops, 259; and W. T. Sherman, 250, 252

Foster's brigade of Vogdes's division, 177
Fowler, James H., 154
Fox, I. I., 190
Frampton, John, 19
France, and cotton, 54, 55
Freehling, William W., 2
Frobel, B. W., 248, 252, 253–54
Frost, Edward, 18, 284
Fuller, Benjamin, 21
Furman, Charles M., 184, 207, 247, 286
Furman Artillery, 218

Gadsden, Christopher, 13
Gaillard, Franklin, 155
Gaillard, Peter C., 286
Gantt, F. Hay, 129
Gartrell, Lucius, 229, 230, 231, 232, 234, 235, 236
Gasden, Christopher S., 299
Geddes, Caroline, 21
Geddings, J. F. M., 110
George W. Williams & Company, 284
Georgia: economy of, 3; railroad investments, 51; railroad miles of, 51; and secession, 43–44, 45; ties to South Carolina, 41, 42–44, 45, 224
Georgia Central Railroad: and C & S Railroad, 15, 16, 17, 20, 22, 32, 40, 47, 72, 167, 169, 242, 274, 304; charter, 3; cooperation with military, 123; cotton shipments, 56, 166; and Hardee, 237; locomotives, 163; rolling stock, 84, 89, 164–65, 257, 259; salaries of workers, 200; and Savannah, 210; and Savannah & Charleston Railroad, 309; and W. T. Sherman, 213, 240, 257, 259; success, 47; and troop transport, 155–56
Georgia legislature, 3, 13, 15
Georgia Light Infantry, 68
Georgia Militia and State Line, 212, 216, 218, 224, 228
Georgia Railroad, 3
German Flying Artillery, 68
German Light Artillery, 236
Gettysburg, Battle of, 151, 152
Gibbes Plantation, 147
Gibbs, H. P., 86
Gillmore, Quincy Adams: and Battery Wagner, 150, 151; and Beauregard, 268; and Broad River, 172; and C & S Railroad, 275, 276–77, 284; and Charleston's defenses, 156, 158; and Department of the South, 144; and Florida, 177; and Higginson's black regiment, 145–48; and James Island,

265, 266; operations against Charleston, 167, 181; portrait of, *144;* and Savannah River Bridge, 76; and Seymour, 175, 176–77, 181; and Terry, 148–49; and Virginia, 188

Gilmer, Jeremy F., 160–61, 171, 172, 177, 188

Gilmore's Brass Band, 33

Gist, William H., 42, 43

Glover, Joseph Edward, 21, 147

Goldsboro, N.C., 122

Gonzales, Ambrosio José, 61, 73, 99, 149, 217, 242–43, 256

Good, Tilghman H., 114

Goodman, Walter, 62

Goodwill Plantation, 82

Gordon's Division, 177

Gourdin, Henry, 27, 58, 86, 184, 201, 207, 295

Gourdin, Robert N., 41, 44, 58

Governor Milton (naval vessel), 145, 146, 147, 148, 192

Graham, R. F., 149, 150

Graham, Thomas Jefferson, 120–21

Grahamville, S.C.: and Charles Jones Colcock, 161, 171, 216, 217–18; and defense of C & S Railroad, 229; and evacuation of Savannah, 256; and Honey Hill, 214, 215, 216, 218, 220, 221; and Lee's troops, 73; and railroad development, 26

Grant, Ulysses S.: and Chattanooga, 160; and command of U.S. Army, 185; and Gillmore, 188; and Lee, 191, 263, 269, 273; and W. T. Sherman, 210–11, 250; in West, 176

Great Britain: and capital for Southern railroads, 50; and cotton, 54, 55; Lee on military aid from, 75; railroad mileage of, 51

Green Island battery, 210

Greenville & Columbia Railroad, 106

Gregg, Maxcy, 81, 85, 90

Gregg, William, 16, 18

Gregorie, Thomas Hutson, 11

Gregorie's Neck, 232, *233,* 234–36, 246, 251, 256–57, 259, 263

Greig & Jones, 27

Guerard, J. J., 154

Gurney, William, 193, 196

H. L. Hunley (naval vessel), 184

Hagood, Johnson: and Beauregard, 111; district of, 107–10; and James Island, 158; and Johns Island, 101; and Petersburg, 188; and Secessionville, 104; and supplies for troops, 164; and William Walker, 112, 171

Haig, Edward M., 21

Haines, Henry Stevens: appointed superintendent of C & S Railroad, 66; and Ashley River Bridge, 208; and Beauregard, 234; cooperation with military, 74, 186, 306; and day-to-day operations of C & S Railroad, 72, 288–89, 296; and Edisto River Bridge, 279; and evacuation of Charleston, 270, 271–72; and evacuation of Savannah, 257; and interline cooperation, 201; and maintenance, 305; and McMicken, 266; and operating expenses, 162, 202, 283; and Plant, 310, 311–12; portrait of, *72;* and postwar assessment, 274, 277, 285, 308; and postwar control, 276; and postwar reconstruction, 289, 291; railroad career of, 66–67, 312, 315; resignation from superintendency of C & S Railroad, 298; and rolling stock, 163, 247, 254, 271, 274, 279, 283; and safety, 163; salary of, 140, 200, 258, 289; and Savannah, Florida and Western Railroad, 310, 312; and Savannah River Bridge, 71; and storage space, 83; and troop transport, 226–27, 228

Hale (naval vessel), 78, 190

Halleck, Henry, 153, 172, 181, 214

Hallowell, Edward Needles, 251, 261, 264

Hamilton, Daniel H., 42

Hamilton, James, Jr., 3

Hamilton, Samuel P., 44

Hamlin, Hannibal, 36

Hammond, James Henry, 45, 51, 54

Hampton, Wade, 45, 190, 265, 269, 271, 272

Hanckel, Charles F., 20

Hanckel, Christian, 21

Hand, Daniel, 284

Hand & Williams Company, 284

Hannon, Patrick, 163

Hardee, William Joseph: background, 205; cooperation with C & S Railroad, 206; and defense of Savannah, 237, 238, 239, 240, 243, 244–45, 247; and evacuation of Charleston, 266, 268, 269–70, 272; and evacuation of Savannah, 248–50, 252–54, 255, 256, 257; and Honey Hill, 221, 228; and Samuel Jones, 211, 229, 235; portrait of, *206;* and railroad iron supply, 206, 257–58, 293; and W. T. Sherman, 214, 237, 248, 250, 251, 261, 262, 264, 265, 266, 272; and Gustavus Smith, 213–14, 228; and troop strength, 211

Hardeeville, S.C.: Drayton stationed at, 72, 88, 101; and evacuation of Savannah,

Hardeeville, S.C. (*continued*) 254–55, 257; keeping C & S Railroad open to, 247; and refugees, 254; and route of C & S Railroad, 21; Union occupation of, 259–60, 267; and Wheeler, 248

Harper's Weekly: engraving of African American soldiers in battle, *143;* engraving of C & S Railroad, *25;* engraving of Evacuation of Savannah, *253;* engraving of Sherman's troops in Pocotaligo, *262;* engraving of Sherman's XV Corps, *265*

Harris, Ira, 177

Harrison, George P., 193, 197, 198–99, 238, 256

Harrison, J. J., 117, 122

Hartridge, Alfred F., 238–39, 240, 245, 256

Hartwell, Alfred S., 214, 215, 219, 220

Hasell, Bentley D.: and accidents, 121; and Ashley River Bridge, 124–25, 126, 131–32; and Augusta railroad convention, 126; and Buckhalter, 118–19; and Columbia railroad convention, 106–7; and finances, 126, 129; as president of C & S Railroad, 120; as president of Shelby & Broad River Railroad, 132, 302; relationship with military, 122, 124, 127–28; relationship with state, 131; and rolling stock, 125–26

Hastie, W. S., 299, 300

Hatch, John Porter: and Ashepoo River, 269; and Battery Pringle, 195–96, 198; and C & S Railroad, 189, 226, 234, 236, 246, 247, 250, 265, 266, 274–77, 283; and Combahee River, 263, 264; and Hardee, 251; and Honey Hill, 214, 215–16, 218, 219, 221, 223, 224, 225; and Johns Island, 192, 194, 197, 199; and occupation of Charleston, 273; and Pocotaligo, 261, 263; portrait of, *189;* and Singletary, 274–75, 283; and troops' experience, 188; and troop strength, 189

Haughton, Nathaniel, 221–22, 223, 236

Haulover Plantation, 180

Hayne, Isaac W., 28, 88, 89, 105–6, 131, 296–97

Hayne Hall Plantation, 134

Hazen, William B., 248

Henderson, D. S., 26

Henry, G. A., 151

Herriott, Robert, 246

Heyward, Charles, 21, 32, 38, 82

Heyward, D. Blake, 93, 95

Heyward, Daniel, 18, 21, 173

Heyward, Duncan Clinch, 33

Heyward, E. B., 234

Heyward, Esther, 53

Heyward, James B., 21

Heyward, Nathaniel Barnwell, 13, 18, 19, 21, 25, 53

Heyward, William C., 67, 73, 142, 154

Heyward, William Henry, 21, 25, 28, 31, 53, 182

Higginson, Thomas Wentworth: as commander of black regiment, 143, 144–45; and Pon Pon River raid, 145–48, 150, 192

Hilton Head Island: and Drayton, 67, 69; Federal troops occupying, 75–76, 91, 99, 102, 104, 110, 119, 122, 171, 188, 214; and Seymour, 181

Hodgson, Margaret Telfair, 11

Hodgson, William Brown: and board of directors of C & S Railroad, 18, 27, 65; and railroad development, 11, 12, 13, 15; and stock in C & S Railroad, 20

Hoke's division, 252

Holcombe's Legion, 105

Holmes, Francis S., 64

Homans, William H., 223

Honey Hill, Battle of, 214–25, *222,* 226, 227, 228

Honour, John Henry, 44, 290

Hood, John Bell, 151, 152, 153, 155, 205, 211, 252, 261, 262, 263, 269

Hooker, Joseph, 138

Hooper, Henry N., 219, 223

Horsey, Samuel Gilman, 147

Housatonic (naval vessel), 184

Howard, Oliver O., 211, 238, 239, 259, 261–62, 263, 264, 272

Howell, Jesse, 32

Hoyt, Henry, 193–94

Hudson, W. S., 29, 31

Huger, Alfred, 41, 130

Huger, Arthur M., 68

Huger, Cleland, 302

Huger, Thomas P., 13

Huger & Hasell, 302

Huguenin, Julius G., 11, 25

Humphrey, M. J., 178

Hunter, David, 99, 102, 104, 137, 144, 145

Hunter, Robert H., 27

Hunter, W. W., 243–44

Hunting Island, 90

Hutchinson, Thomas Leger, 16, 18

Hutchinson Island, 248, 249, 253

Hutson, Thomas Woodward, 114

Hutson, William F., 181, 183–84, 290, 294, 295

hypothecation, 50

industrialization, 3, 281, 303, 305, 313
internal improvement projects, 47, 283
Isaac P. Smith (naval vessel), 137
Isaacs, Alexander, 294, 300
Isundiga (locomotive), 52, 247
Izard, Allen C., 96, 114, 115

J. B. Green and Company, 47
Jackson, Andrew, 11
Jackson, Henry R., 44, 225
Jackson, John H., 90–91
Jackson, N. J., 238, 239, 242
Jackson, Thomas J. "Stonewall," 63
James Adger (locomotive), 52
James Island, S.C.: and Beauregard, 148–49; and Birney, 195, 196; and Colquitt, 179; Confederate troops reinforcing, 150, 151; and Davis, 161; and defense of Charleston, 160; and Foster, 192–93, 199; and Gillmore, 265, 266; and Gonzales, 243; and Manigault, 66, 192, 193; and Robertson, 197, 199; and Gustavus Smith, 255; and Isaac Stevens, 99, 101; and Clement Stevens, 137; and Union forces, 104, 145, 148–49
Jay Cooke & Company, 309
Jeffries Creek Volunteers, 66
Jehossee Island, 190
Jenkins, John, 178, 179, 196–99, 216, 221, 230
Jenkins, Micah, 152, 153, 172, 251
Jennet, James, 178–79
Jesup, Morris K., 289, 302, 310, 311
John Adams (naval vessel), 145, 146, 147
Johns Island, 99, 101, 108, 177–80, 192, 194–97, 198, 199
Johnson, Andrew, 274, 281, 283, 284
Johnson, John, 15
Johnson, Thomas H., 117, 118, 171
Johnson's division, 252
Johnston, A. R., 13
Johnston, Joseph E., 62–63, 149, 175–77, 188, 191, 193, 272–73
Johnston, William, 257
Jones, Charles Colcock, 40–41, 43, 252
Jones, James, 73, 78–79, 90
Jones, Samuel: and Beauregard, 188, 237, 240, 248; and defense of C & S Railroad, 226, 229, 230, 231, 234, 247, 248; and Gregorie's Neck, 234–35, 236; and Hardee, 211, 229, 235; and John Jenkins, 197, 216, 221; military career, 188; and Pringle, 192; and prisoners, 203, 205; and Beverly Robertson, 199; and supplies for troops, 255; and troop strength, 195; and troop transport, 228
Jones, Squire, 82

Kanapaux, J. T., 217, 218
Keitt, Lawrence, 59
Kendrick, J. B., 86
Keokuk (naval vessel), 137
Kershaw's brigade, 155, 156, 172
Kiawah Island, 177
King, Hawkins, 21
King, John Pendleton, 187
Kings Mountain Railroad, 30, 50, 134
Kingville, S.C., 17, 152, 155, 156, 165, 269
Kirk, Manning J., 116
Kirk, R. H., 11
Kirk's squadron, 229
Kirkwood, William, 18, 27

Lackawanna Iron Company, 47
LaFayette Artillery, 112, 118, 218
Lamar, Thomas Gresham, 102
Lane, Joseph, 36
Lartigue, Etienne, 27
Lartigue, Gerald B., 27
Lartigue, John, 14, 19
Lawton, Alexander J., 13
Lawton, Alexander R., 101, 152, 156, 164–67, 267
Lawton, Benjamin W., 13
Lawton, John S., 13
Lawton, Joseph A., 14
Lawtonville, S.C., 12, 15–19, 20, 23
Lazarus, B. D., 86
Leak, W. J., 190
Leake, Walter, 105
Lebby, Robert, 41
LeBleux, L. F., 112, 118
Lee, B. M., 278
Lee, Hutson, 135, 149, 164, 166, 226–27, 229
Lee, Robert Edward: and Ashley River Bridge, 105; and Beauregard, 187, 270; and Bragg, 151; and Burnside, 128; and coastal defense, 80; at Coosawhatchie, 70, 73, 74; and Department of South Carolina, Georgia, and East Florida, 67, 69–70, 80; and Drayton, 302; and earthwork batteries, 76, 114; and Grant, 191, 263, 269, 273; and Honey Hill, 218, 224; and Hooker, 138; legendary horse of, 74–75; and Longstreet, 172; and Meade, 151; and Pemberton, 99; portrait of, *71*; and relations with railroads, 85; at Richmond, 89, 187, 252; and role of railroads,

Lee, Robert Edward (*continued*) 71, 175–76, 177; at Savannah, 81, 89–90; supplies for army of, 201; surrender, 273; and troop strength, 74, 80, 82; and troop transport by rail, 63, 80
Lesesne, Henry D., 64
Leverett, Milton Maxcy, 95
Lewis, John, 21
Lincoln, Abraham: and Charleston area, 161; election of, 40, 41, 42, 43; and Florida, 177, 181; and Fort Sumter, 60, 61; as Republican party candidate, 36, 41; and W. T. Sherman, 211, 256
Llandovery Plantation, 53
Logan, George William, 21
Longstreet, James, 151–53, 156, 165, 167, 173
Louis, D. Daniel, 129
Louisville, Cincinnati & Charleston Railroad, 4, 11, 47
Louisville & Nashville Railroad, 313
Lowndes, Charles Tidyman, 154
Lowndes, W. H., 19
Lowndes Plantation, *266*
Lucas, Robert H., 208
Lumpkin, Wilson, 3

Macbeth, Charles: on board of C & S Railroad, 184; and bondholders of C & S Railroad, 86; and bonds of C & S Railroad, 140; and Cannonsborough Wharf and Mill Company, 207; and cooperation with military, 149, 188; and evacuation of Charleston, 284; and gala for C & S Railroad, 40, 41, 43
Mackay Point Road, *306*
Mackay's Neck, 251, 261
Mackey's Point, 112, 235
Macon (naval vessel), 237, 244
Macon, Ga., 17, 212, 213
Magrath, Andrew Gordon, 41, 42, 44, 45, 88, 263, 274, 302
Magrath, William Joy: and Ashley River railroad bridge, 89; and Battle of Secessionville, 105; on board of C & S Railroad, 65; and military traffic, 91; and Pemberton, 97–98, 101, 106; portrait of, *87*; as president of C & S Railroad, 87–88, 120, 132; as president of South Carolina Railroad, 120, 125, 302–3; and railroad rate convention, 106–7; and rolling stock, 88–89
malaria, 29, 108, 109, 110
Manassas, First Battle of, 188
Manassas, Va., 62

Manassas Gap Railroad, 63
Manigault, Edward M.: appointed superintendent and chief engineer of C & S Railroad, 36; as chief of S. C. ordnance, 66, 67; and gala for C & S Railroad, 6; and James Island, 66, 192, 193; as prisoner of war, 66, 265; and route of C & S Railroad, 20–21, 23; and Savannah River Bridge, 38; on Siege of Charleston, 202; and speed of locomotives, 33
Marion Artillery, 6, 33, 46, 66, 147, 179, 180, 196
Marshall, Edward W., 87, 294, 295
Martin, A. M., 14
Martin, C. C., 38
Martin, William E., 13, 14, 27, 37, 65, 73, 74
Mayflower (naval vessel), 76
Mayor Macbeth (locomotive), 40, 52
May River, 90
McBride, Burwell, 13
McCall, James, 86
McCants, Lawrence W., 26
McClellan, George, 144
McCrady, Edward, 44
McCrady, John, 207
McDonough (naval vessel), 190
McDowell, Irvin, 62
McDowell and Callahan, 38
McDuffie, George, 3
McLaws, Lafayette: and Combahee River, 260; and evacuation of Charleston, 270; and evacuation of Savannah, 253, 256, 257; and Fourth Georgia Cavalry, 197; and Hardee, 212, 269; and James Island, 255; reassignment to Georgia, 151, 155; and Savannah River Bridge, 237; and W. T. Sherman, 261, 262, 263–64, 266
McMicken, M. B., 266
McMillan, W. F., 65
McRae, John, 15–16, 17, 18–19, 20
Meade, George, 151, 152, 172, 176
Meigs, Montgomery, 173–74
Memminger, Christopher G., 18, 27, 60, 65, 311
Memphis & Charleston Railroad, 62
Mercer, George, 124, 150, 160
Mercer, Hugh, 105, 111, 112
Merchants' Exchange, 282, 285
Michigan Central Railroad, 53
Miles, William Porcher, 27, 44, 59, 60, 65, 268
Millen, Ga., 210, 211
Millen, John M., 44
Millen's battalion, 187

Milliken, Edward P., 41, 43
Mills, Otis, 15, 18, 27, 32, 86, 88
Milne, Andrew, 21, 27
Mississippi Central Railroad, 62
Mississippi River, Union control of, 69, 151
Mitchel, Ormsby MacKnight, 110–12, *111*, 119
Mitchell, Charles Taylor, 65, 86, 87, 184, 286
Monteith Swamp, 238–39
Montgomery, James, 142, 145, 189–90, 197
Mordecai, Moses Cohen, 65, 86, 87, 284
Morgan, Joseph H., 93, 95, 114
Morris Island: Confederate prisoners held on, 203–4; Confederate troops on, 151; and evacuation of Charleston, 271; and Gillmore, 144; and Gurney, 193; and Siege of Charleston, 202; Union occupation of, 164, 167, 188, 262–63; and Union campaign to take Charleston, 137, 145, 149–50, 158, 167
Morris's plantation, 145, 147–48
Morrison, Joseph, 193
mosquitoes, 99, 108, 109, 205. *See also* malaria
Mower, Joseph Anthony, 262
Muller, M. P., 32
Murdock, Robert, 147
Myers, Abraham, 107, 135
Myers, Charles, 21

9th Illinois, 272
9th Maine Regiment, 150
9th South Carolina Battalion, 103, 121
9th South Carolina Infantry, 68, 73
9th U.S. Colored Troops, 189, 190
19th Massachusetts, 203
Nashville Manufacturing Company, 54
Nast, Thomas, sketch of African American soldiers in battle, *143*
National Express Company, 279, 289
nationalism: and railroad movement, 2; Southern nationalism, 34
Nelson, Patrick H., 116, 122
Nelson, Scott Reynolds, 50
Nelson, William, 189
Nelson Artillery, 114
Nelson's battalion, 150
Nesbitt, Niles, 193, 231–32
Newcomer, Benjamin, 311
New Ironsides (naval vessel), 137
New Mexico, 35
New Orleans, La., 7, 17–18, 99
New York, and railroad investments, 51
New York, N.Y., and railroad development, 7
New York Engineers, 112, 116, 117

Nims, F., 27
North, Edward, 64
North Carolina Battalion, 239
North Carolina Railroad, 202
Northeastern Railroad: and Ashley River Railroad, 309–10; and Beauregard, 268, 269; and C & S Railroad, 16, 17, 21, 22, 23, 46, 105, 126, 131, 136, 270, 304; C & S Railroad's lease from, 135; and cotton shipments, 165, 166; defense of, 108; and evacuation of Charleston, 271; and Haines, 67; Hatch's control of, 274; mail service, 131; and Plant, 311; postwar restoration of, 275, 283–84; restricted schedules, 201; rolling stock, 279; and supply of firewood to Charleston, 123, 169; and troop transport, 122, 155, 156, 165
North Edisto, 104
Northern capital: and Northern railroads, 51; and postwar investments, 281, 289; and Southern railroads, 47
Northerners: and Charleston's railroad movement, 5; and Charleston's shipping and insurance companies, 2
Northern railroads: capital investment in, 51; government rates paid to, 173–74; maintenance of, 52; rolling stock of, 53; and through traffic, 46
Northern states: and railroad investments, 51; Southern railroads' dependence on supplies from, 54, 84, 305
Northrop, Lucius B., 284
north-south coastline rail network, 7, 10, 18, 23, 46, 58, 301
Norwood, Thomas M., 44
nullification, 1, 5

100th New York Regiment, 78
100th Pennsylvania, 101
102nd U.S. Colored Troops, 214, 223–24, 235, 264
103rd New York Artillery, 195
103rd New York Volunteers, 192
115th New York, 239
127th New York, 193, 214, 216–21, 224, 230, 235, 265
142nd New York, 177–78
144th New York Infantry, 195, 214, 218, 221–22, 230, 235, 265
157th New York, 178, 218, 224, 230, 235, 236
Oak Lawn Plantation, 61
O'Bryan, Lewis, 26
O'Connor, Michael P., 44

O'Hear, James, 21
Ohio Life and Trust Company, 27
Olustee, Battle of, 180
Orr, James, 295, 296
Otey, John M., 254
Ottawa (naval vessel), 78

Page, Powhatan R., 179–80
Page's Point, 234
Palmetto Battalion, 190
Palmetto Guard, 73
Panic of 1857, 27
Panic of 1873, 309, 310
Parker, Charles, 95
Parker, Edward L., 29, 66, 180, 195
Partisan Rangers, 116, 178
Paysinger, T. M., 261
Peake, H. T., 106, 156, 157, 165
Peeples, A. M., 21
Peeples, William B., 217, 224
Pelot, John Francis, 21
Pemberton, John C.: and Ashley River railroad bridge, 89, 105; and Broad River, 81; Charleston headquarters of, 92; and C & S Railroad's steamer service, 124; and cooperation with railroad, 85; defensive strategy of, 91, 99, 101, 102, 104; dispatch of forces, 105; and James Jones, 78–79; and Magrath, 97–98, 101, 106; and Pocotaligo, 82–83, 90, 94, 95; and relationship with railroads, 128; and troop strength, 90, 93; and Vicksburg, 138, 151
Pembina (naval vessel), 78
Pennsylvania, and railroad investments, 51
Pensacola, Fla., 7, 211
Pensacola & Atlantic Railroad, 311
Pensacola & Georgia Railroad, 167, 180
Petersburg, Va., 188, 206, 212, 263, 269, 273
Phelan, William, 27, 29
Phillips's Legion of Georgia Volunteers, 74, 97, 105
Phoenix Iron Works, 54
Pickens, Francis W.: and Fort Sumter, 59–61; and Hayne, 296–97; and Lee, 74; and Pemberton, 90, 106; and Thomas Sherman, 67
picket duty, requirements of, 109–10, 129
Pickett, George, 152
Pinckney, Castle, 59, 137
Plant, Henry B., 289, 310–12
Plant Investment Company, 311
Plant System, 311, 312
Planter (steamship), 92, 116, 118

planters: aristocratic culture, 2; and Charleston's mercantile community, 3; and construction of C & S Railroad, 25, 303; and cotton sales, 55; and Federal blockade, 56; and hiring out of slaves for railroad construction, 24, 29, 143, 162; and Lee's strategy, 80; opposition to C & S Railroad, 26; opposition to railroad movement, 1–2, 47; and postwar reestablishment, 291, 301; protection of, 108; and railroad rights-of-way, 19, 26–27; as stockholders of C & S Railroad, 50
Planters and Mechanics Bank, 294
Plato (naval vessel), 189, 191
Pocahontas (naval vessel), 67
Pocotaligo, First Battle of, 92, 93–97, 99
Pocotaligo, Second Battle of, 111–12, *113*, 114–16, *115*, 119, 153
Pocotaligo, S.C.: and evacuation of Savannah, 256; and Foster, 214, 246; and Samuel Jones, 229, 235, 240; and Union troops, 93–97, 261–62, *262*, 263, 267; and William Walker, 95–97, 108, 110, 169, 171, 172
Poe, Orlando M., 240, 242
Polk, Leonidas, 151
Pon Pon River, 145–48, 150
Pope, George, 223
popular sovereignty doctrine, 35
Port Royal: and Birney, 191; Confederate defense of, 67–69; fall of, 71; and W. T. Sherman, 211; Union victory at, 144, 167; Union occupation of, 99, 136, 137, 160, 169, 171, 173, 177, 188; and Union navy, 70, 75
Port Royal Ferry, 73, 76, 78–80, 93, 96, 97, 109
Port Royal Railroad: and C & S Railroad, 36, 37, 140, 141; and Drayton, 36–37
Potter, Edward E.: and attacks on C & S Railroad, 229–30, 235, 236; and Charleston, 269; and Edisto Island, 264; and Foster, 259; and Honey Hill, 214, 215–16, 218, 219–20, 224; and James Island, 265
Potter, Lorenzo T., 15, 18, 27, 32, 65
Potter & Hunter, 27
Preble, George H., 215, 259
Preston, William Campbell, 1
Prevost, Joseph, 21
Price, Emma, 53
Price, Philip, 53
Pringle, Motte A., 192
prisoners of war: in Andersonville, 203, 204; in Charleston, 203–4, 205; and C & S

Railroad, 203–5, *204;* exchanges of, 203, 206; and James Island, 265; Manigault as, 66, 265; in Savannah, 204; and Sherman's advance to South Carolina, 262
Pritchard, Paul, 11
Provisional Army, and Drayton, 67
public works, southern resistance to Federal investment in, 47
Pynes Plantation, 53

Quick, William, 257
quinine, 29
Quinn, J. M., 63

R. G. Dun & Company, 284
railroad construction: and slave labor, 24, 29, 47, 50, 143, 162, 279–80; and technology, 24–25, 313; and transcontinental railroad, 309. *See also specific railroads*
railroad movement: benefits of railroads, 303–4; Charleston's investments in, 4, 5; cost of right-of-way, 26; and development of railroads, 6–7, 10; in Georgia, 3; and New York venture capitalists, 7; planters' opposition to, 1–2, 47; and promotion of agriculture and commerce, 4–5
railroad roadbeds: durability of, 33; grading of, 25, 27, 29, 47, 51
railroad workers: and conscription laws, 162–63, 188, 268, 305; danger for, 247; and Singletary, 162–63, 188, 200, 201, 202, 207, 247; slaves as, 24–25, 29, 142–43, 202, 313, 315; wages, 200, 202, 207, 278, 283, 296, 309
Randolph, George, 101
Ravenel, Alfred F., 106, 166, 311
Ravenel, William, 21, 41, 64, 86
Read, John P. W., 68
Reagan, John Henninger, 62, 130
Reconstruction, 281, 383, 303
Republican Party, and split in Democratic Party, 36, 40
Resolute (naval vessel), 244
Rhett, Edmund, 13, 14, 27, 42, 58
Rhett, Robert Barnwell, Jr., 1, 34–35, 40, 41, 42, 44
Richards, Frederick, 86, 88, 184, 278, 285
Richardson, John, 11, 13
Richmond, Va.: and Drayton, 105, 107; and Lee, 89, 187, 252; and Pickett, 152; railroad conventions in, 84, 138
Riggs, John S., 41, 44, 86
Riker, R. H., 271

Rikers & Lythgoe, 53
Ripley, Roswell Sabine: and Adams Run, 190; and Ashley River Bridge, 105; and C & S Railroad, 82–83; and Du Pont, 121–22; and earthwork batteries, 76, 114; and evacuation of Sea Islands, 81; and Honey Hill, 218; portrait of, *81;* and Port Royal, 70; and troop strength, 82
Roanoke Island, 82
Robb, James, 309
Robbins, R. E., 289
Robertson, Beverly H., 169, 171, 172, 194, 196–98, 199, 224, 232
Robertson, Jerome, 155, 156, 158
Robinson, James S., 239
Rockwell's battery, 93
Rodman, Daniel, 150
Rogers Locomotive and Machine Works, 47, 299
Rose, William, 228
Rosencrans, William, 151, 156
Roumillat, John L., 71, 272
Russell, William, 63–64
Ruth, A. M., 19
Ruth, Grafton Geddes, 117
Rutledge Mounted Riflemen, 73, 93, 95, 116, 142
Ryan, John S.: and bid for C & S Railroad, 275; as bondholder of C & S Railroad, 65, 87, 182–85, 284, 292, 294; and construction of C & S Railroad, 28–29, 31, 182; and gala for C & S Railroad, 33; and sale of C & S Railroad, 298, 299; and Savannah & Charleston Railroad, 300, 308; and Singletary, 31, 182, 183, 184, 185, 275, 276, 284
Ryan, Thomas, 44, 65, 88

2nd South Carolina Artillery, 193
2nd South Carolina Cavalry, 187, 198
2nd South Carolina Infantry, 155
6th Connecticut, 111, 150
6th South Carolina Cavalry, 108, 147, 178, 187
7th Connecticut, 102, 111, 112, 150
7th Georgia Cavalry, 187, 256
7th Illinois, 272
7th North Carolina Reserve Battalion, 231, 235
7th South Carolina Battalion, 110, 112, 116, 149, 187, 188
7th U.S. Colored Troops, 194
16th South Carolina, 122
16th Tennessee, 92

XVII Corps, 259, 261, 262, 264
17th South Carolina Regiment, 94, 105
61st Ohio Volunteers, 239
63rd Georgia Infantry, 150
74th Pennsylvania, 177
75th Ohio, 178, 194
76th Pennsylvania, 111, 115, 116, 150
79th New York, 93, 101
79th New York Highlanders Regiment, 76, 78
St. Andrews Station: and completion of C & S Railroad, 6, 32, 33; and Davis, 161; storage space at, 83; as terminus for C & S Railroad, 38, 73; and transportation to Charleston, 124
St. Bartholomew Parish, S.C., 14, 25–26
St. John's (steamer), 72
St. Mary's (steamer), 72
St. Peter's Parish, S.C., 12, 13, 14, 15, 19
Salkehatchie Bridge, 21
Sampson (naval vessel), 237, 244
Sass, J. K., 184, 185, 286
Savannah (naval vessel), 252
Savannah, Ga.: beauty of, 210; and completion of C & S Railroad, 6, 40–41; and construction of C & S Railroad, 26; Davis's visit to, 160–61; defenses of, 70, 81, 89–90, 93, 99, 105, 112, 137, 225, 237, 244–45; evacuation of, 248–50, 253–57, *253*; and Hardee, 212, 237, 238, 239, 240, 243, 244–45; Lee in, 81, 89–90; and mail service, 131; prisoners of war in, 204; and railroad development, 11, 12, 15; railroads heading west from, 4; role in Confederate war effort, 210; and route of C & S Railroad, 15, 17, 210; and Thomas Sherman, 76; and W. T. Sherman, 210–11, 212, 226, 237, 242, 243, 247, 250, 256, 259; and terminus of C & S Railroad, 22; and trade, 7, 18, 60; Union and Confederate positions outside, *249*; view of South Carolina's railroad as threat to seaport, 3
Savannah, Albany & Gulf Railroad, 3, 167, 180, 210
Savannah, Florida & Western Railroad, 310, 311, 312
Savannah & Albany Railroad, 3
Savannah & Augusta Railroad, 238
Savannah & Charleston Railroad, 300, 301, 307–10, 311
Savannah River: and Hardee, 212; and route of C & S Railroad, 21; and telegraph line, 153; and trade, 1; and Union army, 91–92, 264
Savannah River Bridge: construction of, 32, 36, 37, 38–39; and Hardee, 212, 243–44;

and McLaws, 237; nightly closing of, 71–72; and Thomas Sherman, 76; and W. T. Sherman, 242–43, 247; and Singletary, 181, 257
Saxton, Rufus, 144, 192, 195, 196, 197, 198
Schimmelfennig, Alexander, 177, 179, 180, 188–89, 192, 193, 195–96, 262–63
Schulz, Frederick C., 145–47
Scott, Winfield, 69
Screven, John E., 11
Screven, John H., 68
Screven's Causeway, 240, 248, 255
Seaboard Air Line, 313
Seaboard Coast Line Railroad Company, 313
Seaboard System, 313
Seabrook, E. M., 86
Seabrook, William, 180
Seabrook Island, 78, 81, 90, 99, 101, 177–78, 195
Seabrook Plantation, 177–78
Sea Island cotton plantations, 69, 144
Sea Islands, 70, 104
Sebring, Edward, 21, 28, 290, 296
secessionists: and board of directors of C & S Railroad, 58; and gala for C & S Railroad, 43–45; and Robert Gourdin, 41; resistance rally, 44–45; and Robert Rhett, 35, 40, 42; and state convention, 45
Secessionville, Battle of, 102–4, *103*, 105, 178
Second Military District of South Carolina, 107–10, 169
sectionalism: and aid from U.S. government, 50; and Democratic Party, 34–35; resistance Northern interference, 2; and Panic of 1857, 27; and public works, 47
Seddon, James A., 160, 171, 201, 208, 257
Seixas, J. M., 165, 166
Selfridge, James, 239
Seneca (naval vessel), 78
Seymour, Truman, 149–50, 175–77, *175*, 180, 181, 203, 210, 264
Sharpe, Thomas, 165
Shaw, Robert Gould, 150, 151
Shelby & Broad River Railroad, 132, 302
Sheridan, Philip, 273
Sherman, Thomas W., 67, 69, 75–76
Sherman, William Tecumseh: and Atlanta, 185, 204, 206, 209, 210; and C & S Railroad, 242, 250, 272; and Columbia, 265, 269, 272; destruction of Southern railroads, 240, *241*; and Hardee, 214, 237, 248, 250, 251, 261, 262, 264, 265, 266, 272; march through Georgia, 211, *211*, 212, 225, 238, 240, 247; "neckties" of, 242; and North

Carolina, 272–73; and Pocotaligo, 262; portrait of, *227*; route from Augusta to Savannah, 231, 238; and Savannah, 210–11, 212, 226, 237, 242, 243, 247, 250, 256, 259; and South Carolina, 259–66, 267, 268, 269, 272
Sickles, Dan, 296
Silliman, William, 235
Simonds, Andrew, 285, 286
Simons, James, 15
Simons, Thomas Grange, 21, 37
Simons, Thomas Y., 43, 44
Simons, W. R., 256
Simonton, Charles A., 296
Simonton and Barker, 298
Sims, Frederick W.: appointed head of the Confederate Quartermaster Department, 139; authority of, 174; and conscription of railroad workers, 163; cooperation with railroads, 164, 165, 166–67, 267–68; and Macon railroad convention, 168; and troop transport, 152, 156
Sineath, Frank, 154
Singletary, Robert Legare: and Ashley River Bridge, 133, 135–36, 181, 207–9; and bill of equity, 296; and blockade-running, 174; on board of C & S Railroad, 88; compared to Drayton, 135; and construction of C & S Railroad, 31–32, 66, 88, 182, 184, 185; and cotton shipments, 166; and creditors, 140, 285–87, 288, 291, 292, 295, 296, 307; criticism of, 291–95, 308; and evacuation of Charleston, 270, 271–72; and evacuation of Savannah, 257; and finances, 133–35, 139–41, 169, 181–84, 201, 209, 258, 267, 277–80, 283, 285, 286–87, 289, 291–93, 307; and Gonzales, 243; and Hatch, 274–75, 283; and interline cooperation, 201–2; and iron, 257, 258–59, 289, 293; and maintenance, 305; military service of, 66, 306; and operating expenses, 162, 202; political career of, 315; and postwar assessment, 274; and postwar control, 275–77; as president of C & S Railroad, 133, 184, 186; and railroad workers, 162–63, 188, 200, 201, 202, 207, 247; and real-estate holdings, 290; and relief bill, 283, 285, 294, 295; and rolling stock, 247, 254, 271; and Ryan, 31, 182, 183, 184, 185, 275, 276, 284; salary of, 200, 258, 278; and sale of C & S Railroad, 298–99, 301, 303; and Savannah River Bridge, 181, 257; and Spring Street depot, 267
Singletary, Sarah Jane Evans, 133
Skinner, R. M., 95

slave code, 34, 35
slaves and slavery: and Battle of Coosawhatchie, 117; and Constitutional Democratic Party, 36; and construction of C & S Railroad, 24–25, 29, 142–43, 313; and Democratic Party, 6; and earthworks, 216–17; and 1860 Association, 41; emancipation of, 280, 281; former slaves as Federal soldiers, 142, 173; and Northern interests, 47; and railroad construction, 24, 29, 47, 50, 143, 162, 279–80; removal from plantations, 81–82; and Second Battle of Pocotaligo, 112; as Confederate laborers, 238, 268; and South Carolina economy, 281; and Southern identity, 34; transporting slaves for war effort, 63
Slocum, Henry W., 211, 238, 251–52, 259, 264
Smalls, Robert, 92, 99, 116
Smart, Henry, 19
Smith, A. D., 103
Smith, Andrew Jackson, *220*
Smith, Derek, 210
Smith, Gustavus Woodson: and evacuation of Savannah, 253; and Hardee, 213–14, 228; and Hartridge, 240; and Honey Hill, 216–17, 218, 221, 224, 225; and James Island, 255; military career, 212; portrait of, *213*
Smith, W. S., 32
Soldiers' Wayside Home, 207
Sol Legare Island, 101, 102
Solomons, S. S., 165
South Atlantic Blockading Squadron, 75, 104, 138
South Carolina: and aid to railroad companies, 50, 88, 89, 131, 280, 283; economy, 47, 69, 281, 294, 309; and interest in C & S Railroad, 287; and Lincoln's election, 42; and postwar recovery, 281; railroad miles, 51; reserve troops, 188; and secession, 42, 43–44, 45, 46, 58, 59; and W. T. Sherman, 259–66, 267, 268, 269, 272; ties to Georgia, 41, 42–44, 45, 224; and U.S. government encroachment, 47; U.S. military control of, 274, 283
South Carolina Executive Council, 88, 89, 125, 126, 131
South Carolina legislature: and charter of C & S Railroad, 15; and funding for C & S Railroad, 20, 27, 30, 31, 47, 50, 283, 285–86, 287, 288; and Midwest trade, 3–4; and railroad movement, 1, 2, 5; and Reconstruction, 283; and route of C & S Railroad, 13, 14, 22; and state convention, 42, 43, 44, 45, 58

South Carolina Light Artillery, 105
South Carolina Military Academy, 226
South Carolina Railroad: accidents, 120; Augusta branch, 15; branches, 7, 19; Camden branch, 4; connection to C & S Railroad, 81, 105, 126, 131, 136; and cotton shipments, 56, 165; defense of, 108, 171; and evacuation of slaves, 82; and Hardee, 265; Hatch's control of, 274; and Honey Hill, 221; and iron supply, 124, 258–59, 289, 293; leadership, 120; William Magrath as president, 120, 125, 302–3; as pioneer railroad, 7; postwar restoration of, 275, 283–84; rolling stock, 53, 84, 89, 164, 279, 297, 308; and route of C & S Railroad, 13, 16, 46, 72–73; and St. Peter's Parish proposal, 12; and sale of C & S Railroad, 299; and Seymour, 176; slaves owned by, 24; and state convention, 46; state support of, 131; stock, 30, 50; and troop transport, 155, 156, 263
South Carolina Siege Train, 61, 66, 73, 149, 180, 193
South Florida Railroad, 312
Southern Express Company, 289, 296, 310
Southern Life Insurance Company, 302
Southern nationalism, 34
Southern railroads: and agriculture, 304; and Atlanta, 209; Augusta convention of, 122, 126–28, 131; and bidirectional traffic, 52; and Bragg, 151; civilian business of, 83; 7th role of, 315; Columbia railroad conventions, 106–7, 124, 186; Confederate coordination of, 61–62, 65, 83, 85, 107, 122–23, 125, 126–28, 138–39, 164, 167, 174, 206, 267–68, 304–5; and conscription of railroad workers, 162–63, 188, 268; and cotton shipments, 56–57; crisis in, 83; debt, 50; and freight tariffs, 107, 127, 187; and government property transport, 106, 107; government rates paid to, 173–74, 186–87, 201, 305–6; and government's role, 168; and in-kind investments, 47, 50; and interstate traffic, 47; lack of organized system, 304; Macon convention of, 168; and mail transport, 62, 130; maintenance, 51–54, 84, 107, 128, 139, 163, 305; map of, *48–49*; meetings of, 62; mileage of, 51; military's abuse of, 85, 122, 128; Montgomery convention of, 130, 135, 306; operations costs, 106; passenger fares, 107, 127, 187; potential advantages of, 61; rails, 52; rate schedules, 106–7; Richmond conventions of, 84, 138; rolling stock, 54, 84, 85, 88–89, 122, 125, 126, 127, 128, 138–39, 152, 156, 163, 167, 187, 240; safety of, 120; and Seymour, 175–76; and W. T. Sherman, 240, *241*; shipping capability of, 123; slaves used in construction of, 142–43; state involvement in, 47, 50–51; supplies for, 83, 84, 88, 127, 128, 138, 162, 187; and tariffs, 127; and through traffic, 46; and troop transport, 62–63, 85, 106, 107, 120, 152, 155–56, 201, 307; U.S. Army threats to, 153; and wartime inflation, 292. *See also specific railroads*
Southern Railway Security Company, 310
Southern States Freight Association, 312
Southern Union Telegraph Company, 289
Southward Ho (locomotive), 52
Southwestern Railroad, 213
Southwestern Railroad Bank, 302
Spartanburg & Union Railroad, 299
Spratt, Leonidas W., 44, 45
Stanton, Edwin M., 177
Star of the West (steamer), 59–60
states' rights, 5, 35–36, 61, 313
steamboat transportation: and Charleston and Savannah, 7; and route of C & S Railroad, 22, 33, 38, 46; steamer service of C & S Railroad, 46, 124, 207, 208, 209, 279, 289
Steinmeyer, John H., 21–23, 38, 45, 88, 184
Stevens, Clement, 137
Stevens, Isaac Ingalls: and Broad River, 76, 78–79; and Fort Walker, 69; and Hilton Head, 76; and James Island, 99, 101; and Pocotaligo, 93, 96, 97; portrait of, *94*; and Secessionville, 102, 103, 104
Stevens, Peter F., 43, 105
Stevens, Reuben, 13
Stevenson, Carter Littlepage, 269
Stoddard, George S., 236
Stokes, William, 81, 94, 153–54, 161
Stone, Edward E., 190
Stone Fleet, 74
Stono (locomotive), 31, 52
Stono River, 92, 99, 137, 149, 192, 195, 196, 197
Stono Scouts, 179, 197
Stowe, Harriet Beecher, 219
Strawbridge, Dewitt C., 116
Strobhart, James A., 11
Strohecker, John, 32, 66, 315
Strong, George, 149
Strong, J. D., 147
Strong, J. J., 11
Stuart, Henry M. "Hal," 112, 118, 243

Sullivan's Cavalry, 179
Sullivan's Island, 59, 60, 193
Swinton, William H.: and freight tariffs, 168; and mail service, 130; resignation as treasurer of C & S Railroad, 278; and Ryan, 184; salary, 140–41, 200, 258; and Singletary, 134, 135; treasurer of C & S Railroad, 66, 131

3rd Battalion South Carolina Infantry, 105
3rd Georgia Reserves, 236
3rd New Hampshire, 90, 103, 111, 150
3rd New York Artillery, 214, 216, 223, 264
3rd Rhode Island Artillery Regiment, 90, 103, 112, 116, 193, 214
3rd South Carolina Cavalry: and Coosawhatchie, 117; and evacuation of Savannah, 255–56; facing Union advance on South Carolina, 261; and Florence, 272; and Haulover Creek, 179; holiday mood of, 173; and Honey Hill, 215, 216, 217, 218, 223; and John Jenkins, 230; and Johns Island, 195, 196; and Samuel Jones, 229; and Whitney, 251
3rd Wisconsin, 239, 244
10th North Carolina Heavy Artillery Battalion, 193
12th Georgia Battalion, 150, 171
12th South Carolina Volunteer Infantry, 68, 73, 90, 187
XX Corps, 238, 242, 259, 262
21st South Carolina, 149, 150, 188
21st U.S. Colored Regiment, 271
23rd South Carolina Regiment, 105, 187
24th South Carolina Regiment, 101, 122, 137
25th North Carolina Volunteer Regiment, 73
25th Ohio: and C & S Railroad, 265; and Combahee River, 264; and Honey Hill, 214, 216, 218–19, 221, 222, 223, 224, 229, 230; and Lowndes Plantation, 266; near Coosawhatchie, 230; and Whale Branch Creek, 229; woodsmen of, 235, 236
25th South Carolina Regiment, 122, 188
26th South Carolina Regiment, 112, 187
26th U.S. Colored Troops, 196, 197–98, 214, 251, 259
26th Virginia Volunteers, 179
27th Georgia Battalion, 238, 240, 256
27th South Carolina Regiment, 188
28th Georgia Battalion, 195
28th Massachusetts, 101, 102
29th Missouri Mounted Infantry, 272
29th South Carolina Regiment, 187
31st Wisconsin, 239
32nd Georgia Regiment: and Adams Run, 190; and Bacon, 231; and Edwards, 235; and Honey Hill, 217, 221, 224; and James Island, 193; and Johns Island, 195, 196, 197, 198; and Samuel Jones, 229; and Pocotaligo, 171; and U.S. Army attack on railroad tracks, 118
32nd U.S. Colored Troops: Confederate skirmish with, 231; and Edisto Island, 264; and Honey Hill, 214, 215, 217, 218–19, 221–23; and James Island, 265
33rd U.S. Colored Troops, 192, 251, 261, 265
34th U.S. Colored Troops, 142, 194, 214
35th U.S. Colored Troops, 194, 214, 219
Tabb, William B., 179
Taliaferro, William Booth, 160, 193
Tate, Samuel, 62
Taylor, James H., 44, 65, 86, 87, 284, 294–96, 298–300, 308
Teaser (steamship), 33
technology, and railroad construction, 24–25, 313
telegraph lines, 153, 160, 234
Tennessee: and Bragg, 151; and cotton shipments, 56; and railroad investments, 51
territories, and slavery, 35
Terry, Alfred, 148–49, 150
Third Military District of South Carolina: and Beauregard, 158, 171; and Charles Colcock, 216; and Drayton, 67; and Gilmer, 172; reinforcements from, 108; reinforcing of, 137; and William Walker, 91, 92, 110, 136, 169
Thomas, Edward J., 254–55
Thomas, George, 211
Thomas, Lorenzo, 76
Thomas Rogers (locomotive), 52, 53
Thornton's Virginia battery, 79
Thurston, H. Lee, 15
Tillinghast, Robert L., 11
Totten, Joseph G., 80
Tower Battery, 102, 103
trade: and Charleston, 6, 7, 18, 22, 60, 283, 296; and feeder lines, 46; and Midwest, 3–4, 6; and Southwest, 6–7
Tredegar Iron Works, 54
Trenholm, George A., 3, 13, 42, 64, 278, 296
Trenholm, William L., 93–95, 96, 142
Trenton Locomotive and Machine Manufacturing Company, 32
Trescot, William H., 42
Trezevant, James Davis, 73

Trowbridge, Charles T., 146, 251
Troy Iron and Nail Foundry, 47
Tubman, Harriet, 142
Tupper, Samuel Y., 41
Turkey Roost Swamp, 239
Turner, George E., 157, 175, 307
Turner, Joseph W., 73
Twiggs, John D., 190
Tybee Island, 76

Union blockade: and cotton, 55, 56, 174, 305; Davis's policy toward, 174–75; and *Planter*, 92; and Southern railroads, 83, 84, 164, 165; tightening of, 72; and George Williams, 284
unionists of South Carolina, 45
U.S. Army: attacks on Charleston, 99, 101–4, 158, 160; attacks on C & S Railroad, 110–12, 114–18, 121, 145–49, *148*, 194, 229–37, 246–47, 251; black regiments of, 142, *143*, 144, 150, 177, 188; and Pocotaligo, 93–97, 261–62, *262*, 263, 267; and Port Royal, 99, 136, 137, 160, 169, 171, 173, 177; and Savannah River, 91–92; and Thomas Sherman, 69; and troop strength, 75; and weather, 99, 101–2, 195, 199, 259–60. *See also specific units*
U.S. Arsenal, Charleston, 59
U.S. Congress: and slavery, 35; and South Carolina railroad movement, 1, 2
U.S. Marines, 235, 236
U.S. Navy: attack on Charleston harbor, 136, 137–38, 139; blockading forces, 69; and Broad River, 214, 215; and Coosawhatchie River, 73; effectiveness of, 70; and Gregorie's Neck, *233*; gunboats, 68, 70, 73, 76, 78, 79, 80, 97, 99, 101, 102, 103, 104, 109, 112, 116, 118, 119, 137, 149, 150, 189, 194, 199, 225; South Atlantic Blockading Squadron, 75; Stone Fleet, 74; and Stono River, 99, 103
U.S. Postal Service, 62
Utah, 35

Vicksburg, Miss., 138, 151, 175
Virginia: railroad companies of, 84; railroad miles of, 51
Virginia Light Artillery, 105
Virginia & Tennessee Railroad, 152
Vixen (naval vessel), 190
Von Gilsa, Leopold, 177

Wadley, William M., 125, 126–28, 138, 139, 164

Wagner, Theodore D., 184, 284
Wagner, Thomas M., 65, 86
Walker, Leo D., 74
Walker, Leroy P., 62
Walker, Robert J., 91
Walker, Robert S., 162
Walker, W. H. T., 172
Walker, William Stephen: and Beauregard, 112, 158, 169, 171; and Davis, 161; and defense of Charleston, 158, 160, 169, 171; and defense of C & S Railroad, 111, 112, 114, 115, 116, 119, 151, 160, 169; and Drayton, 67, 68–69, 94; military career of, 91; and Pocotaligo, 95–97, 108, 110, 169, 171, 172; portrait of, *96*; reinforcements for, 173; and Third Military District command, 91, 92, 110, 136, 169; troops commanded by, 136; and Wilmington, 187
Walter, George H., 108
Walters, W. T., 311
Walter's Light Battery, 147
Walters syndicate, 311
War Between the States, role of railroads in, 52, 61. *See also specific battles and generals*
Ward, W. T., 259
Warley, Charles, 21
Washington, D. C., and C & S Railroad, 17–18
Washington Artillery, 43, 108, 147, 197, 243
Washington Light Infantry, 43
Wassaw Island, 210
Waud, Alfred R.: Green Pond Drive station sketch, *31*; train transporting prisoners sketch, *204*
Waycross Short Line, 310, 311
Wayne, Robert A., 197, 256
weather: and camp life, 110; and evacuation of Savannah, 256; and postwar repairs, 279; and sickly season, 29, 105, 108, 110, 277; and Union Army, 99, 101–2, 195, 199, 259–60
Webb, John H., 11
Webb, Thomas, 202
Webb's Battery, 197
Webster, Daniel, 2
Weehawken (naval vessel), 137
Welles, Gideon, 137–38
West, Francis H., 239
West, James B., 146
Western & Atlantic Railroad, 3, 156, 289
Western Coalition, 151
Wharton & Petsch, 53, 89
Wheeler, Joseph, 237, 248, 252, 255, 257, 260
Whippy Swamp Guards, 117, 119

White, E. John, 162
White, James B., 226, 230, 232
White, Thomas G., 146–47
White Hall Plantation, 53, 134, 201
Whitemarsh Island, 210
Whiting, William Henry Chase, 122, 164, 166, 193
Whitney, H. A., 251
Wilderness Campaign, 181, 188
Wilkes, John, 21
Willcoxon, John B., 90
Williams, George Walton, 86, 88, 278, 284–85, 295–96, 298–99, 309
Williams, James S., 20–22, 23
Williams, Reuben, 272
Williman, A. B., 21
Williman's Island, 153, 154
Willingham, B. L., 19
Willis, Francis T., 299, 300
Wilmington, Columbia & Augusta Railroad, 311
Wilmington, N.C., 4, 7, 15, 161, 164–65, 187
Wilmington & Manchester Railroad: accidents, 120; and cotton shipments, 165–66; and Drane, 136, 303; and Haines, 66–67; Hatch's control of, 274; locomotives, 163; postwar condition, 276; and right-of-way costs, 37; rolling stock, 279; and route of C & S Railroad, 14, 15; and South Carolina Railroad, 4; stock, 30, 50; and troop transport, 155; U.S. Army raids of, 153; Whiting on, 122
Wilmington & Weldon Railroad, 311
Wilson's brigade, 122
Wise, Henry A., 153, 156, 160, 171, 179–80, 188
Wise, Stephen, 80
Woodford, Stewart L., 219–20, 230, 235
Woods Road, 218–19, 221, 223, 224
Wright, Ambrose R., 253, 266, 269
Wright, Horatio G., 69, 99, 101, 102, 104
Wyman, B. F., 96, 112, 114, 117

Yancey, William L., 34, 35, 40
Yates, Joseph A., 194
Yeadon, Richard, 86, 184, 207, 286, 289
yellow fever, 29, 112, 119
Young, Pierce M. B., 247, 248–49, 256

ABOUT THE AUTHOR

H. David Stone, Jr. is a physician in Florence, South Carolina. He is a graduate of Furman University and the University of South Carolina School of Medicine.